Fundamentals of
Queueing Theory

Fundamentals of Queueing Theory

Third Edition

DONALD GROSS
Research Professor of Operations Research
George Mason University
Professor Emeritus
The George Washington University

CARL M. HARRIS
BDM Professor of Operations Research
George Mason University

A Wiley-Interscience Publication
JOHN WILEY & SONS, INC.
New York • Chichester • Weinheim • Brisbane • Singapore • Toronto

Copyright © 1998 by John Wiley & Sons, Inc. All rights reserved.

Published simultaneously in Canada.

Library of Congress Cataloging-in-Publication Data:
Gross, Donald.
    Fundamentals of queueing theory / Donald Gross, Carl M. Harris. – 3rd ed.
      p.   cm.
    Includes index.
    ISBN 0-471-17083-6 (cloth : alk. paper)
    1. Queueing theory. I. Harris, Carl M., 1940–   II. Title.
  IN PROCESS
  519.8'2–dc21   97-13171

Contents

Preface

The changes in this third edition reflect the more than 20 years of experience we have had using the prior versions as a text in teaching queueing theory and as a reference work, plus numerous comments we have had from colleagues since the first and second editions appeared. The most important modifications in this new edition relate to our incorporation of spreadsheet-based software in recognition of the incredible strides made in personal computing in the short 12 years since the second edition appeared. An abbreviated listing follows of the major technical and editorial changes we have made in this revision.

SPECIFIC MAJOR CHANGES

1. The old Section 1.7 on deterministic models has been streamlined and combined with material on general results previously scattered throughout the book and with the data bookkeeping presentation previously found in the discussion of simulation in the old Chapter 8. We have introduced more complete discussions of queue sample paths and the development of Little's formulas as a result. At the end of Chapter 1, we have moved up the presentation of the basic results on birth–death processes from Section 2.1 into Section 1.10, following the discussion of Markov processes in Section 1.9.

2. Less the material on birth–death processes moved to Section 1.10, Chapter 2 remains much as it was previously. The narrative has been changed somewhat, but the order of material is the same. In light of the relative simplicity of the formulations in this chapter, virtually all the models discussed here have been included in the software as distinct modules.

3. There are some additional results added in Chapter 3 on the classical Markovian bulk and Erlang models, including more details on the $E_j/E_k/1$ queue. The inclusion of these sorts of models in the software is especially important given their inherent numerical complexity.

4. In the second edition, we introduced an entirely new Chapter 4 on the important topic of queueing networks. This time, we have smoothed out the narrative, included new material on reversibility, and added a section on mean-value analysis (MVA).

5. Chapter 5's first nine sections on $M/G/1$ queues and variations are largely unchanged, but Section 5.10 has been expanded to combine departure-point state dependence with the concepts of decomposition and server vacations. Section 5.2 contains a new proof of Erlang's loss formula built around the reversibility of Markov processes, first discussed by us in Chapter 4. The material following in the chapter on $G/M/c$ has also been expanded to include a more complete discussion on the necessary rootfinding involved and the extension from the single-server problem to multiple servers. This chapter is also very well covered in the software.

6. Chapter 6 contains new material on the $G/E_k/1$ queue, including the extension of simple rootfinding into the complex domain. This material is combined with expanded discussions on matrix geometric solutions and quasi-birth–death processes. The discussion on the $G/G/1$ problem has thus been moved to Section 6.2 and remains largely as it was. The solution to the problem following this—the $M/D/c$ queue—is then connected back to the same sort of complex rootfinding problem as for the $G/E_k/1$. The sections in this chapter on Markov renewal processes, alternative disciplines, design and control, and statistical inference have all been updated and expanded.

7. The new Chapter 7 combines the important topics of bounds, approximations, numerics, and simulation. The discussions on bounds and approximations have been updated and slightly expanded from the previous edition. But our coverage of simulation has been reduced from a full chapter (8) to just a section and moved into Chapter 7. In view of the subject's explosive growth since the second edition, we felt that we could not do justice to the subject in a limited number of pages, and have instead merely summarized the most general elements of simulation modeling.

COURSE COVERAGE

Although the amount of material that can be covered in a given time will depend on the background of the students (again, this edition assumes only a knowledge of undergraduate differential and integral calculus, elements of differential equations and matrix manipulations, and a calculus-based probability and statistics course), the following table gives our suggestion on possible course coverages:

Course length	Recommended Text Material
One quarter	Chapters 1, 2 (through 2.9), 3 (omit 3.4), 4 (omit 4.5, 4.6), 5 (5.1.1–5.1.3, 5.1.7, 5.3.1)
One semester	Chapters 1, 2 (through 2.10.1), 3 (omit 3.4), 4, 5 (5.1.1–5.1.3, 5.1.7, 5.3.1), 6 (6.1, 6.2), 7 (7.1–7.3)
Two quarters	Chapters 1, 2, 3 (omit 3.4), 4, 5 (5.1.1–5.1.3, 5.1.7, 5.1.10, 5.2.2, 5.3.1), 6 (6.1–6.5), 7
Two semesters	Entire text

ACKNOWLEDGMENTS

We are grateful for the assistance given to us over the years by many professional colleagues and students, whose numerous comments and suggestions have been so helpful in improving our work. With heartfelt thanks, we extend special appreciation once more to our families for their unlimited and continuing encouragement and to all the people at John Wiley who have been wonderfully supportive of us through these three editions.

DONALD GROSS
CARL M. HARRIS

Fairfax, Virginia
November 1997

Fundamentals of
Queueing Theory

CHAPTER 1

Introduction

All of us have experienced the annoyance of having to wait in line. Unfortunately, this phenomenon continues to be common in congested, urbanized, "high-tech" societies. We wait in line in our cars in traffic jams or at toll booths; we wait on hold for an operator to pick up our telephone calls; we wait in line at supermarkets to check out; we wait in line at fast-food restaurants; and we wait in line at banks and post offices. We, as customers, do not generally like these waits, and the managers of the establishments at which we wait also do not like us to wait, since it may cost them business. Why then is there waiting?

The answer is simple: there is more demand for service than there is facility for service available. Why is this so? There may be many reasons; for example, there may be a shortage of available servers, it may be infeasible economically for a business to provide the level of service necessary to prevent waiting, or there may be a space limit to the amount of service that can be provided. Generally these limitations can be removed with the expenditure of capital, and to know how much service should then be made available, one would need to know answers to such questions as, "How long must a customer wait?" and "How many people will form in the line?" Queueing theory attempts (and in many cases succeeds) to answer these questions through detailed mathematical analysis. The word "queue" is in more common usage in Great Britain and other countries than in the United States, but it is rapidly gaining acceptance in this country, although it must be admitted that it is just as displeasing to spend time in a queue as in a waiting line.

1.1 DESCRIPTION OF THE QUEUEING PROBLEM

A queueing system can be described as customers arriving for service, waiting for service if it is not immediate, and if having waited for service, leaving the system after being served. The term "customer" is used in a general sense and does not imply necessarily a human customer. For example, a

1

Fig. 1.1   A typical queueing process.

customer could be a ball bearing waiting to be polished, an airplane waiting in line to take off, or a computer program waiting to be run. Such a basic system can be schematically shown as in Figure 1.1. Although any queueing system may be diagrammed in this manner, it should be clear that a reasonably accurate representation of such a system would require a detailed characterization of the underlying processes.

Queueing theory was developed to provide models to predict the behavior of systems that attempt to provide service for randomly arising demands; not unnaturally, then, the earliest problems studied where those of telephone traffic congestion. The pioneer investigator was the Danish mathematician A. K. Erlang, who, in 1909, published "The Theory of Probabilities and Telephone Conversations". In later works he observed that a telephone system was generally characterized by either (1) Poisson input, exponential holding (service) times, and multiple channels (servers), or (2) Poisson input, constant holding times, and a single channel. Erlang was also responsible for the notion of stationary equilibrium, for the introduction of the so-called balance-of-state equations, and for the first consideration of the optimization of a queueing system.

Work on the application of the theory to telephony continued after Erlang. In 1927, E. C. Molina published his paper "Application of the Theory of Probability to Telephone Trunking Problems", which was followed one year later by Thornton Fry's *Probability and Its Engineering Uses*, which expanded much of Erlang's earlier work. In the early 1930s, Felix Pollaczek did some further pioneering work on Poisson input, arbitrary output, and single- and multiple-channel problems. Additional work was done at that time in Russia by Kolmogorov and Khintchine, in France by Crommelin, and in Sweden by Palm. The work in queueing theory picked up momentum rather slowly in its early days, but accelerated in the 1950s, and there has been a great deal of work in the area since then.

There are many valuable applications of the theory, most of which have been well documented in the literature of probability, operations research, management science, and industrial engineering. Some examples are traffic flow (vehicles, aircraft, people, communications), scheduling (patients in hospitals, jobs on machines, programs on a computer), and facility design (banks, post offices, amusement parks, fast-food restaurants).

Queueing theory originated as a very practical subject, but much of the literature up to the middle 1980s was of little direct practical value. However, queueing theorists have once again become concerned about the application of the sophisticated theory that has largely arisen since the close of World War II. The emphasis in the literature on the exact solution of queueing problems with clever mathematical tricks is now becoming secondary to model building and the direct use of these techniques in management decisionmaking. Most real problems do not correspond exactly to a mathematical model, and increasing attention is being paid to complex computational analysis, approximate solutions, sensitivity analyses, and the like. The development of the practice of queueing theory must not be restricted by a lack of closed-form solutions, and problem solvers must be able to put the developed theory to good use. These points should be kept in mind by the reader, and we attempt to illustrate them whenever possible throughout this text.

1.2   CHARACTERISTICS OF QUEUEING PROCESSES

In most cases, six basic characteristics of queueing processes provide an adequate description of a queueing system: (1) arrival pattern of customers, (2) service pattern of servers, (3) queue discipline, (4) system capacity, (5) number of service channels, and (6) number of service stages.

1.2.1   Arrival Pattern of Customers

In usual queueing situations, the process of arrivals is stochastic, and it is thus necessary to know the probability distribution describing the times between successive customer arrivals (interarrival times). It is also necessary to know whether customers can arrive simultaneously (batch or bulk arrivals), and if so, the probability distribution describing the size of the batch.

It is also necessary to know the reaction of a customer upon entering the system. A customer may decide to wait no matter how long the queue becomes, or, on the other hand, if the queue is too long, the customer may decide not to enter the system. If a customer decides not to enter the queue upon arrival, the customer is said to have *balked*. A customer may enter the queue, but after a time lose patience and decide to leave. In this case, the customer is said to have *reneged*. In the event that there are two or more parallel waiting lines, customers may switch from one to another, that is, *jockey* for position. These three situations are all examples of queues with *impatient customers*.

One final factor to be considered regarding the arrival pattern is the manner in which the pattern changes with time. An arrival pattern that does not change with time (i.e., the probability distribution describing the input process is time-independent) is called a *stationary* arrival pattern. One that is not time-independent is called *nonstationary*.

1.2.2 Service Patterns

Much of the discussion above concerning the arrival pattern is appropriate in discussing service. Most importantly, a probability distribution is needed to describe the sequence of customer service times. Service may also be single or batch. One generally thinks of one customer being served at a time by a given server, but there are many situations where customers may be served simultaneously by the same server, such as a computer with parallel processing, sightseers on a guided tour, or people boarding a train.

The service process may depend on the number of customers waiting for service. A server may work faster if the queue is building up or, on the contrary, may get flustered and become less efficient. The situation in which service depends on the number of customers waiting is referred to as *state-dependent* service. Although this term was not used in discussing arrival patterns, the problems of customer impatience can be looked upon as ones of state-dependent arrivals, since the arrival behavior depends on the amount of congestion in the system.

Service, like arrivals, can be stationary or nonstationary with respect to time. For example, learning may take place, so that service becomes more efficient as experience is gained. The dependence on time is not to be confused with dependence on state. The former does not depend on the number of customers in the system, but rather on how long it has been in operation. The latter does not depend on how long the system has been in operation, but only on the state of the system at a given time, that is, on how many customers are currently in the system. Of course, a queueing system can be both nonstationary and state-dependent.

Even if the service rate is high, it is very likely that some customers will be delayed by waiting in the line. In general, customers arrive and depart at irregular intervals; hence the queue length will assume no definitive pattern unless arrivals and service are deterministic. Thus it follows that a probability distribution for queue lengths will be the result of two separate processes— arrivals and services—which are generally, though not universally, assumed mutually independent.

1.2.3 Queue Discipline

Queue discipline refers to the manner in which customers are selected for service when a queue has formed. The most common discipline that can be observed in everyday life is first come, first served (FCFS). However, this is certainly not the only possible queue discipline. Some others in common usage are last come, first served (LCFS), which is applicable to many inventory systems when there is no obsolescence of stored units, as it is easier to reach the nearest items, which are the last in; selection for service in random order independent of the time of arrival to the queue (RSS); and a variety of *priority* schemes, where customers are given priorities upon entering the system, the ones with higher priorities to be selected for service ahead of those with lower priorities, regardless of their time of arrival to the system.

There are two general situations in priority disciplines. In the first, which is called *preemptive*, the customer with the highest priority is allowed to enter service immediately even if a customer with lower priority is already in service when the higher-priority customer enters the system; that is, the lower-priority customer in service is preempted, its service stopped, to be resumed again after the higher-priority customer is served. There are two possible additional variations: the preempted customer's service when resumed can either continue from the point of preemption or start anew. In the second general priority situation, called the *nonpreemptive* case, the highest-priority customer goes to the head of the queue but cannot get into service until the customer presently in service is completed, even though this customer has a lower priority.

1.2.4 System Capacity

In some queueing processes there is a physical limitation to the amount of waiting room, so that when the line reaches a certain length, no further customers are allowed to enter until space becomes available as the result of a service completion. These are referred to as finite queueing situations; that is, there is a finite limit to the maximum system size. A queue with limited waiting room can be viewed as one with forced balking where a customer is forced to balk if it arrives when the queue size is at its limit. This is a simple case, since it is known exactly under what circumstances arriving customers must balk.

1.2.5 Number of Service Channels

As we shortly explain in more detail, it is generally preferable to design multiserver queueing systems to be fed by a single line. Thus, when we specify the number of service channels, we are typically referring to the number of parallel service stations which can serve customers simultaneously. Figure 1.1 depicts an illustrative single-channel system, while Figure 1.2 shows two variations of multichannel systems. The two multichannel systems differ in that the first has a single queue, while the second allows a queue for each channel. A hair-styling salon with many chairs is an example of the first type of multichannel system (assuming no customer is waiting for any particular stylist), while a supermarket or fast-food restaurant might fit the second description. It is generally assumed that the service mechanisms of parallel channels operate independently of each other.

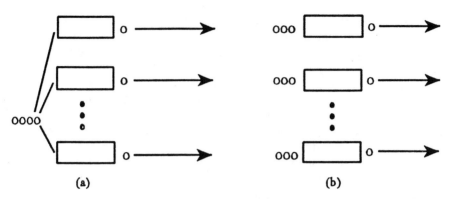

Fig. 1.2 Multichannel queueing systems.

1.2.6 Stages of Service

A queueing system may have only a single stage of service, as in the hair-styling salon, or it may have several stages. An example of a multistage queueing system would be a physical examination procedure, where each patient must proceed through several stages, such as medical history; ear, nose, and throat examination; blood tests; electrocardiogram; eye examination; and so on. In some multistage queueing processes recycling or feedback may occur. Recycling is common in manufacturing processes, where quality-control inspections are performed after certain stages, and parts that do not meet quality standards are sent back for reprocessing. Similarly, a telecommunications network may process messages through a randomly selected sequence of nodes, with the possibility that some messages will require rerouting on occasion through the same stage. A multistage queueing system with some feedback is depicted in Figure 1.3.

The six characteristics of queueing systems discussed in this section are generally sufficient to completely describe a process under study. Clearly, a wide variety of queueing systems can be encountered. Before performing any mathematical analyses, however, it is absolutely necessary to describe adequately the process being modeled. Knowledge of the basic six characteristics is essential in this task.

It is extremely important to use the correct model or at least the model

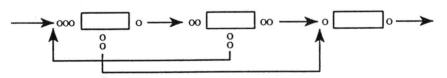

Fig 1.3 A multistage queueing system with feedback.

that best describes the real situation being studied. A great deal of thought is often required in this model selection procedure. For example, let us reconsider the supermarket mentioned previously. Suppose there are c checkout counters. If customers choose a checkout counter on a purely random basis (without regard to the queue length in front of each counter) and never switch lines (no jockeying), then we truly have c independent single-channel models. If, on the other hand, there is a single waiting line and when a checker becomes idle, the customer at the head of the line (or with the lowest number if numbers are given out) enters service, we have a c-channel model. Neither, of course, is generally the case in most supermarkets. What usually happens is that queues form in front of each counter, but new customers enter the queue which is the shortest (or has shopping carts which are lightly loaded). Also, there is a great deal of jockeying between lines. Now the question becomes which choice of models (c independent single channels or a single c-channel) is more appropriate. If there were complete jockeying, the single-c-channel model would be quite appropriate, since even though in reality there are c lines, there is little difference, when jockeying is present, between these two cases. This is so because no servers will be idle as long as customers are waiting for service, which would not be the case with c truly independent single channels. As jockeying is rather easy to accomplish in supermarkets, the c-channel model will be more appropriate and realistic than the c-single-channels model, which one might have been tempted to choose initially prior to giving much thought to the process. Thus it is important not to jump to hasty conclusions but to select carefully the most appropriate model.

1.3 NOTATION

As a shorthand for describing queueing processes, a notation has evolved, due for the most part to Kendall (1953), which is now rather standard throughout the queueing literature. A queueing process is described by a series of symbols and slashes such as $A/B/X/Y/Z$, where A indicates in some way the interarrival-time distribution, B the service pattern as described by the probability distribution for service time, X the number of parallel service channels, Y the restriction on system capacity, and Z the queue discipline (Appendix 1 contains a dictionary of symbols used throughout this text). Some standard symbols for these characteristics are presented in Table 1.1. For example, the notation $M/D/2/\infty/FCFS$ indicates a queueing process with exponential interarrival times, deterministic service times, two parallel servers, no restriction on the maximum number allowed in the system, and first-come, first-served queue discipline.

In many situations only the first three symbols are used. Current practice is to omit the service-capacity symbol if no restriction is imposed ($Y = \infty$)

Table 1.1 Queueing Notation *A/B/X/Y/Z*

Characteristic	Symbol	Explanation
Interarrival-time distribution (*A*) Service-time distribution (*B*)	*M*	Exponential
	D	Deterministic
	E_k	Erlang type k ($k = 1, 2, \ldots$)
	H_k	Mixture of k exponentials
	PH	Phase type
	G	General
Number of parallel servers (*X*)	$1, 2, \ldots, \infty$	
Restriction on system capacity (*Y*)	$1, 2, \ldots, \infty$	
Queue discipline (*Z*)	FCFS	First come, first served
	LCFS	Last come, first served
	RSS	Random selection for service
	PR	Priority
	GD	General discipline

and to omit the queue discipline if it is first come, first served (*Z* = FCFS). Thus *M/D/2* would be a queueing system with exponential input, deterministic service, two servers, no limit on system capacity, and first-come, first-served discipline.

The symbols in Table 1.1 are, for the most part, self-explanatory; however, a few require further comment. The symbol *G* represents a general probability distribution; that is, no assumption is made as to the precise form of the distribution. Results in these cases are applicable to any probability distribution. These general-time distributions, however, are required to represent independent and identically distributed random variables.

It may also appear strange that the symbol *M* is used for exponential. The use of the symbol *E*, as one might expect, would be too easily confused with E_k, which is used for the type-k Erlang distribution (a gamma with an integer shape parameter). So *M* is used instead; it stands for the Markovian or memoryless property of the exponential, which is developed in some detail in Section 1.9.

The reader may have noticed that the list of symbols is not complete. For example, there is no indication of a symbol to represent bulk arrivals, to represent series queues, to denote any state dependence, and so on. If a suitable notation does exist for any previously unmentioned model, it is indicated when that particular model is brought up in the text. However, there still remain models for which no symbolism has either been developed or accepted as standard, and this is generally true for those models less frequently analyzed in the literature.

1.4 MEASURING SYSTEM PERFORMANCE

Up to now the concentration has been on the physical description of queueing processes. What, then, might one like to know about the effectiveness of a queueing system? Generally there are three types of system responses of interest. These are: (1) some measure of the waiting time that a typical customer might be forced to endure; (2) an indication of the manner in which customers may accumulate; and (3) a measure of the idle time of the servers. Since most queueing systems have stochastic elements, these measures are often random variables and their probability distributions, or at the very least their expected values, are desired.

There are two types of customer waiting times, the time a customer spends in the queue and the total time a customer spends in the system (queue plus service). Depending on the system being studied, one may be of more interest than the other. For example, if we are studying an amusement park, it is the time waiting in the queue that makes the customer unhappy. On the other hand, if we are dealing with machines that require repair, then it is the total down time (queue wait plus repair time) that we wish to keep as small as possible. Correspondingly, there are two customer accumulation measures as well: the number of customers in the queue and the total number of customers in the system. The former would be of interest if we desire to determine a design for waiting space (say the number of seats to have for customers waiting in a hair-styling salon), while the latter may be of interest for knowing how many of our machines may be unavailable for use. Idle-service measures can include the percentage of time any particular server may be idle, or the time the entire system is devoid of customers.

The task of the queueing analyst is generally one of two things. He or she is either to determine the values of appropriate measures of effectiveness for a given process, or to design an "optimal" (according to some criterion) system. To do the former, one must relate waiting delays, queue lengths, and such to the given properties of the input stream and the service procedures. On the other hand, for the design of a system the analyst might want to balance customer waiting time against the idle time of servers according to some inherent cost structure. If the costs of waiting and idle service can be obtained directly, they can be used to determine the optimum number of channels to maintain and the service rates at which to operate these channels. Also, to design the waiting facility it is necessary to have information regarding the possible size of the queue to plan for waiting room. There may also be a space cost which should be considered along with customer-waiting and idle-server costs to obtain the optimal system design. In any case, the analyst will strive to solve this problem by analytical means; however, if these fail, he or she must resort to simulation. Ultimately, the issue generally comes down to a trade-off of better customer service versus the expense of providing more service capability, that is, determining the increase in investment of service for a corresponding decrease in customer delay.

1.5   SOME GENERAL RESULTS

We present some general results and relationships for $G/G/1$ and $G/G/c$ queues in this section, prior to specific model development. These results will prove useful in many of the following sections and chapters, as well as providing some insight at this early stage.

Denoting the average rate of customers entering the queueing system as λ and the average rate of serving customers as μ, a measure of traffic congestion for c-server systems is $\rho \equiv \lambda/c\mu$. When $\rho > 1$ ($\lambda > c\mu$), the average number of arrivals into the system exceeds the maximum average service rate of the system, and we would expect, as time goes on, the queue to get bigger and bigger, unless, at some point, customers were not allowed to join. If we are interested in steady-state conditions (the state of the system after it has been in operation a long time), when $\rho > 1$, the queue size never settles down (assuming customers are not prevented from entering the system) and there is no steady state. It turns out that for steady-state results to exist, ρ must be strictly less than 1 (again, assuming no denial of customer entry). When $\rho = 1$, unless arrivals and service are deterministic and perfectly scheduled, no steady state exists, since randomness will prevent the queue from ever emptying out and allowing the servers to catch up, thus causing the queue to grow without bound. Therefore, if one knows the average arrival rate and average service rate, the minimum number of parallel servers required to guarantee a steady-state solution can be immediately calculated by finding the smallest c such that $\lambda/c\mu < 1$.

What we most often desire in solving queueing models is to find the probability distribution for the total number of customers in the system at time t, $N(t)$, which is made up of those waiting in queue, $N_q(t)$, plus those in service, $N_s(t)$. Let $p_n(t) = \Pr\{N(t) = n\}$, and $p_n = \Pr\{N = n\}$ in the steady state. Considering c-server queues in steady state, two expected-value measures of major interest are the mean number in the system,

$$L = \mathrm{E}[N] = \sum_{n=0}^{\infty} np_n \, ,$$

and the expected number in queue,

$$L_q = \mathrm{E}[N_q] = \sum_{n=c+1}^{\infty} (n - c)p_n.$$

1.5.1   Little's Formulas

One of the most powerful relationships in queueing theory was developed by John D. C. Little in the early 1960s (see Little, 1961, for the original proof—a host of papers refining the proof followed in the ensuing decades).

Little related the steady-state mean system sizes to the steady-state average customer waiting times as follows. Letting T_q represent the time a customer (transaction) spends waiting in the queue prior to entering service and T represent the total time a customer spends in the system ($T = T_q + S$, where S is the service time, and T, T_q, and S are random variables), two often used measures of system performance with respect to customers are $W_q = E[T_q]$ and $W = E[T]$, the mean waiting time in queue and the mean waiting time in the system, respectively. Little's formulas are

$$L = \lambda W \tag{1.1a}$$

and

$$L_q = \lambda W_q. \tag{1.1b}$$

Thus it is necessary to find only one of the four expected value measures, in view of Little's formulas and the fact that $E[T] = E[T_q] + E[S]$, or, equivalently, $W = W_q + 1/\mu$, where μ, as before, is the mean service rate.

Although the following does not constitute a proof, we illustrate the concept of Little's formulas by considering a sample path of one busy period (time from when a customer enters an empty system until it next empties out again). Consider the illustration in Figure 1.4, where the number of customers (say N_c), that arrive over the time period $(0, T)$ is 4.

The calculations for L and W are

$$L = [1(t_2 - t_1) + 2(t_3 - t_2) + 1(t_4 - t_3) + 2(t_5 - t_4)$$
$$+ 3(t_6 - t_5) + 2(t_7 - t_6) + 1(T - t_7)]/T$$
$$= (\text{area under curve})/T$$
$$= (T + t_7 + t_6 - t_5 - t_4 + t_3 - t_2 - t_1)/T \tag{1.2a}$$

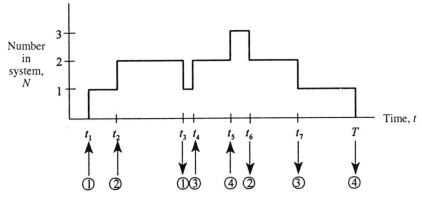

Fig. 1.4 Busy-period sample path.

and

$$W = [(t_3 - t_1) + (t_6 - t_2) + (t_7 - t_4) + (T - t_5)]/4$$
$$= (T + t_7 + t_6 - t_5 - t_4 + t_3 - t_2 - t_1)/4$$
$$= (\text{area under curve})/N_c. \qquad (1.2b)$$

Thus we see from Equations (1.2a) and (1.2b) that the area under curve is $LT = WN_c$, which yields $L = WN_c/T$. The fraction N_c/T is the number of customers arriving over the time T and is, for this period, the arrival rate λ, so that $L = \lambda W$. A similar argument would hold for a picture of the number in the queue N_q over the period $(0, T)$, yielding $L_q = \lambda W_q$. While this is not a proof (since it needs to be shown that these relationships hold in the limit over many busy periods as time goes to infinity), one can see the idea behind the relationships.

An interesting result that can be derived from Little's formulas [Equations (1.1a) and (1.1b)] and the relation between W and W_q is

$$L - L_q = \lambda(W - W_q) = \lambda(1/\mu) = \lambda/\mu. \qquad (1.3)$$

But $L - L_q = \mathrm{E}[N] - \mathrm{E}[N_q] = \mathrm{E}[N - N_q] = \mathrm{E}[N_s]$, so that the expected number of customers in service in the steady state is λ/μ, which we will denote by r. Note for a single-server system that $r = \rho$ and it also follows from simple algebra that

$$L - L_q = \sum_{n=1}^{\infty} np_n - \sum_{n=1}^{\infty} (n-1)p_n = \sum_{n=1}^{\infty} p_n = 1 - p_0 .$$

From this, we can easily derive the probability that any given server is busy in a multiserver system in the steady state. We denote this probability by p_b. Since we have just shown that the expected number present in service at any instant in the steady state is r, it follows from the symmetry of the c servers that the expected number present at one server is r/c. Then, by a simple expected-value argument, we can show that $p_b = \rho$, since

$$r/c = \rho = 0 \cdot (1 - p_b) + 1 \cdot p_b.$$

For a single-server queue $(G/G/1)$, the probability of the system being idle $(N = 0)$ is the same as the probability of a server being idle. Thus, $p_0 = 1 - p_b$ in this case, and $p_0 = 1 - \rho = 1 - r = 1 - \lambda/\mu$. The quantity $r = \lambda/\mu$, the expected number of customers in service, has another interesting connotation. It is sometimes also referred to as the *offered load*, since, on average, each customer requires $1/\mu$ time units of service and the average number of customers arriving per unit time is λ, so that the product $\lambda(1/\mu)$ is the amount of work arriving to the system per unit time. Dividing this by the number of

Table 1.2 Summary of General Results for *G/G/c* Queues

$\rho = \lambda/c\mu$	Traffic intensity; offered work load rate to a server
$L = \lambda W$	Little's formula
$L_q = \lambda W_q$	Little's formula
$W = W_q + 1/\mu$	Expected-value argument
$p_b = \lambda/c\mu = \rho$	Busy probability for an arbitrary server
$r = \lambda/\mu$	Expected number of customers in service; offered work load rate
$L = L_q + r$	Combined result--Equation (1.3)
$p_0 = 1 - \rho$	*G/G/*1 empty-system probability
$L = L_q + (1 - p_0)$	Combined result for *G/G/*1

servers *c* (which yields ρ) gives the average amount of work coming to each server per unit time.

Table 1.2 summarizes the results of this section.

1.6 SIMPLE DATA BOOKKEEPING FOR QUEUES

At this point, it might be useful to use a table format to show how the random events of arrivals and service completions interact for a sample single-server system to form a queue. In the following, we begin at time 0 with a first arrival and then update the system state when events (arrivals or departures) occur—thus, the name *event-oriented bookkeeping* is used for this sort of table.

Consider the elementary case of a constant rate of arrivals to a single channel which possesses a constant service rate. (Figure 1.5 is an illustration of this with interarrival times of 3 and serve times of 5.) These regularly

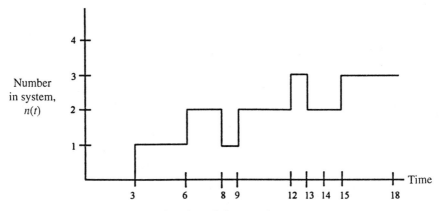

Fig. 1.5 Sample path for queueing process.

spaced arrivals are to be served first come, first served (FCFS). Let it also be assumed that at time $t = 0$ there are no customers waiting and that the channel is empty. Let λ be defined as the number of arrivals per unit time, and $1/\lambda$ then will be the constant time between successive arrivals. The particular unit of time (minutes, hours, etc.) is up to the choice of the analyst. However, consistency must be adhered to once the unit is chosen so that the same basic unit is used throughout the analysis. Similarly, if μ is to be the rate of service in terms of completions per unit time when the server is busy, then $1/\mu$ is the constant service time. We would like to calculate the number in the system at an arbitrary time t, say $n(t)$, and the time the nth arriving customer must wait in the queue to obtain service, say $W_q^{(n)}$. From these, it then becomes easy to compute the major measures of effectiveness.

Under the assumption that as soon as a service is completed another is begun, the number in the system (including the customer in service) at time t is determined by the equation

$$n(t) = \{\text{number of arrivals in } (0, t]\}$$
$$- \{\text{number of services completed in } (0, t]\}. \qquad (1.4)$$

It should be pointed out that there are usually three waiting times of interest—the time spent by the nth customer waiting for service (or line delay), which we write here as $W_q^{(n)}$; the time the nth customer spent in the system, which we shall call $W^{(n)}$; and what is called the virtual line wait $V(t)$, namely, the wait a fictitious arrival would have to endure if it arrived at time t. The reader is cautioned that various authors are not consistent and each of these quantities is sometimes referred to simply as the waiting time.

To find the waiting times in queue until service begins, we observe that the line waits $W_q^{(n)}$ and $W_q^{(n+1)}$ of two successive customers in *any* single-server queue (deterministic or otherwise) are related by the simple recurrence relation

$$W_q^{(n+1)} = \begin{cases} W_q^{(n)} + S^{(n)} - T^{(n)} & (W_q^{(n)} + S^{(n)} - T^{(n)} > 0), \\ 0 & (W_q^{(n)} + S^{(n)} - T^{(n)} \le 0), \end{cases} \qquad (1.5)$$

where $S^{(n)}$ is the service time of the nth customer and $T^{(n)}$ is the interarrival time between the nth and $(n + 1)$st customers. This can be seen by a simple diagram as shown in Figure 1.6. (This is an important general relation which is also utilized in later portions of the text.)

Bookkeeping has to do with updating the system status when events occur, recording items of interest, and calculating measures of effectiveness. Event-oriented bookkeeping updates the system state only when events (arrivals or departures) occur. Since there is not necessarily an event every basic time unit, in next-event bookkeeping the master clock is increased by a variable

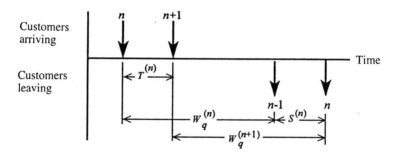

Fig. 1.6 Successive $G/G/1$ waiting times.

Table 1.3 Input Data

i	1	2	3	4	5	6	7	8	9	10	11	12
Interarrival time between customers $i+1$ and i	2	1	3	1	1	4	2	5	1	4	2	—
Service time of customer i	1	3	6	2	1	1	4	2	5	1	1	3

amount each time, rather than a fixed amount as it would be in time-oriented bookkeeping. The event-oriented approach will be illustrated here by an example, using the arrival and service data given in Table 1.3.

We see from simple averaging calculations for columns (5) and (6) in Table 1.4 that the mean line delay of the 12 customers was $40/12 = \frac{10}{3}$, while their mean system waiting time turned out to be $70/12 = \frac{35}{6}$. Further, we observe that we can estimate the mean arrival rate as $\frac{12}{31}$ customers per unit time, since there were 12 arrivals over the 31-time-unit observation horizon. Thus, the application of Little's law to these numbers tells us that the average system size L over the full time horizon was

$$L = \lambda W = \frac{70/12}{31/12} = \frac{70}{31}.$$

The mean queue size can be computed similarly.

Table 1.4 Event-Oriented Bookkeeping

(1) Master Clock Time	(2) Arrival/ Departure Customer i	(3) Time Arrival i Enters Service	(4) Time Arrival i Leaves Service	(5) Time in Queue	(6) Time in System	(7) No. in Queue Just After Master Clock Time	(8) No. in System Just After Master Clock Time
0	1-A	0	1	0	1	0	1
1	1-D					0	0
2	2-A	2	5	0	3	0	1
3	3-A	5	11	2	8	1	2
5	2-D					0	1
6	4-A	11	13	5	7	1	2
7	5-A	13	14	6	7	2	3
8	6-A	14	15	6	7	3	4
11	3-D					2	3
12	7-A	15	19	3	7	3	4
13	4-D					2	3
14	8-A;5-D	19	21	5	7	2	3
15	6-D					1	2
19	9-A;7-D	21	26	2	7	1	2
20	10-A	26	27	6	7	2	3
21	8-D					1	2
24	11-A	27	28	3	4	2	3
26	12-A;9-D	28	31	2	5	2	3
27	10-D					1	2
28	11-D					0	1
31	12-D					0	0

1.7 POISSON PROCESS AND THE EXPONENTIAL DISTRIBUTION

The most common stochastic queueing models assume that interarrival times and service times obey the exponential distribution or, equivalently, that the arrival rate and service rate follow a Poisson distribution. In this section we will derive the Poisson distribution and show that assuming the number of occurrences in some time interval to be a Poisson random variable is equivalent to assuming the time between successive occurrences to be an exponentially distributed random variable.

We consider an arrival counting process $\{N(t), t \geq 0\}$, where $N(t)$ denotes the total number of arrivals up to time t, with $N(0) = 0$, and which satisfies the following three assumptions:

i. The probability that an arrival occurs between time t and time $t + \Delta t$ is equal to $\lambda \Delta t + o(\Delta t)$. We write this as Pr{arrival occurs between t and $t + \Delta t$} $= \lambda \Delta t + o(\Delta t)$, where λ is a constant independent of $N(t)$, Δt is an incremental element, and $o(\Delta t)$ denotes a quantity that becomes negligible when compared to Δt as $\Delta t \to 0$, that is,

$$\lim_{\Delta t \to 0} \frac{o(\Delta t)}{\Delta t} = 0.$$

ii. Pr{more than one arrival between t and $t + \Delta t$} $= o(\Delta t)$.
iii. The numbers of arrivals in nonoverlapping intervals are statistically independent; that is, the process has independent increments.

We wish to calculate $p_n(t)$, the probability of n arrivals in a time interval of length t, n being an integer ≥ 0. We will do this by first developing differential–difference equations for the arrival process. For $n \geq 1$ we have

$$p_n(t + \Delta t) = \text{Pr}\{n \text{ arrivals in } t \text{ and none in } \Delta t\}$$
$$+ \text{Pr}\{n - 1 \text{ arrivals in } t \text{ and one in } \Delta t\}$$
$$+ \text{Pr}\{n - 2 \text{ arrivals in } t \text{ and two in } \Delta t\} + \cdots$$
$$+ \text{Pr}\{\text{no arrivals in } t \text{ and } n \text{ in } \Delta t\}. \tag{1.6}$$

Using assumptions i, ii, and iii, Equation (1.6) becomes

$$p_n(t + \Delta t) = p_n(t)[1 - \lambda \Delta t - o(\Delta t)] + p_{n-1}(t)[\lambda \Delta t + o(\Delta t)] + o(\Delta t), \tag{1.7}$$

where the last term, $o(\Delta t)$, represents the terms Pr{$n - j$ arrivals in t and j in Δt; $2 \leq j \leq n$}.
For the case $n = 0$, we have

$$p_0(t + \Delta t) = p_0(t)[1 - \lambda \Delta t - o(\Delta t)]. \tag{1.8}$$

Rewriting (1.7) and (1.8) and combining all $o(\Delta t)$ terms, we have

$$p_0(t + \Delta t) - p_0(t) = -\lambda \Delta t \, p_0(t) + o(\Delta t) \tag{1.9}$$

and

$$p_n(t + \Delta t) - p_n(t) = -\lambda \Delta t \, p_n(t) + \lambda \Delta t \, p_{n-1}(t) + o(\Delta t) \quad (n \geq 1). \tag{1.10}$$

We divide (1.9) and (1.10) by Δt, take the limit as $\Delta t \to 0$, and obtain the differential–difference equations

$$\lim_{\Delta t \to 0} \left[\frac{p_0(t + \Delta t) - p_0(t)}{\Delta t} = -\lambda p_0(t) + \frac{o(\Delta t)}{\Delta t} \right],$$

$$\lim_{\Delta t \to 0} \left[\frac{p_n(t + \Delta t) - p_n(t)}{\Delta t} = -\lambda p_n(t) + \lambda p_{n-1}(t) + \frac{o(\Delta t)}{\Delta t} \right], \qquad (n \geq 1),$$

which reduce to

$$\frac{dp_0(t)}{dt} = -\lambda p_0(t) \tag{1.11}$$

and

$$\frac{dp_n(t)}{dt} = -\lambda p_n(t) + \lambda p_{n-1}(t) \qquad (n \geq 1). \tag{1.12}$$

We now have an infinite set of linear, first-order ordinary differential equations to solve. Equation (1.11) clearly has the general solution $p_0(t) = Ce^{-\lambda t}$, where the constant C is easily determined to be equal to 1, since $p_0(0) = 1$. Next, let $n = 1$ in (1.12), and we find that

$$\frac{dp_1(t)}{dt} = -\lambda p_1(t) + \lambda p_0(t),$$

or

$$\frac{dp_1(t)}{dt} + \lambda p_1(t) = \lambda p_0(t) = \lambda e^{-\lambda t}.$$

The solution to this equation is

$$p_1(t) = Ce^{-\lambda t} + \lambda t e^{-\lambda t}.$$

Use of the boundary condition $p_n(0) = 0$ for all $n > 0$ yields $C = 0$ and gives

$$p_1(t) = \lambda t e^{-\lambda t}.$$

Continuing sequentially to $n = 2, 3, \ldots$ in (1.12) and proceeding similarly, we find that

$$p_2(t) = \frac{(\lambda t)^2}{2} e^{-\lambda t}, \qquad p_3(t) = \frac{(\lambda t)^3}{3!} e^{-\lambda t}, \ldots \qquad (1.13)$$

From (1.13), we conjecture that the general formula is

$$p_n(t) = \frac{(\lambda t)^n}{n!} e^{-\lambda t}. \qquad (1.14)$$

It is left as an exercise (see Problem 1.8) to use mathematical induction to verify (1.14), which is the well-known formula for a Poisson probability distribution with mean λt. Thus if we consider the random variable defined as the number of arrivals to a queueing system by time t, this random variable has the Poisson distribution given by (1.14) with a mean of λt arrivals, or a mean arrival rate (arrivals per unit time) of λ.

Poisson processes have a number of interesting additional properties. One of the most important is that the numbers of occurrences in intervals of equal width are identically distributed (stationary increments). In particular, for $t > s$, the difference $N(t) - N(s)$ is identically distributed as $N(t + h) - N(s + h)$, with probability function

$$p_n(t - s) = \frac{[\lambda(t - s)]^n}{n!} e^{-\lambda(t-s)}.$$

This can easily be seen by the following argument. Since the Poisson has independent increments (assumption iii), there is no loss of generality if $N(s)$ and $N(s + h)$ are assumed to be zero. Then if the Poisson derivation is carried out for both $N(t)$ and $N(t + h)$ under assumptions i, ii, and iii, the foregoing formula results for each (see Problem 1.11).

We now show that if the arrival process is Poisson, an associated random variable defined as the time between successive arrivals (interarrival time) follows the exponential distribution. Let T be the random variable "time between successive arrivals"; then

$$\Pr\{T \geq t\} = \Pr\{\text{no arrivals in time } t\} = p_0(t) = e^{-\lambda t}.$$

Therefore, we see that the cumulative distribution function of T can be written as

$$A(t) = \Pr\{T \leq t\} = 1 - e^{-\lambda t},$$

with corresponding density function

$$a(t) = \frac{dA(t)}{dt} = \lambda e^{-\lambda t}.$$

Thus T has the exponential distribution with mean $1/\lambda$. We would intuitively expect the *mean* time between arrivals to be $1/\lambda$ if the *mean* arrival rate is λ. Our analysis substantiates this. It can also be shown that if the interarrival times are independent and have the same exponential distribution, then the arrival rate follows the Poisson distribution, and a proof of this assertion follows.

To begin, let the cumulative distribution function (CDF) of the arrival counting process, $\Pr\{N(t) \le n\}$, be denoted by $P_n(t)$. Then it follows that

$$p_n(t) = \Pr\{N(t) = n\}$$
$$= P_n(t) - P_{n-1}(t).$$

But

$$P_n(t) = \Pr\{(\text{sum of } n + 1 \text{ interarrival times}) > t\}.$$

However, the sum of independent and identically distributed exponential random variables has an Erlang distribution (which is a special type of gamma distribution); hence

$$P_n(t) = \int_t^\infty \frac{\lambda(\lambda x)^n}{n!} e^{-\lambda x} \, dx. \tag{1.15}$$

The transformation of variables $u = x - t$ gives

$$P_n(t) = \int_0^\infty \frac{\lambda^{n+1}(u + t)^n}{n!} e^{-\lambda t} e^{-\lambda u} \, du$$
$$= \int_0^\infty \frac{\lambda^{n+1} e^{-\lambda t} e^{-\lambda u}}{n!} \sum_{i=0}^n u^{n-i} t^i \frac{n!}{(n - i)! i!} \, du,$$

from the binomial theorem. The Σ and $\int$ may be switched to give

$$P_n(t) = \sum_{i=0}^n \frac{\lambda^{n+1} e^{-\lambda t} t^i}{(n - i)! i!} \int_0^\infty e^{-\lambda u} u^{n-i} \, du.$$

But the integral in the above equation is essentially the well-known gamma function and equals $(n - i)!/\lambda^{n-i+1}$. So

$$P_n(t) = \sum_{i=0}^n \frac{(\lambda t)^i e^{-\lambda t}}{i!},$$

which is clearly recognizable as the CDF of the Poisson process.

The Poisson–exponential arrival process derived here is sometimes referred to as completely random arrivals. Although the reader might think that completely random would allude to some sort of haphazard arrival process or a uniform distribution for interarrival times, when encountered in queueing literature it specifically refers to the Poisson-arrival-rate–exponential-interarrival-time pattern. This can be explained in light of the following characteristic of a Poisson process. Given that k arrivals have occurred in an interval $[0, T]$, the k times $\tau_1 < \tau_2 < \cdots < \tau_k$ at which the arrivals occurred are distributed as the order statistics of k uniform random variables on $[0, T]$. Note that it is not the interarrival times, but rather the times at which the arrivals occurred, that are uniformly distributed. This can be proven as follows.

The differential element of the conditional probability density may be written as

$$f_\tau(t|k)\, dt \equiv f(t_1, t_2, \ldots, t_k | k \text{ arrivals in } [0, T])\, dt_1\, dt_2 \cdots dt_k$$

$$\doteq \Pr\{t_1 \le \tau_1 \le t_1 + dt_1, \ldots, t_k \le \tau_k \le t_k + dt_k | k \text{ arrivals in } [0, T]\}.$$

Using the definition of conditional probability gives

$$f_\tau(t|k)\, dt =$$

$$\frac{\Pr\{t_1 \le \tau_1 \le t_1 + dt_1, \ldots, t_k \le \tau_k \le t_k + dt_k \text{ and } k \text{ arrivals in } [0, T]\}}{\Pr\{k \text{ arrivals in } [0, T]\}}$$

The numerator of the right-hand side above can be found by making direct use of the Poisson probability function and its properties, since we wish to find the probability that exactly one event occurs in each of the k time intervals $(t_i, t_i + dt_i)$, and no events occur elsewhere, that is, in $T - dt_1 - dt_2 - \cdots - dt_k$. Therefore, since the probability of k occurrences in a time t is Poisson, we have

$$f_\tau(t|k)\, dt \doteq \frac{\lambda\, dt_1\, e^{-\lambda dt_1} \lambda\, dt_2\, e^{-\lambda dt_2} \cdots \lambda\, dt_k e^{-\lambda dt_k} e^{-\lambda(T - dt_1 - dt_2 - \cdots - dt_k)}}{(\lambda T)^k e^{-\lambda T}/k!}$$

$$= \frac{k!}{T^k}\, dt_1\, dt_2 \cdots dt_k.$$

Hence

$$f_\tau(t_1, t_2, \ldots, t_k | k \text{ arrivals in } [0, T]) = \frac{k!}{T^k}, \tag{1.16}$$

which is identical to the joint density of the order statistics of k random variables uniform on $[0, T]$.

One important consequence of the uniform property of the Poisson process is that the outcomes of random observations of a stochastic process $X(t)$ have the same probabilities as if the scans were taken at Poisson-selected points. When $X(t)$ is a queue, this property is called *PASTA*, for "Poisson arrivals see time averages."

Making similar assumptions to those above for arrivals, one could utilize the same type of process to describe the service pattern. If we change the three assumptions in the beginning of this section slightly by using the word service instead of arrival and condition the probability statements by requiring the system to be nonempty, we obtain a Poisson service rate and an exponential service-time distribution for describing the service pattern. In the following section, we prove an important property of the exponential distribution which aids in a relatively simple analysis of queueing problems when arrival and service patterns exhibit the Poisson–exponential characteristics as derived in this section.

1.8 MARKOVIAN PROPERTY OF THE EXPONENTIAL DISTRIBUTION

We will now prove the Markovian or memoryless property of the exponential distribution. To explain this property in words, suppose service times are exponentially distributed. This property states that the probability that a customer currently in service has t units of remaining service is independent of how long it has already been in service. Thus we wish to prove that

$$\Pr\{T \le t_1 \mid T \ge t_0\} = \Pr\{0 \le T \le t_1 - t_0\}. \qquad (1.17)$$

The proof is relatively straightforward and proceeds as follows. From the definition of conditional probability we have

$$\Pr\{T \le t_1 \mid T \ge t_0\} = \frac{\Pr\{(T \le t_1) \text{ and } (T \ge t_0)\}}{\Pr\{T \ge t_0\}}$$

$$= \frac{e^{-\lambda t_0} - e^{-\lambda t_1}}{e^{-\lambda t_0}}$$

$$= 1 - e^{-\lambda(t_1 - t_0)} = \Pr\{0 \le T \le t_1 - t_0\}.$$

It is also true that the exponential distribution is the only continuous distribution which exhibits this memoryless property. (The only other distribution to exhibit this property is the geometric, which is the discrete analog of the exponential.) The proof of this assertion rests on the fact that the only continuous function solution of the equation

$$g(s + t) = g(s) + g(t)$$

is the linear form

$$g(y) = Cy, \qquad (1.18)$$

here C is an arbitrary constant. This rather intuitive result turns out not to be a trivial matter to prove. However, the proof is well documented in the literature (e.g., see Parzen, 1962), and, for the purposes of this text, this additional detail is not necessary. Then, under the assumption of this result, we proceed as follows. We wish to show that if the memorylessness of (1.17) holds, then it follows that

$$\Pr\{T \leq t\} = F(t) = 1 - e^{Ct}.$$

Now subtract both sides of (1.17) from 1 and denote the complementary CDF by $\tilde{F}$. Thus

$$\tilde{F}(t_1 | T \geq t_0) = \tilde{F}(t_1 - t_0). \qquad (1.19)$$

From the laws of conditional probability, we can write (1.19) as

$$\frac{\tilde{F}(t_1 \text{ and } T \geq t_0)}{\tilde{F}(t_0)} = \frac{\tilde{F}(t_1)}{\tilde{F}(t_0)} = \tilde{F}(t_1 - t_0),$$

or

$$\tilde{F}(t_1) = \tilde{F}(t_0)\tilde{F}(t_1 - t_0).$$

Letting $t = t_1 - t_0$ yields

$$\tilde{F}(t + t_0) = \tilde{F}(t_0)\tilde{F}(t),$$

and when natural logarithms are taken of both sides, it is found that

$$\ln \tilde{F}(t + t_0) = \ln \tilde{F}(t_0) + \ln \tilde{F}(t).$$

It thus follows from (1.18) that

$$\ln \tilde{F}(t) = Ct, \qquad \text{or} \qquad \tilde{F}(t) = e^{Ct}.$$

Thus $F(t) = 1 - e^{Ct}$.

There are many possible and well-known generalizations of the Poisson–exponential process, most of which have rather obvious applications to queues and are taken up in greater detail later in the text. The most obvious of the generalizations is a truncation of the infinite domain, that is, omission of some of the nonnegative integers from the range of possible values. This is done whenever the removed values are either theoretically meaningless or practically unobservable. An example of this occurs in the $M/M/c/c$ queue and gives rise to *Erlang's loss* or *B formula*. The only change to be made here, with caution, is the rescaling of the respective probabilities because the Poisson terms no longer sum to one.

Another generalization arises if we go back to the axiomatic derivation and no longer permit λ to be a constant independent of time. If instead the functional relationship is denoted by $\lambda(t)$, then the probability of one occurrence in a small time increment is rewritten as $\lambda(t)\,\Delta t + o(\Delta t)$, and it turns out that the resulting distribution of the counting process is the so-called nonhomogeneous Poisson given by

$$p_n(t) = e^{-m(t)}\frac{[m(t)]^n}{n!}, \quad m(t) = \int_0^t \lambda(s)\,ds \qquad (n \ge 0).$$

A third, and very common, generalization occurs when one relaxes the Poisson assumption that more than one occurrence in Δt has probability $o(\Delta t)$. Instead, let

$$\Pr\{i \text{ occurrences in } (t, t + \Delta t)\} = \lambda_i\,\Delta t + o(\Delta t) \qquad (i = 1, 2, \ldots, n)$$

with

$$\sum_{i=1}^{n} \lambda_i = \lambda.$$

It should be immediately clear now that this is equivalent to allowing the event of i simultaneous occurrences in Δt with probability $\lambda_i\Delta t + o(\Delta t)$, and each individual stream of occurrences of the same batch size (i) itself forms a Poisson process. If these substreams are denoted by $N_i(t)$, then it should also be clear that the total process is $N(t) = \Sigma_i\, iN_i(t)$, with probability function

$$p_n(t) = \Pr\{n \text{ occurrences in } [0, t]\}$$

$$= \sum_{i=0}^{n} e^{-\lambda t}\frac{(\lambda t)^i}{i!}c_n^{(i)} \qquad (c_0^{(0)} \equiv 1),$$

where $c_n^{(i)}$ is the probability that i occurrences give a grand total of n (i.e., the probability associated with the i-fold convolution of the batch-size probabilities $\{\lambda_i/\lambda\}$). The process $N(t)$ is known as the *multiple Poisson* and clearly also has the stationary and independent-increment properties.

The foregoing probability function $p_n(t)$ has an alternative derivation as a compound distribution, since it admits of a random sum interpretation as follows. Consider the process $N(t)$ to be defined by

$$N(t) = \sum_{n=1}^{M(t)} Y_n \,,$$

where $M(t)$ is a regular Poisson process and $\{Y_n\}$ is a sequence of independent and identically distributed (IID) discrete random variables with probabilities

$$c_j = \Pr\{Y_n = j\} \qquad \text{(for all } n)$$
$$= \frac{\lambda_j}{\Sigma_i \lambda_i} \,;$$

that is, occurrences happen according to a Poisson process $\{M(t)\}$ but are not necessarily singles in that their size is j with probability c_j. Then, by the laws of probability,

$$\Pr\{N(t) = m\} = \sum_{k=0}^{m} \left[\Pr\{M(t) = k\} \cdot \Pr\left\{ \sum_{n=1}^{k} Y_n = m \right\} \right]$$
$$= \sum_{k=0}^{m} e^{-\lambda t} \frac{(\lambda t)^k}{k!} c_m^{(k)} \,.$$

In other words, the compound approach looks at the process as one Poisson stream with a randomly varying batch size, whereas the multiple approach looks at the process as the sum of Poisson streams, each with a constant batch size.

Poisson–exponential streams are special cases of a larger class of problems called *renewal processes*. An ordinary renewal process arises from any sequence of nonnegative IID random variables. Many of the properties that we have derived for Poisson–exponential sequences can also be derived in a renewal context. Some other specific results will be needed later in the text, particularly when the input is an arbitrary G stream, but these will be derived as needed by direct probabilistic arguments. The reader particularly interested in renewal theory is referred to Chapter 3 of Ross (1996), Chapter 9 of Çinlar (1975), or Chapter 5 of Heyman and Sobel (1982).

In subsequent chapters of the book, the Poisson process and its associated characteristics will play a key role in the development of many queueing models. This is true not only because of the many mathematically agreeable properties of the Poisson–exponential, but also because many real-life situations in fact do obey the appropriate requirements. Though it may seem at first glance that the demands of exponential interoccurrence times are rather stringent, this is not the case.

A strong argument in favor of exponential inputs is the one that often occurs in the context of reliability. It is the result of the well-known fact that the limit of a binomial distribution is Poisson, which says that if a mechanism consists of many parts, each of which can fail with only a small probability, and if the failures for the different parts are mutually independent and identical, then the total flow of failures can be considered Poisson. Another view that favors the exponential comes from the theory of extreme values. Here the exponential appears quite frequently as the limiting distribution of the (normalized) first-order statistic of random samples drawn from continuous populations (see Problem 1.19 for one such example).

There is also an additional argument that comes out of information theory. It is that the exponential distribution is the one that provides the least information, where information content or negative entropy of the distribution $f(x)$ is defined as $\int f(x) \log f(x)\, dx$. It can easily be shown that the exponential distribution has least information or highest entropy, and is therefore the most random law that can be used, and thus certainly provides a reasonably conservative approach. We treat the topic of choosing the appropriate probability model in more detail in Chapters 6 and 7.

1.9   STOCHASTIC PROCESSES AND MARKOV CHAINS

A stochastic process is the mathematical abstraction of an empirical process whose development is governed by probabilistic laws (the Poisson process is one example). From the point of view of the mathematical theory of probability, a stochastic process is best defined as a family of random variables, $\{X(t), t \in T\}$, defined over some index set or parameter space T. The set T is sometimes also called the time range, and $X(t)$ denotes the state of the process at time t. Depending upon the nature of the time range, the process is classified as a discrete-parameter or continuous-parameter process as follows:

1. If T is a countable sequence, for example, $T = \{0, \pm 1, \pm 2, \ldots\}$ or $T = \{0, 1, 2, \ldots\}$, then the stochastic process $\{X(t), t \in T\}$ is said to be a discrete-parameter process defined on the index set T.

2. If T is an interval or an algebraic combination of intervals, for example, $T = \{t : -\infty < t < +\infty\}$ or $T = \{t : 0 < t < +\infty\}$, then the stochastic process $\{X(t), t \in T\}$ is called a continuous-parameter process defined on the index set T.

1.9.1   Markov Process

A discrete-parameter stochastic process $\{X(t), t = 0, 1, 2, \ldots)$ or a continuous-parameter stochastic process $\{X(t), t > 0)$ is said to be a Markov process if, for any set of n time points $t_1 < t_2 < \cdots < t_n$ in the index set or time range of the process, the conditional distribution of $X(t_n)$, given the values of $X(t_1), X(t_2), X(t_3), \ldots, X(t_{n-1})$, depends only on the immediately preceding value, $X(t_{n-1})$; more precisely, for any real numbers $x_1, x_2, \ldots, x_n$,

$$\Pr\{X(t_n) \le x_n \mid X(t_1) = x_1, \ldots, X(t_{n-1}) = x_{n-1}\}$$
$$= \Pr\{X(t_n) \le x_n \mid X(t_{n-1}) = x_{n-1}\}.$$

In nonmathematical language one says that, given the "present" condition of the process, the "future" is independent of the "past," and the process is thus "memoryless."

Markov processes are classified according to:

1. the nature of the index set of the process (whether discrete or continuous parameter), and
2. the nature of state space of the process (whether discrete or continuous parameter).

A real number x is said to be a state of a stochastic process $\{X(t), t \in T\}$ if there exists a time point t in T such that the $\Pr\{x - h < X(t) < x + h\}$ is positive for every $h > 0$. The set of possible states constitutes the state space of the process. If the state space is discrete, the Markov process is generally called a Markov chain, although some authors reserve the term "chain" for only those Markov processes with both discrete state space and discrete parameter space. In this text, we shall say that a discrete-parameter Markov process with discrete state space is a plain Markov chain, and that a continuous-parameter Markov process with discrete state space is a continuous-time Markov chain. (Multivariate extensions can be nicely formulated for vector states x.)

A Markov chain is finite if the state space is finite; otherwise it is a denumerable or infinite Markov chain. Since a discrete-parameter process is observed at a countable number of time points, let the successive observations be denoted by $X_0, X_1, X_2, \ldots, X_n, \ldots$ where X_n is the random variable whose value represents the state of the system at the nth time point. An arbitrary sequence of random variables $\{X_n\}$ is thus a Markov chain if each random variable X_n is discrete and the following holds: for any integer $m > 2$ and any set of m points $n_1 < n_2 < \cdots < n_m$, the conditional distribution of X_{n_m}, given values of $X_{n_1}, X_{n_2}, \ldots, X_{n_{m-1}}$ depends only on $X_{n_{m-1}}$, the immediately preceding value; that is,

$$\Pr\{X_{n_m} = x_{n_m} \mid X_1 = x_{n_1}, \ldots, X_{n_{m-1}} = x_{n_{m-1}}\}$$
$$= \Pr\{X_{n_m} = x_{n_m} \mid X_{n_{m-1}} = x_{n_{m-1}}\}.$$

When a Markov process has a continuous state space and discrete parameter space, we call it a discrete-parameter Markov process. If both the state space and parameter space are continuous, it is called a continuous-

Table 1.5   Classification of Markov Processes

State Space	Type of Parameter	
	Discrete	Continuous
Discrete	(Discrete-parameter) Markov chain	Continuous-parameter Markov chain
Continuous	Discrete-parameter Markov process	Continuous-parameter Markov process

parameter Markov process. Table 1.5 summarizes our classification scheme for Markov processes.

An important generalization of the Markov chain which is very useful in queueing is the semi-Markov process (SMP) or Markov renewal process (MRP). The state transitions in an SMP form a discrete Markov chain, but the times between successive transitions are random variables. If these random variables are distributed exponentially for a continuous-parameter case or geometrically for a discrete-parameter case, with mean dependent on the current state only, the SMP reduces to a Markov process because of the memoryless property of these random variables.

1.9.2   Discrete-Parameter Markov Chains

Consider a sequence of random variables, $\{X_n, n = 0, 1, 2, \ldots | X_n = 0, 1, 2, \ldots\}$, which forms a Markov chain with discrete parameter space; that is, for all n,

$$\Pr\{X_n = j | X_0 = i_0, X_1 = i_1, \ldots, X_{n-1} = i_{n-1}\} = \Pr\{X_n = j | X_{n-1} = i_{n-1}\}.$$

If the value of the random variable X_n is j, then the system is said to be in state j after n steps or transitions. The conditional probabilities $\Pr\{X_n = j | X_{n-1} = i\}$ are called the *single-step transition probabilities* or just *transition probabilities*. If these probabilities are independent of n, then the chain is said to be *homogeneous* and the probabilities $\Pr\{X_n = j | X_{n-1} = i\}$ can be written as p_{ij}. The matrix formed by placing p_{ij} in the (i, j) location is known as the *transition matrix* or *chain matrix* (call it P). For homogeneous chains, the m-step transition probabilities

$$\Pr\{X_{n+m} = j | X_n = i\} = p_{ij}^{(m)}$$

are also independent of n. The unconditional probability of state j at the nth trial will be written as

$$\Pr\{X_n = j\} = \pi_j^{(n)},$$

so that the initial distribution is given by $\pi_j^{(0)}$.

From the basic laws of probability, one can easily show that the matrix formed by the elements $\{p_{ij}^{(m)}\}$, say $P^{(m)}$, can be found simply by multiplying $P^{(m-k)}$ by $P^{(k)}$ for any value of k, $0 < k < m$. This is the matrix equivalent of the well-known Chapman–Kolmogorov (CK) equations for this Markov process, namely,

$$p_{ij}^{(m)} = \sum_r p_{ir}^{(m-k)} p_{rj}^{(k)} \qquad (0 < k < m)$$

or, in matrix notation,

$$P^{(m)} = P^{(m-k)} P^{(k)}. \qquad (1.20)$$

Letting $k = m - 1$ in (1.20) gives

$$P^{(m)} = P \cdot P^{(m-1)}, \qquad (1.21)$$

and, continuing this procedure recursively, we can easily show that

$$P^{(m)} = P \cdot P \cdots P = P^m.$$

Hence $P^{(m)}$ can be obtained by multiplying the matrix P by itself m times.

Often one is interested in the probabilities of being in state j after m transitions regardless of the starting state. If the vector $\boldsymbol{\pi}^{(m)}$ is created from the probabilities $\{\pi^{(m)}\}$, then

$$\boldsymbol{\pi}^{(m)} = \boldsymbol{\pi}^{(m-1)} P, \qquad (1.22)$$

which, when used recursively, gives

$$\boldsymbol{\pi}^{(m)} = \boldsymbol{\pi}^{(0)} P^m \qquad (1.23)$$

for the initial state vector $\boldsymbol{\pi}^{(0)}$.

Defining the matrix Q as $P - I$, where I is the identity matrix, and subtracting $\boldsymbol{\pi}^{(m-1)}$ from both sides of (1.22) gives

$$\boldsymbol{\pi}^{(m)} - \boldsymbol{\pi}^{(m-1)} = \boldsymbol{\pi}^{(m-1)} Q. \qquad (1.24)$$

Note that P is always a stochastic matrix (i.e., its rows sum to one), while Q has rows that sum to zero.

1.9.3 Continuous-Parameter Markov Chains

We now consider a continuous-parameter Markov chain $\{X(t), t \in T)$ for $T = \{t : 0 \le t < \infty\}$. Consider any times $s > t > u \ge 0$ and states i and j; then

$$p_{ij}(u, s) = \sum_r p_{ir}(u, t)p_{rj}(t, s), \tag{1.25}$$

where $p_{ij}(u, s)$ is the probability of moving from state i to j in the time beginning at u and ending at s, and the summation is over all states of the chain. This result should be fairly intuitive and says that the chain can reach state j at time s by starting from state i at time u and stopping off at time t at any other possible state r. This is the CK equation for the continuous process [analogous to (1.20) for the discrete process]. In matrix notation, Equation (1.25) becomes

$$P(u, s) = P(u, t)P(t, s).$$

Letting $u = 0$ and $s = t + \Delta t$ in (1.25) gives

$$p_{ij}(0, t + \Delta t) = \sum_r p_{ir}(0, t)p_{rj}(t, t + \Delta t).$$

Defining $p_i(0)$ as the probability that the chain starts in state i at time 0 and $p_j(t)$ as the unconditional probability that the chain is in state j at time t regardless of starting state, we multiply the above equation by $p_i(0)$ and sum over all states i to get

$$\sum_i p_i(0)p_{ij}(0, t + \Delta t) = \sum_r \sum_i p_{ir}(0, t)p_i(0)p_{rj}(t, t + \Delta t),$$

or

$$p_j(t + \Delta t) = \sum_r p_r(t)p_{rj}(t, t + \Delta t). \tag{1.26}$$

For the Poisson process treated earlier,

$$p_{rj}(t, t + \Delta t) = \begin{cases} \lambda \Delta t + o(\Delta t) & (r = j - 1, j \ge 1), \\ 1 - \lambda \Delta t + o(\Delta t) & (r = j) \\ o(\Delta t) & (\text{elsewhere}). \end{cases}$$

Substituting this into (1.26) gives

$$p_j(t + \Delta t) = [\lambda \Delta t + o(\Delta t)]p_{j-1}(t)$$
$$+ [1 - \lambda \Delta t + o(\Delta t)]p_j(t) + o(\Delta t) \qquad (j \ge 1).$$

which is Equation (1.7). What we did in Section 1.7 was to derive the CK equation for the Poisson process from scratch, appealing to the same basic probability arguments which yield the general CK equations (1.25) and (1.26).

There is an additional theory which takes one from the CK equation to two differential equations, which are called the forward and backward equations. If the transition probability functions $p(u, s)$ of the chain have the additional properties that there exist continuous functions $q_i(t)$ and $q_{ij}(t)$ such that

$$\Pr\{\text{a change of state in } (t, t + \Delta t)\} = 1 - p_{ii}(t, t + \Delta t) = q_i(t)\, \Delta t + o(\Delta t),$$
(1.27a)

$$p_{ij}(t, t + \Delta t) = q_{ij}(t)\, \Delta t + o(\Delta t),$$
(1.27b)

then under some mild regularity conditions (e.g., see Ross, 1996), Equation (1.25) leads to (see Problem 1.15)

$$\frac{\partial}{\partial t} p_{ij}(u, t) = -q_j(t) p_{ij}(u, t) + \sum_{r \neq j} p_{ir}(u, t) q_{rj}(t)$$
(1.28a)

and

$$\frac{\partial}{\partial u} p_{ij}(u, t) = q_i(u) p_{ij}(u, t) - \sum_{r \neq i} q_{ir}(u) p_{rj}(u, t)$$
(1.28b)

These two differential equations are known, respectively, as Kolmogorov's forward and backward equations. Consider Equation (1.28a) further. Let $u = 0$, and assume a homogeneous process so that $q_i(t) = q_i$ and $q_{ij}(t) = q_{ij}$ for all t. We then obtain

$$\frac{dp_{ij}(0, t)}{dt} = -q_j p_{ij}(0, t) + \sum_{r \neq j} p_{ir}(0, t) q_{rj}.$$

Multiplying both sides of the above equation by $p_i(0)$ and summing over all i yields

$$\frac{dp_j(t)}{dt} = -q_j p_j(t) + \sum_{r \neq j} p_r(t) q_{rj}.$$

In matrix notation, this equation can be written as

$$\boldsymbol{p}'(t) = \boldsymbol{p}(t)\boldsymbol{Q},$$
(1.29)

where $p(t)$ is the vector $(p_0(t), p_1(t), p_2(t), \ldots)$, $p'(t)$ is the vector of its derivatives, and

$$
Q = \begin{pmatrix}
-q_0 & q_{01} & q_{02} & q_{03} \cdots \\
q_{10} & -q_1 & q_{12} & q_{13} \cdots \\
q_{20} & q_{21} & -q_2 & q_{23} \cdots \\
\cdot & \cdot & \cdot & \cdot \\
\cdot & \cdot & \cdot & \cdot \\
\cdot & \cdot & \cdot & \cdot
\end{pmatrix}.
$$

Note from (1.27) that $q_i = \Sigma_{j \neq i} q_{ij}$, since

$$
\sum_j p_{ij}(t, t + \Delta t) = 1,
$$

which implies that

$$
1 - q_i \, \Delta t + o(\Delta t) + \sum_{j \neq i} [q_{ij} \, \Delta t + o(\Delta t)] = 1,
$$

or

$$
-q_i \, \Delta t + o(\Delta t) = -\sum_{j \neq i} [q_{ij} \, \Delta t + o(\Delta t)],
$$

so that $q_i = \Sigma_{j \neq i} q_{ij}$.

The matrix Q can also be looked at as

$$
Q = \lim_{\Delta t \to 0} \frac{P(t, t + \Delta t) - I}{\Delta t},
$$

where

$$
P(t, t + \Delta t) = \{p_{ij}(t, t + \Delta t)\}.
$$

Thus Q plays a similar role for continuous-parameter Markov chains to the one $Q = P - I$ played for discrete-parameter Markov chains. The matrix Q is often called the *intensity matrix* (or *infinitesimal generator*) of the continuous-parameter Markov chain.

Again referring to the Poisson process, we can use (1.29) to get (1.12) directly by noting that $q_j = \lambda$, $q_{rj} = \lambda$ for $r = j - 1$ and $j \geq 1$, and $q_{ij} = 0$ elsewhere, so that

$$
\frac{dp_j(t)}{dt} = -\lambda p_j(t) + \lambda p_{j-1}(t).
$$

Equation (1.11) can be obtained similarly.

Since the state space of any queue is composed of nonnegative integers (representing the number of customers present), a large percentage of queueing problems can be categorized as continuous-parameter (time) Markov chains. Many such models have the additional *birth–death* property that the net change across an infinitesimal time interval can never be other than -1, 0, or $+1$, and that

$$\Pr\{\text{increase } n \to n + 1 \text{ in } (t, t + \Delta t)\} = \lambda_n \, \Delta t + o(\Delta t) \qquad (n \ge 0),$$

$$\Pr\{\text{decrease } n \to n - 1 \text{ in } (t, t + \Delta t)\} = \mu_n \, \Delta t + o(\Delta t) \qquad (n \ge 1).$$

Hence

$$\Pr\{\text{no change in } (t, t + \Delta t)\} = 1 - (\lambda_n + \mu_n) \, \Delta t + o(\Delta t)$$

and

$$
\begin{aligned}
q_{n,n+1} &= \lambda_n, \\
q_{n,n-1} &= \mu_n & (\mu_n \ne 0), \\
q_{rj} &= 0 & \text{(elsewhere)}, \\
q_n &= \lambda_n + \mu_n & (q_0 = \lambda_0),
\end{aligned}
$$

so that it is possible to get either a forward or a backward Kolmogorov equation.

Substituting for q_i and q_{ij}, the matrix Q is

$$
Q = \begin{pmatrix}
-\lambda_0 & \lambda_0 & 0 & 0 \cdots \\
\mu_1 & -(\lambda_1 + \mu_1) & \lambda_1 & 0 \cdots \\
0 & \mu_2 & -(\lambda_2 + \mu_2) & \lambda_2 \cdots \\
\cdot & \cdot & \cdot & \cdot \\
\cdot & \cdot & \cdot & \cdot \\
\cdot & \cdot & \cdot & \cdot
\end{pmatrix},
$$

and using the Kolmogorov forward Equation (1.29), we obtain a set of differential–difference equations for the birth–death process, namely,

$$\frac{dp_j(t)}{dt} = -(\lambda_j + \mu_j)p_j(t) + \lambda_{j-1}p_{j-1}(t) + \mu_{j+1}p_{j+1}(t) \qquad (j \ge 1),$$

$$\frac{dp_0(t)}{dt} = -\lambda_0 p_0(t) + \mu_1 p_1(t)$$

$$(1.30)$$

Note that setting $\mu = 0$ and $\lambda_i = \lambda$ for all i also gives the Poisson equations (1.11) and (1.12) of the previous section. The Poisson process is thus often called the *pure birth* process (see Problem 1.16).

1.9.4 Imbedded Markov Chains

In many of the situations in this text requiring the use of a continuous-parameter queueing model, we can often get quite satisfactory results by instead looking at the state of the queue only at certain selected times, leading to an *imbedded* discrete-parameter Markov chain. To illustrate this, consider the birth–death process *only* at transition times; that is, we create an imbedded stochastic process which turns out to be a discrete-parameter Markov chain. The transition probability matrix $P = \{p_{ij}\}$, where p_{ij} is the probability that, given the process is in state i and a transition occurs, it goes next to state j, can be shown to be (see Problem 1.18)

$$p_{ij} = \begin{cases} \dfrac{\lambda_i}{\lambda_i + \mu_i} & (j = i + 1, i \geq 1), \\[2mm] \dfrac{\mu_i}{\lambda_i + \mu_i} & (j = i - 1, i \geq 1), \\[2mm] 1 & (i = 0, j = 1), \\[1mm] 0 & \text{otherwise.} \end{cases} \qquad (1.31)$$

Another way to view a continuous-time Markov chain is as a process which traverses from state to state in continuous time, the holding times being exponential (required to satisfy the Markov property) random variables with mean $1/q_i$ [in the birth–death example, $1/q_i = 1/(\lambda_i + \mu_i)$] for holding in state i. At times of transition, we have a discrete-parameter Markov chain with transition probabilities $p_{ij} = q_{ij}/q_i$ [in the birth–death example, p_{ij} is given above by (1.31)].

1.9.5 Long-Run Behavior of Markov Processes

We are often interested in the behavior of a Markov process after a long period of time, particularly in whether its behavior "settles down" probabilistically. We discuss three related concepts having to do with long-run behavior, namely, *limiting distributions*, *stationary distributions*, and *ergodicity*.

Consider a discrete-parameter Markov chain, and suppose that

$$\lim_{m \to \infty} p_{ij}^{(m)} = \pi_j \qquad \text{(for all } i\text{)};$$

that is, after a long time, the probability that the process is in state j given that it started in state i is independent of the starting state i. This means that P^m approaches a limit as m goes to infinity, namely, that all rows of P^m become identical. We call the $\{\pi_j\}$ the *limiting* or *steady-state* probabilities of the Markov chain.

Consider now the unconditional state probabilities after m steps, as given by $\pi^{(m)} = \pi^{(0)} P^{(m)}$; that is,

$$\pi_j^{(m)} = \sum_i \pi_i^{(0)} p_{ij}^{(m)} \; .$$

Then

$$\lim_{m\to\infty} \pi_j^{(m)} = \lim_{m\to\infty} \sum_i \pi_i^{(0)} p_{ij}^{(m)} = \sum_i \pi_i^{(0)} \lim_{m\to\infty} p_{ij}^{(m)}$$

$$= \sum_i \pi_i^{(0)} \pi_j = \pi_j \sum_i \pi_i^{(0)} = \pi_j \, ,$$

and hence $\pi_j^{(m)}$ goes to the same limit as $p_{ij}^{(m)}$ and is independent of starting-state probabilities and the time parameter m. When these unconditional limiting probabilities exist, they can be found as follows. From (1.22),

$$\lim_{m\to\infty} \boldsymbol{\pi}^{(m)} = \lim_{m\to\infty} \boldsymbol{\pi}^{(m-1)} \boldsymbol{P}.$$

Letting $\boldsymbol{\pi} = (\pi_0, \pi_1, \dots)$ represent the limiting probability vector, we have

$$\lim_{m\to\infty} \boldsymbol{\pi}^{(m)} = \lim_{m\to\infty} \boldsymbol{\pi}^{(m-1)} = \boldsymbol{\pi},$$

so that

$$\boxed{\boldsymbol{\pi} = \boldsymbol{\pi}\boldsymbol{P}, \quad \text{or} \quad \boldsymbol{0} = \boldsymbol{\pi}\boldsymbol{Q}.} \tag{1.32}$$

From this, together with the *boundary condition* that $\Sigma_j \pi_j = 1$, we can obtain the $\{\pi_j\}$. These well-known equations are called the *stationary equations* of the Markov chain, and their solution is called the *stationary distribution*. The equations (1.32) will play a major role in the solution to some of the more advanced queueing models treated in Chapter 5. Note that the boundary condition may be written in vector notation as $\boldsymbol{\pi}\boldsymbol{e} = 1$, for $\boldsymbol{e}$ a column vector with all elements equal to one.

It is possible in some cases to get solutions to (1.32) even when no limiting distribution exists. Thus when a limiting distribution exists, this implies a solution to (1.32), and the resulting stationary distribution is the limiting distribution. But the converse is not true; that is, a solution to (1.32) does not imply the existence of a limiting distribution. We illustrate this with the following examples.

Example 1.1
Consider a degenerate discrete-parameter stochastic process which sequentially alternates between two states, -1 and $+1$. If we call -1 state 0 and

+1 state 1, the transition probability matrix is

$$P = \begin{pmatrix} 0 & 1 \\ 1 & 0 \end{pmatrix}.$$

We can solve Equation (1.32) and get the stationary distribution vector by

$$(\pi_0, \pi_1) = (\pi_0, \pi_1)\begin{pmatrix} 0 & 1 \\ 1 & 0 \end{pmatrix},$$

which gives $\pi_0 = \pi_1$. Using the fact that $\pi_0 + \pi_1 = 1$ (for a valid probability distribution), we have the stationary probability distribution $\pi_0 = \pi_1 = \frac{1}{2}$. But is this also the limiting distribution? Intuitively we would think not, since the process keeps alternating forever and at any time in the future, the probability that it will be found in a particular state (say -1) is either one or zero, depending on the particular value of m chosen. This is easily verified by successive multiplication of the matrix P, yielding

$$P^{(m)} = \begin{cases} \begin{pmatrix} 1 & 0 \\ 0 & 1 \end{pmatrix}, & m \text{ even,} \\ \\ \begin{pmatrix} 0 & 1 \\ 1 & 0 \end{pmatrix}, & m \text{ odd.} \end{cases}$$

Thus even though a solution to $\pi = \pi P$ exists, there is no $\lim_{m \to \infty} p_{ij}^{(m)}$. Furthermore, $\lim_{m \to \infty} p_{ij}^{(m)}$ does not exist unless $\pi^{(0)} = (\frac{1}{2}, \frac{1}{2})$, that is, unless the initial probability distribution is set equal to the stationary distribution. For this $\pi^{(0)}$ (and only this $\pi^{(0)}$),

$$\pi^{(1)} = (\tfrac{1}{2}, \tfrac{1}{2})\begin{pmatrix} 0 & 1 \\ 1 & 0 \end{pmatrix} = (\tfrac{1}{2}, \tfrac{1}{2}),$$

and hence

$$\pi^{(m)} = (\tfrac{1}{2}, \tfrac{1}{2}) \qquad \text{for all } m.$$

What we have done by using $\pi^{(0)} = (\frac{1}{2}, \frac{1}{2})$, the stationary solution, in the above example is to make this stochastic process strictly stationary. A strictly stationary stochastic process is defined as follows: for all k and h, the joint probability distribution of $X(t_1)$, $X(t_2)$, ..., $X(t_k)$ (called a *finite-dimensional distribution* of order k) is equal to the joint probability distribution of $X(t_1 + h)$, $X(t_2 + h)$, ..., $X(t_k + h)$. Note that stationary processes therefore possess time-independent distribution functions. The solution to $\pi = \pi P$ does not imply strict stationarity, but strict stationarity does imply that $\pi^{(m)}$ is time-independent. The process in Example 1.1 is not in general strictly stationary, but can be made so by using an initial probability vector equal to the stationary vector $(\frac{1}{2}, \frac{1}{2})$. While strict stationarity guarantees the time independence of $\pi^{(m)}$, the process still does not possess a steady state, since $p_{ij}^{(m)}$ never becomes independent of the starting state i or the time parameter m.

Example 1.2
Consider now a discrete-parameter Markov chain similar to the one of Example 1.1 but with two possible states, 0 and 1, and transition probability matrix

$$P = \begin{pmatrix} \frac{1}{2} & \frac{1}{2} \\ \frac{1}{2} & \frac{1}{2} \end{pmatrix}.$$

Note that at each transition, there is a probability $\frac{1}{2}$ of staying in the current state and a probability $\frac{1}{2}$ that the chain will change state. Again, it is easy to get the stationary distribution by $\pi = \pi P$ and $\pi e = 1$, which yields $\pi_0 = \pi_1 = \frac{1}{2}$.

Now let us see if a limiting distribution exists. Successive multiplication of P yields

$$P^{(m)} = \begin{pmatrix} \frac{1}{2} & \frac{1}{2} \\ \frac{1}{2} & \frac{1}{2} \end{pmatrix} \qquad \text{for all } m.$$

Thus $\pi = (\frac{1}{2}, \frac{1}{2})$ is the limiting (steady-state) distribution vector, namely,

$$\lim_{m \to \infty} p_{i0}^{(m)} = \tfrac{1}{2} \qquad (i = 0, 1),$$

$$\lim_{m \to \infty} p_{i1}^{(m)} = \tfrac{1}{2} \qquad (i = 0, 1).$$

This process is also not, in general, stationary unless $\pi^{(0)} = (\frac{1}{2}, \frac{1}{2})$, but it does, as we have shown, possess a steady state, since $\lim_{m \to \infty} p_{ij}^{(m)} = \pi_j$, regardless of which $\pi^{(0)}$ is used.

Let us complicate the example slightly by changing P to

$$P = \begin{pmatrix} \frac{1}{3} & \frac{2}{3} \\ \frac{2}{3} & \frac{1}{3} \end{pmatrix}.$$

The stationary solution is still $\boldsymbol{\pi} = (\frac{1}{2}, \frac{1}{2})$, which is easy to verify by solving $\boldsymbol{\pi} = \boldsymbol{\pi}P$ and $\boldsymbol{\pi}e = 1$. We must now see whether $\boldsymbol{P}^{(m)}$ goes to a limit; that is, whether $\lim_{m \to \infty} p_{ij}^{(m)}$ exists. Successive multiplication of $\boldsymbol{P}$ here by itself yields

$$\boldsymbol{P} \cdot \boldsymbol{P} = \boldsymbol{P}^2 = \begin{pmatrix} \frac{1}{3} & \frac{2}{3} \\ \frac{2}{3} & \frac{1}{3} \end{pmatrix} \begin{pmatrix} \frac{1}{3} & \frac{2}{3} \\ \frac{2}{3} & \frac{1}{3} \end{pmatrix} = \begin{pmatrix} \frac{5}{9} & \frac{4}{9} \\ \frac{4}{9} & \frac{5}{9} \end{pmatrix}.$$

Now

$$\boldsymbol{P} \cdot \boldsymbol{P}^2 = \boldsymbol{P}^3 = \begin{pmatrix} \frac{1}{3} & \frac{2}{3} \\ \frac{2}{3} & \frac{1}{3} \end{pmatrix} \begin{pmatrix} \frac{5}{9} & \frac{4}{9} \\ \frac{4}{9} & \frac{5}{9} \end{pmatrix} = \begin{pmatrix} \frac{13}{27} & \frac{14}{27} \\ \frac{14}{27} & \frac{13}{27} \end{pmatrix},$$

and continuing on with this procedure, we see that $\boldsymbol{P}^{(m)}$ converges to a matrix all of whose entries are $\frac{1}{2}$. Thus, the stationary distribution is indeed the steady-state distribution. Once again, we can make the process completely stationary by letting $\boldsymbol{\pi}^{(0)} = (\frac{1}{2}, \frac{1}{2})$, since then $\boldsymbol{\pi}^{(0)}\boldsymbol{P} = (\frac{1}{2}, \frac{1}{2})$ and $\boldsymbol{\pi}^{(m)} = (\frac{1}{2}, \frac{1}{2})$ for all m. If we were to use any other $\boldsymbol{\pi}^{(0)}$, we would find that $\boldsymbol{\pi}^{(m)}$ equals $(\frac{1}{2}, \frac{1}{2})$ only in the limit.

For continuous-parameter processes, the stationary solution can be obtained from

$$0 = pQ, \tag{1.33}$$

where Q is the intensity matrix as defined in Equation (1.29). The same concepts of stationarity and steady state apply for the continuous-parameter case, with t replacing m in the limiting process. Direct determination of the existence of a steady-state solution is more difficult here, since it would involve obtaining the solution to the differential equations of (1.29) and then taking the $\lim_{t \to \infty} p'(t)$. If it exists, it will, of course, equal the p obtained from (1.33). This is a considerably more difficult task than successive multiplication of the transition probability matrix P required for the discrete-parameter case.

At the end of this section, we shall present some theorems that will allow us to determine the existence of a limiting distribution. Once we know a limiting distribution exists, it can be obtained from $\boldsymbol{\pi} = \boldsymbol{\pi}P$ or $0 = pQ$, and the respective boundary conditions $\boldsymbol{\pi}e = 1$ or $pe = 1$.

1.9.6 Ergodicity

Closely associated with the concepts of limiting and stationary distributions is the idea of *ergodicity*, which has to do with the information contained in one infinitely long sample path of a process (e.g., see Papoulis, 1991).

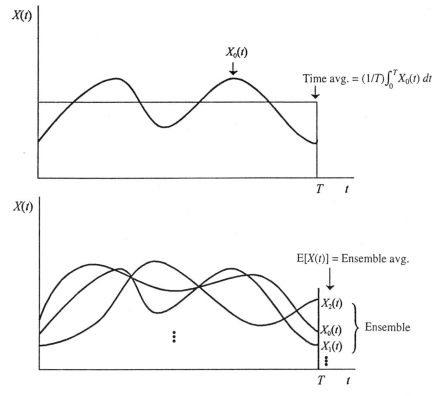

Fig. 1.7 Ergodicity.

Ergodicity is important in that it deals with the problem of determining measures of a stochastic process $X(t)$ from a single realization, as is often done in analyzing simulation output. That is to say, $X(t)$ is ergodic in the most general sense if, with probability one, all its "measures" can be determined or well approximated from a single realization, $x_0(t)$, of the process. Since statistical measures of the process are usually expressed as time averages, this is often stated as follows: $X(t)$ is ergodic if time averages equal ensemble averages, that is, expected values (see Figure 1.7). One may not always be interested in all, but sometimes only in certain measures (or moments) of a process. We can then define ergodicity with respect to these moments, and a process might thus be ergodic for certain moments, but not for others. However, in queueing theory, we are typically interested in fully ergodic processes, that is, processes which are ergodic with respect to all moments.

When dealing with stationary processes, statistical averages are time-independent; hence concern with respect to ergodicity centers only on convergence of time averages (e.g., see Karlin and Taylor, 1975, pp. 474 ff., or Heyman and Sobel, 1982, pp. 366 ff.). In queueing theory, the processes of interest are in general not stationary, and thus our interest in ergodicity involves the convergence of both time and ensemble averages.

Mathematically now, the time average of the square of a realization of a process, for example, would be written as

$$\overline{x_T^2} = \frac{1}{T} \int_0^T [x_0(t)]^2 \, dt. \tag{1.34}$$

This is then the second moment of the sample function $x_0(t)$. The ensemble average at time t would be denoted by

$$E[\{X(t)\}]^2 = \lim_{n \to \infty} \frac{\sum_{i=0}^n \{x_i(t)\}^2}{n} = m_2(t) \tag{1.35}$$

and is the population second moment. To then say that the process is ergodic with respect to its second moment is equivalent to requiring that

$$\lim_{T \to \infty} \overline{x_T^2} = \lim_{t \to \infty} m_2(t) = \overline{x^2} < \infty. \tag{1.36}$$

We thus say that a process is ergodic if this property holds for all its moments (ergodic in distribution function), and thus that each moment possesses a limit (becomes free of "time"). Note that ergodicity means that moments become time-independent, but not necessarily independent of the initial state. This independence of the process from its actual starting state (existence of a limiting distribution) is a somewhat stronger condition than ergodicity, though the two often go together. Clearly, processes need not be ergodic with respect to any specific number of moments, and the conditions under which any moment is ergodic are the subject of a large class of theorems commonly referred to as ergodic theorems, some of which will be presented later in this section.

We now go back to our previous examples to illustrate the concept of ergodicity. We shall show that the process of Example 1.1 is not ergodic by showing that the relationship (1.36) does not hold for the first moment (this is all that is required to prove nonergodicity, since for a process to be ergodic it must be ergodic in *all* its moments).

Consider the equivalent of (1.34) for the first moment of our alternating process. The average value of the process after $m - 1$ transitions [assume it starts in state $+1$, i.e., $\boldsymbol{\pi}^{(0)} = (1, 0)$] is

$$\overline{x_m} = \frac{1}{m} \sum_{i=0}^{m-1} x_i$$

$$= \begin{cases} 0 & (m \text{ even}), \\ 1/m & (m \text{ odd}). \end{cases}$$

The limit of x_m as m goes to infinity is clearly zero.

Since each path must be identical, the ensemble average after m transitions is

$$m_1(m) = E[X_m] = \begin{cases} +1 & (m \text{ even}), \\ -1 & (m \text{ odd}). \end{cases}$$

The limit of the ensemble average as m goes to infinity does not equal zero; in fact, it does not even exist. Hence, the process is not ergodic in its first moment and is thus not ergodic in any way.

If we use $\pi^{(0)} = (\frac{1}{2}, \frac{1}{2})$, however, then the limit of $m_1(m)$ as m goes to infinity is also zero and the process is indeed ergodic even though, as we have previously shown, no limiting distribution exists. We note that stationary processes [e.g., our problem when $\pi^{(0)} = (\frac{1}{2}, \frac{1}{2})$] are ergodic. But we shall see shortly that ergodic processes need not be stationary.

Now let us see if the process of Example 1.2 is ergodic. We first consider the time average of the kth power of the process. We need to calculate

$$\lim_{m \to \infty} \overline{x_m^k} = \lim_{m \to \infty} \frac{\sum_{i=0}^{m-1} x_i^k}{m}.$$

Now x_i^k, the value of the kth power of the process at the ith transition, is an observation from a random variable taking on values 0 or 1, each with probability $\frac{1}{2}$. Thus, ignoring the starting state (it washes out in the limit, since x_0^k is either zero or one and $\lim_{m \to \infty} x/m = 0$), we have that $\lim_{m \to \infty} \overline{x_m^k}$ is the limit of the average value of $m - 1$ IID Bernoulli random variables with parameter $\frac{1}{2}$. By the strong law of large numbers, this quantity goes to the Bernoulli mean of $\frac{1}{2}$.

Let us now consider the ensemble average at time m, $E[X_m^k] = m_k(m)$, which is the kth population moment of a Bernoulli random variable with parameter $\frac{1}{2}$ and can be calculated as

$$E[X_m^k] = \sum_{i=0}^{1} x_i^k p_i = 0^k \left(\tfrac{1}{2}\right) + 1^k \left(\tfrac{1}{2}\right) = \tfrac{1}{2} \qquad (k = 1, 2, 3, \ldots).$$

The limit of $E[X_m^k]$ as m goes to infinity is also $\frac{1}{2}$; hence the process is ergodic. Note that ergodicity of this process holds regardless of $\pi^{(0)}$, so that even though in general the process is not stationary, it is nevertheless ergodic.

It is generally not easy to show ergodicity or nonergodicity by direct methods, as in the foregoing examples, and the theorems to follow will aid in this task. Prior to stating the key theorems which will enable us to determine when a solution to the stationary equation exists, if a limiting distribution exists, and whether or not the process is ergodic, we first present some definitions required for characterizating discrete-parameter Markov chains.

Two states, i and j, are said to *communicate* ($i \leftrightarrow j$) if i is accessible from j ($j \rightarrow i$) and j is accessible from i ($i \rightarrow j$). A chain is said to be *irreducible* if all of its states communicate, that is, if there exists an n such that $p_{ij}^{(n)} > 0$ for all pairs (i, j).

The period of a return state k of a chain is defined as the greatest common divisor (GCD) of the set of integers $\{n\}$ for which $p_{kk}^{(n)} > 0$. A state is said to be aperiodic if this GCD is 1, that is, if it has period 1. A chain is said to be aperiodic if each of its states is aperiodic. For example, if a process can return to a certain state in 5 or 7 time units, this state is aperiodic, since possible return times are 5, 7, 10, 12, 14, 15, 17, 19, 20, 21, 22, 24, 25, 26, 27, . . . , and this sequence has greatest divisor equal to one. On the other hand, if the return times are 3 and 6, then we get a sequence of possible returns of 3, 6, 9, 12, . . . , meaning that the state is periodic with period equal to three.

Define $f_{ij}^{(n)}$ as the probability that a chain starting at state j returns for the first time to j in n transitions. Hence the probability that the chain ever returns to j is

$$f_{jj} = \sum_{n=1}^{\infty} f_{jj}^{(n)}.$$

If $f_{jj} = 1$, then j is said to be a *recurrent* state; if $f_{jj} < 1$, then j is a *transient* state. When $f_{jj} = 1$,

$$m_{jj} = \sum_{n=1}^{\infty} n f_{jj}^{(n)}$$

is the *mean recurrence time*. If $m_{jj} < \infty$, then j is known as a *positive recurrent state*, while if $m_{jj} = \infty$, then j is a *null recurrent state*.

To illustrate, consider the following transition probability matrix, with six states numbered $0 \rightarrow 5$:

$$
\begin{pmatrix}
\frac{1}{6} & 0 & 0 & \frac{1}{2} & \frac{1}{3} & 0 \\
0 & 1 & 0 & 0 & 0 & 0 \\
\frac{1}{5} & 0 & \frac{2}{5} & \frac{1}{5} & 0 & \frac{1}{5} \\
\frac{1}{4} & 0 & 0 & \frac{3}{4} & 0 & 0 \\
\frac{1}{2} & 0 & 0 & \frac{1}{2} & 0 & 0 \\
0 & \frac{1}{4} & \frac{1}{4} & 0 & 0 & \frac{1}{2}
\end{pmatrix}.
$$

Now state 1 is recurrent, since if the process starts there, it stays there forever. State 5 is transient (can visit a few times and then never return), as is state 2. States 0, 3, 4 are recurrent—once one is entered the process moves among these forever. These states form a recurrent *class*.

As another example, consider the imbedded birth–death transition probability matrix given by (1.31). We see that it is irreducible, may be recurrent (e.g., if $\mu_i \geq \lambda_i$ for all $i \geq$ some value n), and has period two.

Define $f_{ij}^{(n)}$, $i \neq j$, as the probability that the first passage from state i to state j occurs in exactly n steps. Then the probability that state j is ever reached from i is

$$
f_{ij} = \sum_{n=1}^{\infty} f_{ij}^{(n)}.
$$

The expected value of the sequence $\{f_{ij}^{(n)}, n = 1, 2 \ldots\}$ of first-passage probabilities for a fixed pair (i, j), $i \neq j$, is denoted by m_{ij} and is called the *mean first-passage time*; that is,

$$
m_{ij} = \sum_{n=1}^{\infty} n f_{ij}^{(n)} \qquad (i \neq j).
$$

There are an extensive number of theorems in the literature which permit one to determine the presence of recurrence in a Markov chain and to calculate the mean recurrence time whenever appropriate (e.g., see Çinlar, 1975). The following theorems, which we state without proofs, tie much of the theory together and relate the concepts of ergodicity, limiting probabilities, and stationary probabilities.

Theorem 1.1.

(a) *In an irreducible and positive recurrent discrete-parameter Markov chain, a nondegenerate solution to the stationary equations*

$$
\pi = \pi P, \qquad \pi e = 1
$$

always exists, where the solution vector $\boldsymbol{\pi} = \{\pi_j\}$ is such that $\pi_j = 1/m_{jj}$.

Furthermore, when all moments of this stationary distribution are finite, we have the following:

(b) *If the starting probability vector $\boldsymbol{\pi}^{(0)}$ is set equal to the stationary probability vector $\boldsymbol{\pi}$, the above chain becomes a stationary stochastic process and hence ergodic.*

(c) *If the above chain is aperiodic as well as irreducible and positive recurrent, the process is ergodic and has a limiting probability distribution equal to the stationary distribution.*

Note that existence of a limiting distribution is the strongest condition, ergodicity is somewhat weaker, and a nondegenerate solution to the stationary equations is the weakest of the three conditions.

Theorem 1.2. *An irreducible, aperiodic chain is positive recurrent if there exists a nonnegative solution of the system*

$$\sum_{j=0}^{\infty} p_{ij}x_j \le x_i - 1 \qquad (i \neq 0)$$

such that

$$\sum_{j=0}^{\infty} p_{0j}x_j < \infty.$$

Theorem 1.3. *For a continuous-parameter Markov chain, the imbedded chain (at points of transition) need not be aperiodic as long as the mean holding times in all states are bounded, for Theorem 1.1 to be valid.*

The reason aperiodicity is not required for the imbedded Markov chain is that since the process is continuous in time, the time between transitions varies continuously, and as long as it is bounded, the transition times "wash out" any periodicity that may come from the imbedded process.

Returning to the birth–death process and considering its imbedded Markov chain, we see that the matrix P as given by (1.31) is irreducible and positive recurrent as long as certain conditions hold (for example, $\mu_i > \lambda_i$ for all $i \ge n$). Further, if the mean state holding times, $1/(\lambda_i + \mu_i)$, are bounded (λ_i and/or $\mu_i \ge \epsilon > 0$ for all i), then Theorem 1.3 is satisfied and the process is ergodic and possesses a limiting (steady-state) probability distribution. Such processes are sometimes also referred to as achieving *statistical equilibrium* (see Cooper, 1981).

When dealing with Markov chains, some authors (e.g., Feller, 1957, pp. 353 ff.; Heyman and Sobel, 1982, pp. 230 ff.) define a positive recurrent

Table 1.6 Long-Run Behavior Concepts

Concept	Definition	Relationship/Interpretation
Stationary distribution (SD)	Solution to $\pi = \pi P$, $\pi e = 1$ ($0 = pQ$, $pe = 1$).	If all states communicate, π_j can be interpreted as the long-run percentage of time the process spends in state j.
Stationary stochastic process (SSP)	State distributions invariant over time.	If all states communicate, a solution to $\pi = \pi P$ exists and the process is ergodic.
Ergodic process (EP)	Time averages equal ensemble averages in limit.	A solution to $\pi = \pi P$ exists. EPs become SSPs in limit as time goes to infinity.
Limiting or steady-state or statistical equilibrium distribution	Limiting state probability distribution independent of time and initial state.	Process is also ergodic. Stationarity and independence of starting state achieved in limit. The limiting distribution is the SD.

aperiodic state as ergodic, and thus the necessary conditions for Theorem 1.1 to hold, that is, irreducibility and positive recurrence, can be stated as requiring all states of a Markov chain to be ergodic. While it is certainly true that such chains are ergodic as defined by Equation (1.36) and the text immediately following it, we believe it is somewhat confusing to define ergodicity as synonymous with the existence of limiting distributions, since it is a somewhat weaker condition, as we have previously mentioned.

To summarize the long-run behavior concepts of Markov chains studied in this section, we recap the preceding definitions in Table 1.6.

1.10 STEADY-STATE BIRTH–DEATH PROCESSES

With the proper conditions on λ_i and μ_i as mentioned in the previous section, a steady-state solution exists and can be determined from $0 = pQ$. Referring to the end of Section 1.9.3, these equations are found by setting $dp_j(t)/dt = 0$ and $p_j(t) = p_j$ in Equation (1.30):

$$0 = -(\lambda_j + \mu_j)p_j + \lambda_{j-1}p_{j-1} + \mu_{j+1}p_{j+1} \qquad (j \geq 1),$$
$$0 = -\lambda_0 p_0 + \mu_1 p_1. \qquad\qquad\qquad (1.37)$$

Rewriting, we get

$$p_{j+1} = \frac{\lambda_j + \mu_j}{\mu_{j+1}} p_j - \frac{\lambda_{j-1}}{\mu_{j+1}} p_{j-1} \qquad (j \geq 1),$$

$$p_1 = \frac{\lambda_0}{\mu_1} p_0.$$

(1.38)

If follows that

$$p_2 = \frac{\lambda_1 + \mu_1}{\mu_2} p_1 - \frac{\lambda_0}{\mu_2} p_0$$

$$= \frac{\lambda_1 + \mu_1}{\mu_2} \frac{\lambda_0}{\mu_1} p_0 - \frac{\lambda_0}{\mu_2} p_0$$

$$= \frac{\lambda_1 \lambda_0}{\mu_2 \mu_1} p_0,$$

$$p_3 = \frac{\lambda_2 + \mu_2}{\mu_3} p_2 - \frac{\lambda_1}{\mu_3} p_1$$

$$= \frac{\lambda_2 + \mu_2}{\mu_3} \frac{\lambda_1 \lambda_0}{\mu_2 \mu_1} p_0 - \frac{\lambda_1 \lambda_0}{\mu_3 \mu_1} p_0$$

$$= \frac{\lambda_2 \lambda_1 \lambda_0}{\mu_3 \mu_2 \mu_1} p_0,$$

.

.

.

The pattern which appears to be emerging is that

$$p_n = \frac{\lambda_{n-1} \lambda_{n-2} \cdots \lambda_0}{\mu_n \mu_{n-1} \cdots \mu_1} p_0 \qquad (n \geq 1)$$

$$= p_0 \prod_{i=1}^{n} \frac{\lambda_{i-1}}{\mu_i},$$

(1.39)

and we guess that this is the correct formula. To verify that (1.39) is, in fact, proper, mathematical induction will be used on Equation (1.38).

First, the formula is clearly correct for $n = 0$. We have also shown it to be valid for $n = 1, 2,$ and 3. If it is assumed to be true now for $n = k$ (≥ 0), then it is seen that

$$p_{k+1} = \frac{\lambda_k + \mu_k}{\mu_{k+1}} p_0 \prod_{i=1}^{k} \frac{\lambda_{i-1}}{\mu_i} - \frac{\lambda_{k-1}}{\mu_{k+1}} p_0 \prod_{i=1}^{k-1} \frac{\lambda_{i-1}}{\mu_i}$$

$$= \frac{p_0 \lambda_k}{\mu_{k+1}} \prod_{i=1}^{k} \frac{\lambda_{i-1}}{\mu_i} + \frac{p_0 \mu_k}{\mu_{k+1}} \prod_{i=1}^{k} \frac{\lambda_{i-1}}{\mu_i} - \frac{p_0 \mu_k}{\mu_{k+1}} \prod_{i=1}^{k} \frac{\lambda_{i-1}}{\mu_i}$$

$$= p_0 \prod_{i=1}^{k+1} \frac{\lambda_{i-1}}{\mu_i},$$

and the proof by induction is complete.

Since probabilities must sum to 1, it follows that

$$\boxed{p_0 = \left(1 + \sum_{n=1}^{\infty} \prod_{i=1}^{n} \frac{\lambda_{i-1}}{\mu_i} \right)^{-1}.} \tag{1.40}$$

From (1.40), we thus see that a necessary and sufficient condition for the existence of a steady-state solution is the convergence of the infinite series

$$1 + \sum_{n=1}^{\infty} \prod_{i=1}^{n} \frac{\lambda_{i-1}}{\mu_i}.$$

Equations (1.39) and (1.40) are extremely useful in generating a variety of queueing models. For example, we can see that by letting $\lambda_n = \lambda$ and $\mu_n = \mu$ for all n, we obtain the $M/M/1$ queue.

The transient solutions to birth–death processes are, for the most part, extremely difficult to obtain. There are a few isolated cases for which tractable solutions do exist, and we discuss them in the context of the particular model involved, rather than at this juncture.

PROBLEMS

*Whenever a problem is best solved by the use of this book's accompanying software, we have added a boldface letter **C** to the right of the problem number.*

1.1. Discuss the following queueing situations in terms of the characteristics given in Section 1.2.

(a) Aircraft landing at an airport.

(b) Supermarket checkout procedures.

(c) Post-office or bank customer windows.

(d) Toll booths on a bridge or highway.

(e) Gasoline station with several pump islands.

(f) Automatic car wash facility.

(g) Telephone calls coming into a customer information system.

(h) Appointment patients coming into a doctor's office.

(i) Tourists wishing a guided tour of the White House.

(j) Electronic components on an assembly line consisting of three operations and one inspection at end of line.

(k) Processing of programs coming from a number of independent sources on a local area network into a central computer.

1.2. Give three examples of a queueing situation other than those in Problem 1.1, and discuss them in terms of the basic characteristics of Section 1.2.

1.3. The Carry Out Curry House, a fast-food Indian restaurant, must decide on how many parallel service channels to provide. They estimate that, during the rush hour, the average number of arrivals per hour will be approximately 40. They also estimate that, on the average, a server will take about 5.5 min to serve a typical customer. Using only this information, how many service channels will you recommend they install?

1.4. Fluffy Air, a small local feeder airline, needs to know how many slots to provide for telephone callers to be placed on hold. They plan to have enough answerers so that the average waiting time on hold for a caller will be 75 seconds during the busiest period of the day. They estimate the average call-in rate to be 3 per minute. What would you advise?

1.5. The Happidaiz frozen-yogurt stand does a thriving business on warm summer evenings. Even so, there is only a single person on duty at all times. It is known that service time (dishing out the yogurt and collecting the money) is normally distributed with mean 2.5 min and standard deviation 0.5 min. (Note that although the normal distribution allows for negative values and these make no sense when using a normal to describe service times, the standard deviation with respect to the mean is small so that negative values are more than 4 standard deviations below the mean and the probability of negative values is essentially zero.) You arrive on a particular evening to get your favorite crunchy chocolate yogurt cone and find 8 people ahead of you. Estimate the average time until you get the first lick. What is the probability that you will have to wait more than $\frac{1}{2}$ hour? [*Hint*: Remember that the sum of normal random variables is itself normally distributed.]

1.6C. The following observations have been made regarding the time between successive arrivals to a single-server, FCFS queueing process, and the actual time to serve them:

Customer	Interarrival Time	Service Time
1	1	3
2	9	7
3	6	9
4	4	9
5	7	10
6	9	4
7	5	8
8	8	5
9	4	5
10	10	3
11	6	6
12	12	3
13	6	5
14	8	4
15	9	9
16	5	9
17	7	8
18	8	6
19	8	8
20	7	3

(a) Compute the average line and system waiting times.

(b) Calculate the average system waiting time of those customers who had to wait for service (i.e., exclude those who were immediately taken into service), the average length of the queue, the average number in the system, and the fraction of idle time of the server.

1.7C. Items arrive at an initially unoccupied inspection station at a uniform rate of one every 5 min. With the time of the first arrival set equal to 5, the chronological times for inspection completion of the first 10 items were observed to be 7, 17, 23, 29, 35, 38, 39, 44, 46, and 60, respectively. By manual simulation of the operation for 60 min, using these data, develop sample results for (a) mean number in system and (b) percentage idle time experienced.

1.8. Derive the equations (1.13) of Section 1.7 by the sequential use of Equation (1.12); then employ mathematical induction to prove (1.14).

1.9. Given the probability function found for the Poisson process in Equation (1.14), find its moment generating function, $M_{N(t)}(\theta)$, that is, the expected value of $e^{\theta N(t)}$. Then use this MGF to show that the mean and variance both equal λt.

1.10. Derive the Poisson process by using the third assumption that the numbers of arrivals in nonoverlapping intervals are statistically independent and then applying the binomial distribution.

1.11. Prove that the Poisson process has stationary increments.

1.12. By the use of the arguments of Sections 1.7 and 1.8, find the distribution of the counting process associated with IID Erlang interoccurrence times.

1.13. Assume that arrivals can occur singly or in batches of two, with the batch size following the probability distribution

$$f(1) = p, \quad f(2) = 1 - p \quad (0 < p < 1)$$

and with the time between successive batches following the exponential probability distribution

$$a(t) = \lambda e^{-\lambda t} \quad (t > 0).$$

Show that the probability distribution for the number of arrivals in time t is the compound Poisson distribution given by

$$p_n(t) = e^{-\lambda t} \sum_{k=0}^{[n/2]} \frac{p^{n-2k}(1 - p)^k (\lambda t)^{n-k}}{(n - 2k)! \, k!},$$

where $[n/2]$ is the greatest integer $\le n/2$.

1.14. (a) You are given two Poisson processes with intensities λ_1 and λ_2. Find the probability that there is an occurrence of the first stream before the second, starting at time $t = 0$.

(b) A queueing system is being observed. We see that all m identical, exponential servers are busy, with n more customers waiting, and decide to shut off the arrival stream. On the average, how long will it take for the system to empty completely?

1.15. Verify Equations (1.28a) and (1.28b) by using (1.27) in (1.25).

[*Hint*: To obtain (1.28a), let $s = t + \Delta t$. What is required to obtain (1.28b)?]

1.16. Derive the Poisson equations (1.11) and (1.12) by considering the Poisson process as a pure birth process and by using the Kolmogorov forward equation (1.28a) with the proper $q_k(t)$ and $q_{i,k}(t)$, that is, put the proper matrix Q into Equation (1.30).

1.17C. (a) Compute the stationary probability distribution for Examples 1.1 and 1.2.

(b) Compute the stationary probability distribution for a Markov chain with the following single-step transition probability matrix:

$$\begin{pmatrix} .25 & .20 & .12 & .43 \\ .25 & .20 & .12 & .43 \\ 0 & .25 & .20 & .55 \\ 0 & 0 & .25 & .75 \end{pmatrix}.$$

1.18. Derive the imbedded transition probabilities for the birth–death process as given by Equation (1.31). [*Hint*: Use results of Problem 1.14(a) or the law of conditional probability: $\Pr\{A|B,C\} = \Pr\{A, B|C\}/\Pr\{B|C\}$].

1.19. Consider the first order statistic (call it $T_{(1)}$) of a random sample of size n drawn from a uniform $(0, 1)$ population. Its CDF is found to be

$$F(t) = 1 - (1 - t)^n.$$

Show that the random variable $nT_{(1)}$ converges in law to an exponential as $n \to \infty$.

1.20C. You are told that a small single-server, birth–death-type queue with finite capacity cannot hold more than three customers. The three arrival or birth rates are $(\lambda_0, \lambda_1, \lambda_2) = (3, 2, 1)$, while the service or death rates are $(\mu_1, \mu_2, \mu_3) = (1, 2, 2)$. Find the steady-state probabilities $\{p_i, i = 0, 1, 2, 3\}$ and L. Then determine the average or effective arrival rate $\lambda_{\text{eff}} = \Sigma \lambda_i p_i$, and the expected system waiting time W.

1.21C. The finite-capacity constraint of Problem 1.20 has been pushed up to 10 now, with the arrival rates known to be $(4, 3, 2, 2, 3, 1, 2, 1, 2, 1)$ and service rates to be $(1, 1, 1, 2, 2, 2, 3, 3, 3, 4)$. Do as before and find p_i, $i = 0, 1, 2, \ldots, 10$, the mean system size L, the effective arrival rate $\lambda_{\text{eff}} = \Sigma \lambda_i p_i$, and the expected system waiting time W.

1.22. Suppose that you have learned that an $M/G/1/K$ has a blocking probability of $p_K = .1$, that $\lambda = \mu = 1$, and that $L = 5$. Find λ_{eff}, W, W_q, p_0, and ρ_{eff}.

1.23C. In choosing the proper distributions to represent interarrival and service times, the *coefficient of variation* (CV) can often be useful. The CV is defined as the ratio of the standard deviation to the mean, and provides a measure of the relative spread of a distribution. For example, service consisting of routine tasks should have a relatively small spread around the mean (CV ≤ 1), whereas service consisting of diverse tasks (some quick, some time-consuming) should have a relatively large spread around the mean (CV ≥ 1). The exponential distribution, widely used in queue modeling, has CV $= 1$ (standard deviation $=$ mean). Two other distributions often employed in queueing are the *Erlang* distribution and the *mixed-exponential* distribution (the *hyperexponential* distribution is a special case). The Erlang is a two-parameter distribution, having a type or shape parameter k (an integer ≥ 1) and a scale parameter, which we shall denote by β. The mean of the Erlang is the product $k\beta$, and its standard deviation is the product $\beta\sqrt{k}$. The CV for an Erlang is then $1/\sqrt{k} \leq 1$. When $k = 1$, the Erlang reduces to the exponential distribution. The mixed-exponential distribution function is a convex linear combination of exponential distributions, mixed according to some probability distribution (e.g., we select from one exponential population, mean μ_1, with probability p, and from a second exponential, mean μ_2, with probability $1 - p$). The CV for a mixed exponential distribution can be shown to be always >1. Using the software, solve the following problems:

(a) Data taken on a server who provides espresso to customers at the Betterbean Boutique show that the mean time *to* serve a customer is 2.25 min with a standard deviation of 1.6 min. What is the probability that service takes more than 5 min? [*Hint:* Find the closest integer value for k and then solve for β.]

(b) Data collected at a small post office in the rural town of Arling-rock reveal that the clerk has basically two types of customers—those who desire to purchase stamps only and those who require other more complicated functions. The distributions of service times for each type of customer can be well approximated by the exponential distribution; the stamp-only customers take on average 1.06 min, while the nonstamp customers take on the average 3.8 min. What is the probability that a stamp customer requires more than 5 min? What is the probability that a non-stamp customer takes more than 5 min? If 15% of all arrivals are stamp-only customers, what is the probability that the next customer in line requires more than 5 min?

CHAPTER 2

Simple Markovian Birth–Death Queueing Models

The purpose of this chapter is to develop a broad class of queueing models by the direct application of the results obtained in Chapter 1 for the birth–death process, which we showed to be a special type of continuous-parameter Markov chain. We now exploit the results of Section 1.10 to allow us to model a large number of queueing problems, beginning with the classical Poisson-input, exponential-service, single-server $M/M/1$ queue.

2.1 STEADY-STATE SOLUTION FOR THE $M/M/1$ MODEL

The density functions for the interarrival times and service times for the $M/M/1$ queue are given, respectively, as

$$a(t) = \lambda e^{-\lambda t},$$
$$b(t) = \mu e^{-\mu t},$$

where $1/\lambda$ then is the mean interarrival time and $1/\mu$ the mean service time. Interarrival times, as well as service times, are assumed to be statistically independent. Since interarrival and service times are exponential, and arrival and conditional service rates Poisson, we have

Pr{an arrival occurs in an infinitesimal interval of length Δt} =
$\lambda \Delta t + o(\Delta t)$;
Pr{more than one arrival occurs in Δt} = $o(\Delta t)$;
Pr{a service completion in Δt, given system is not empty} = $\mu \Delta t + o(\Delta t)$;
and
Pr{more than one service completion in Δt, given more than one in system} = $o(\Delta t)$.

Therefore, the $M/M/1$ problem is a birth–death process with $\lambda_n = \lambda$ and $\mu_n = \mu$, for all n. Arrivals can be considered as "births" to the system, since if the system is in state n (we consider system state as the number in the system) and an arrival occurs, the state is changed to $n + 1$. On the other hand, a departure occurring while the system is in state n sends the system down one to $n - 1$ and can be looked upon as a "death." Hence, the steady-state equations are found from Equations (1.37) and (1.38) as

$$0 = -(\lambda + \mu)p_n + \mu p_{n+1} + \lambda p_{n-1} \qquad (n \ge 1),$$
$$0 = -\lambda p_0 + \mu p_1, \tag{2.1a}$$

or

$$p_{n+1} = \frac{\lambda + \mu}{\mu} p_n - \frac{\lambda}{\mu} p_{n-1} \qquad (n \ge 1),$$
$$p_1 = \frac{\lambda}{\mu} p_0. \tag{2.1b}$$

While the $M/M/1$ queue is a special case of the $M/M/c$ queue, it is instructive to first look in detail at the $M/M/1$ system before solving for the more general $M/M/c$. Several techniques to be used in the following sections and chapters can be more simply explained looking first at the $M/M/1$.

2.1.1 Stochastic Balance

A shortcut method for directly obtaining the stationary equations (2.1) is by means of a *flow* balance procedure called *stochastic balance*. Essentially, the analysis looks at a given state and requires that average total flow into the state be equal to average total flow out of the state if steady-state conditions exist. To begin, consider a state n ($n \ge 1$) in an $M/M/1$ queue. Then a rate transition diagram can be constructed to appear as shown in Figure 2.1 from the following considerations. From state n, the system goes to $n - 1$ if a service is completed or to $n + 1$ if an arrival occurs. The system can go to state n from $n - 1$ if an arrival comes, or to state n from $n + 1$ if a service

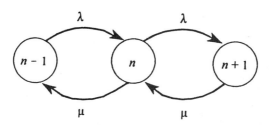

Fig. 2.1 Rate transition diagram—$M/M/1$.

is completed. The flow rates are all in terms of λ and μ such that the mean total flow out of state n equals $\lambda p_n + \mu p_n$ while the mean total flow into state n equals $\mu p_{n+1} + \lambda p_{n-1}$.

Intuitively, we explain the mean total flow being the product of the steady-state probabilities and flow rates (λ and μ) as follows. We can look at p_n as the percentage of time in the steady state that the system is in state n. The average arrival rate is λ units per unit time. Since the Poisson has the conditional uniform property, if the system is in state n a fraction p_n of the time (look at this as if a gate were open p_n of the time), then λp_n is the mean rate of flow (through the gate) upward to $n + 1$. Likewise, μp_n is the mean rate of flow in the other direction (downward) to state $n - 1$.

Equating flow out to flow in gives

$$0 = -(\lambda p_n + \mu p_n) + \mu p_{n+1} + \lambda p_{n-1} \qquad (n \geq 1)$$

or

$$\lambda p_n + \mu p_n = \mu p_{n+1} + \lambda p_{n-1} \qquad (n \geq 1),$$

which is the top line of (2.1a). Changing the state diagram of Figure 2.1 to reflect a barrier at state 0 (it is not possible to go below state 0) will yield the second line of (2.1a) (see Problem 2.1). Actually, the Kolmogorov differential equation can also be gotten in this manner, since if we are not in steady state, the flow rates are not equal and the difference in the rates of flow is the rate of change of the system state with respect to time, namely, $p_n'(t)$.

Equations (2.1) are called *global* balance equations, since they equate the total mean flow into each state with the total mean flow out of that state. Further analysis yields what we shall refer to as a *detailed* balance equation relating the mean flow into and out of adjacent states. To illustrate, place an artificial line in Figure 2.1 between states n and $n + 1$. Mean flows across this barrier must also be equal to zero for steady-state conditions, that is, flow to the right must equal flow to the left. This yields the equation $\lambda p_n = \mu p_{n+1}$. If we consider a barrier between states $n - 1$ and n, we get $\lambda p_{n-1} = \mu p_n$. Note that adding these two equations yields the global balance equations of (2.1a).

Sometimes it may be advantageous to work with the detailed (rather than the global) balance equations in solving for the steady-state probabilities. The equating of these adjacent flows relates to something called *reversibility*, a concept which becomes particularly useful later in our work on queueing networks.

This flow balance analysis between two states works in this case because of the birth–death characteristic that only adjacent states can directly communicate, with constant transition rates (see Figure 2.1). For more general Markovian models, this is not necessarily true, so that a detailed balance analysis may become considerably more complicated and not always particularly fruitful. However, for all Markovian models, equating the total flow out of a state with the total flow into the state always yields the global balance equations, from which the $\{p_n\}$ can be determined.

2.2 METHODS OF SOLVING STEADY-STATE DIFFERENCE EQUATIONS

We now return to procedures for solving Equations (2.1). We present three methods for doing this. The first and probably the most straightforward is an iterative procedure, the second involves generating functions, and the third involves the concept of linear operators and is analogous to methods used for differential equations. The reason for presenting the three methods of solution is that one may be more successful than the others, depending on the particular model at hand. For the model under study here, all three methods work equally well, and we take this opportunity to illustrate their use.

2.2.1 Iterative Method of Solving the Steady-State Difference Equations for $\{p_n\}$

By this, we mean that we iteratively use the global balance equations given by Equations (2.1) or equivalently the local balance equation $\lambda p_n = \mu p_{n+1}$ to obtain a sequence of state probabilities, $p_1, p_2, p_3, \ldots$, each in terms of p_0. When we believe that we have enough information about the form of these state probabilities, we make a conjecture on their general form for all states n, and then attempt to verify that our surmise is indeed correct, using mathematical induction or the like. But this is precisely the sort of thing that we did for the general birth–death process in Chapter 1, where we verified that Equation (1.39) is the appropriate formula for the steady-state probability p_n for any birth–death process with birth rates $\{\lambda_n, n = 0, 1, 2, \ldots\}$ and death rates $\{\mu_n, n = 1, 2, 3, \ldots\}$. Since the $M/M/1$ system is indeed a birth–death process with constant birth and death rates, we can directly apply (1.39) with $\lambda_n = \lambda$ and $\mu_n = \mu$ for all n. It follows that

$$p_n = p_0 \prod_{i=1}^{n} \left(\frac{\lambda}{\mu}\right) \qquad (n \geq 1)$$

$$= p_0 (\lambda/\mu)^n.$$

To get p_0 now, we utilize the fact that probabilities must sum to 1 and it follows that

$$1 = \sum_{n=0}^{\infty} \left(\frac{\lambda}{\mu}\right)^n p_0.$$

Recall our earlier definition in Section 1.5 of the traffic intensity or utilization rate ρ as the ratio λ/μ for single-server queues. Rewriting thus gives

$$p_0 = \frac{1}{\sum_{n=0}^{\infty} \rho^n}.$$

Now $\sum_{n=0}^{\infty} \rho^n$ is the geometric series $1 + \rho + \rho^2 + \rho^3 + \cdots$ and converges if and only if $\rho < 1$. Thus for the existence of a steady-state solution, $\rho = \lambda/\mu$ must be less than 1, or equivalently, λ less than μ. This makes intuitive sense, for if $\lambda > \mu$, the mean arrival rate is greater than the mean service rate, and the server will get further and further behind. That is to say, the system size will keep building up without limit. It is not as intuitive, however, why no steady-state solution exists when $\lambda = \mu$. One possible way to explain infinite build up when $\lambda = \mu$ is that as the queue grows, it is more and more difficult for the server to decrease the queue because the average service rate is no higher than the average arrival rate.

Making use of the well-known expression for the sum of the terms of a geometric progression,

$$\sum_{n=0}^{\infty} \rho^n = \frac{1}{1 - \rho} \qquad (\rho < 1),$$

we have

$$p_0 = 1 - \rho \qquad (\rho = \lambda/\mu < 1), \tag{2.2}$$

which confirms the general result for p_0 we derived previously in Section 1.5 for all $G/G/1$ queues. Thus the full steady-state solution for the $M/M/1$ system is the *geometric* probability function

$$\boxed{p_n = (1 - \rho)\rho^n \qquad (\rho = \lambda/\mu < 1).} \tag{2.3}$$

2.2.2 Solving Steady-State Difference Equations for $\{p_n\}$ by Generating Functions

The probability generating function, $P(z) = \sum_{n=0}^{\infty} p_n z^n$ (z complex with $|z| \leq 1$), can be utilized to find the $\{p_n\}$. The procedure involves finding a closed expression for $P(z)$ from Equation (2.1a) and then finding the power series expansion to "pick off" the $\{p_n\}$ which are the coefficients. For some models, it is relatively easy to find a closed expression for $P(z)$, but quite difficult to find its series expansion to obtain the $\{p_n\}$. However, even if the series expansion cannot be found, $P(z)$ still provides useful information. For example, $dP(z)/dz$ evaluated at $z = 1$ gives the expected number in the system, $L = \Sigma\, np_n$. For the $M/M/1$ model considered here, we can completely solve for the $\{p_n\}$ using $P(z)$.

To do so, begin by rewriting (2.1b) in terms of ρ and obtain

$$p_{n+1} = (\rho + 1)p_n - \rho p_{n-1} \qquad (n \geq 1),$$
$$p_1 = \rho p_0. \tag{2.4}$$

When both sides of the first line of (2.4) are multiplied by z^n we find

$$p_{n+1}z^n = (\rho + 1)p_n z^n - \rho p_{n-1}z^n,$$

or

$$z^{-1}p_{n+1}z^{n+1} = (\rho + 1)p_n z^n - \rho z p_{n-1}z^{n-1}.$$

When both sides of the above equation are summed from $n = 1$ to ∞, it is found that

$$z^{-1} \sum_{n=1}^{\infty} p_{n+1}z^{n+1} = (\rho + 1) \sum_{n=1}^{\infty} p_n z^n - \rho z \sum_{n=1}^{\infty} p_{n-1}z^{n-1},$$

or

$$z^{-1}[P(z) - p_1 z - p_0] = (\rho + 1)[P(z) - p_0] - \rho z P(z). \tag{2.5}$$

We know from (2.4) that $p_1 = \rho p_0$, and hence

$$z^{-1}[P(z) - (\rho z + 1)p_0] = (\rho + 1)[P(z) - p_0] - \rho z P(z).$$

Solving for $P(z)$ we finally have

$$P(z) = \frac{p_0}{1 - z\rho}. \tag{2.6}$$

To find p_0 now, we use the boundary condition that probabilities sum to 1 in the following way. Consider $P(1)$, which can be seen to be

$$P(1) = \sum_{n=0}^{\infty} p_n 1^n = \sum_{n=0}^{\infty} p_n = 1.$$

Thus from (2.6), we have

$$P(1) = 1 = \frac{p_0}{1 - \rho},\qquad(2.7)$$

so that $p_0 = 1 - \rho$.

Because the $\{p_n\}$ are probabilities, $P(z) > 0$ for $z > 0$; hence $P(1) > 0$. From (2.7) we see that $P(1) = p_0/(1 - \rho) > 0$; therefore ρ must be <1, since p_0 is a probability and is >0. Thus

$$\boxed{P(z) = \frac{1 - \rho}{1 - \rho z} \qquad (\rho < 1,\ \ |z| \le 1).}\qquad(2.8)$$

It is a easy to expand (2.8) as a power series by simple long division or to recognize it as the sum of a geometric series, since $|\rho z| < 1$. So doing yields

$$\frac{1}{1 - \rho z} = 1 + \rho z + (\rho z)^2 + (\rho z)^3 + \cdots,$$

and thus the probability generating function is

$$P(z) = \sum_{n=0}^{\infty} (1 - \rho)\rho^n z^n.\qquad(2.9)$$

The coefficient of z^n, p_n, is therefore given by

$$p_n = (1 - \rho)\rho^n \qquad (\rho = \lambda/\mu < 1),$$

which is what was previously obtained in Equation (2.3).

We make a number of concluding observations about the algebraic form of the generating function given in (2.8). First, we note that this expression is the quotient of two (simple) polynomials (i.e., it is a rational function), the numerator being a constant, while the denominator is the linear form $1 - \rho z$. It thus follows that the denominator has the single zero of $1/\rho$ (>1), which is clearly just the reciprocal of the traffic intensity. These points are important, because (2.8) is the first of many generating functions we shall be seeing as we develop more queueing models in the text.

2.2.3 Solving Steady-State Difference Equations for $\{p_n\}$ by the Use of Operators

We begin here by considering a linear operator D defined on the sequence

$\{a_0, a_1, a_2, \ldots\}$, such that

$$Da_n = a_{n+1} \quad \text{(for all } n\text{)}.$$

Then the general linear difference equation with constant coefficients

$$C_n a_n + C_{n+1} a_{n+1} + \cdots + C_{n+k} a_{n+k} = \sum_{i=n}^{n+k} C_i a_i = 0 \qquad (2.10)$$

may be written as

$$\left(\sum_{i=n}^{n+k} C_i D^{i-n} \right) a_n = 0,$$

since

$$D^m a_n = a_{n+m} \quad \text{(for all } n \text{ and } m\text{)}.$$

For example, in the event that (2.10) is of the form

$$c_2 a_{n+2} + c_1 a_{n+1} + c_0 a_n = 0, \qquad (2.11)$$

then

$$(c_2 D^2 + c_1 D + c_0) a_n = 0, \qquad (2.12)$$

and if the quadratic in D has the real roots r_1 and r_2, then it is also true that

$$(D - r_1)(D - r_2) a_n = 0.$$

Since r_1 and r_2 are roots of Equation (2.12), $d_1 r_1^n$ and $d_2 r_2^n$ are solutions to (2.11), where d_1 and d_2 are arbitrary constants. This can be verified by the substitution of $d_1 r_1^n$ and $d_2 r_2^n$ into (2.11), where $a_n = d_1 r_1^n$ or $d_2 r_2^n$. For example, letting $a_n = d_1 r_1^n$ we have, upon substitution in (2.11),

$$c_2 d_1 r_1^{n+2} + c_1 d_1 r_1^{n+1} + c_0 d_1 r_1^n = 0,$$

or

$$d_1 r_1^n (c_2 r_1^2 + c_1 r_1 + c_0) = 0,$$

which is of the form of (2.12). Similarly, one can show that $a_n = d_2 r_2^n$ is a solution and hence that their sum, $d_1 r_1^n + d_2 r_2^n$, is also a solution. It can be shown in a manner similar to that used for ordinary linear differential equations that this sum is the most general solution.

This approach is directly applicable to the solution of the steady-state difference equations (2.1a), since these are linear difference equations of a form like (2.11). If D is defined by the operation $Dp_n = p_{n+1}$ and if (2.1a) is rewritten as

$$\mu p_{n+2} - (\lambda + \mu)p_{n+1} + \lambda p_n = 0 \qquad (n \geq 0),$$

then these limiting probabilities are the solution to

$$[\mu D^2 - (\lambda + \mu)D + \lambda]p_n = 0 \tag{2.13}$$

subject to the boundary conditions

$$p_1 = \frac{\lambda}{\mu}p_0 = \rho p_0,$$

$$\sum_{n=0}^{\infty} p_n = 1.$$

This quadratic in D factors easily to give

$$(D - 1)(\mu D - \lambda)p_n = 0,$$

and hence

$$p_n = d_1(1)^n + d_2(\lambda/\mu)^n = d_1 + d_2\rho^n, \tag{2.14}$$

where d_1 and d_2 are to be found with the use of the boundary conditions. We see from (2.14) that

$$p_1 = d_1 + d_2\rho \quad \text{and} \quad p_2 = d_1 + d_2\rho^2.$$

But d_1 must be zero to allow the probabilities to sum to one, so that $p_1 = d_2\rho$. Now we also know from (2.4) that $p_1 = \rho p_0$. Thus $d_2 = p_0$ and

$$p_n = \rho^n p_0.$$

The empty probability p_0 is now found as in the first method by summing the $\{p_n\}$ over all n and is

$$p_0 = 1 - \rho \qquad (\rho < 1).$$

2.2.4   Measures of Effectiveness

The steady-state probability distribution for the system size allows us to calculate the system's measures of effectiveness. Two of immediate interest are the expected number in the system and the expected number in the queue at steady state. To derive these, let N represent the random variable "number of customers in the system in steady state" and L represent its expected value. We can then write

$$L = E[N] = \sum_{n=0}^{\infty} np_n$$

$$= (1 - \rho) \sum_{n=0}^{\infty} n\rho^n. \tag{2.15}$$

Consider the summation

$$\sum_{n=0}^{\infty} n\rho^n = \rho + 2\rho^2 + 3\rho^3 + \cdots$$

$$= \rho(1 + 2\rho + 3\rho^2 + \cdots)$$

$$= \rho \sum_{n=1}^{\infty} n\rho^{n-1}.$$

We observe that $\sum_{n=1}^{\infty} n\rho^{n-1}$ is simply the derivative of $\sum_{n=0}^{\infty} \rho^n$ with respect to ρ, since the summation and differentiation operations may be interchanged, as the functions are sufficiently well behaved. Since $\rho < 1$,

$$\sum_{n=0}^{\infty} \rho^n = \frac{1}{1 - \rho};$$

hence

$$\sum_{n=1}^{\infty} n\rho^{n-1} = \frac{d[1/(1 - \rho)]}{d\rho}$$

$$= \frac{1}{(1 - \rho)^2}.$$

So the expected number in the system at steady state is then

$$L = \frac{\rho(1 - \rho)}{(1 - \rho)^2},$$

or simply

$$L = \frac{\rho}{1 - \rho} = \frac{\lambda}{\mu - \lambda}. \tag{2.16}$$

If the random variable "number in queue in steady state" is denoted by N_q and its expected value by L_q, then we have

$$L_q = \sum_{n=1}^{\infty} (n-1)p_n = \sum_{n=1}^{\infty} np_n - \sum_{n=1}^{\infty} p_n$$

$$= L - (1 - p_0) = \frac{\rho}{1 - \rho} - \rho.$$

[Note that $L_q = L - (1 - p_0)$ holds for all single-channel, one-at-a-time service queues, since no assumptions were made in the derivation as to the input and service distributions; this can also be seen from (1.3).] Thus the mean queue length is

$$L_q = \frac{\rho^2}{1 - \rho} = \frac{\lambda^2}{\mu(\mu - \lambda)}. \tag{2.17}$$

We might also be interested in the expected queue size of nonempty queues, which we denote by L'_q; that is, we wish to ignore the cases where the queue is empty. Another way of looking at this measure is to view it as the expected size of the queues which form from time to time. We can write

$$L'_q = E[N_q \mid N_q \neq 0]$$

$$= \sum_{n=1}^{\infty} (n-1)p'_n = \sum_{n=2}^{\infty} (n-1)p'_n,$$

where p'_n is the conditional probability distribution of n in the system given the queue is not empty, or $p'_n = \Pr\{n \text{ in system} \mid n \geq 2\}$. From the laws of conditional probability,

$$p'_n = \frac{\Pr\{n \text{ in system and } n \geq 2\}}{\Pr\{n \geq 2\}}$$

$$= \frac{p_n}{\sum_{n=2}^{\infty} p_n} \qquad (n \geq 2)$$

$$= \frac{p_n}{1 - (1 - \rho) - (1 - \rho)\rho}$$

$$= \frac{p_n}{\rho^2}.$$

The probability distribution $\{p'_n\}$ is the distribution $\{p_n\}$ normalized by omitting the cases $n = 0$ and 1. Thus

$$L'_q = \sum_{n=2}^{\infty} (n-1)\frac{p_n}{\rho^2}$$
$$= \frac{L - p_1 - (1 - p_0 - p_1)}{\rho^2}.$$

Hence

$$\boxed{L'_q = \frac{1}{1-\rho} = \frac{\mu}{\mu - \lambda}.}$$ (2.18)

As a side observation, it is not at all by coincidence that it turned out that

$$\Pr\{n \text{ in system} \geq 2\} = \rho^2,$$

because it can easily be established for all n that

$$\Pr\{N \geq n\} = \rho^n.$$

The proof is as follows:

$$\Pr\{N \geq n\} = \sum_{k=n}^{\infty} (1 - \rho)\rho^k$$
$$= (1 - \rho)\rho^n \sum_{k=n}^{\infty} \rho^{k-n}$$
$$= \frac{(1 - \rho)\rho^n}{1 - \rho} = \rho^n.$$

To complete the basic part of this presentation, we recall from Chapter 1 that the expected steady-state system waiting time W and line delay W_q can be found easily from L and L_q by using Little's formulas, $L = \lambda W$ and $L_q = \lambda W_q$. In the case of the $M/M/1$ queue, it thus follows from (2.16) and (2.17) that

$$\boxed{W = \frac{L}{\lambda} = \frac{\rho}{\lambda(1-\rho)} = \frac{1}{\mu - \lambda}}$$ (2.19)

and

$$W_q = \frac{L_q}{\lambda} = \frac{\rho^2}{\lambda(1-\rho)} = \frac{\rho}{\mu - \lambda}. \qquad (2.20)$$

All of these measures of effectiveness are illustrated by the following.

Example 2.1
Ms. H. R. Cutt runs a one-person, unisex hair salon. She does not make appointments, but runs the salon on a first-come, first-served basis. She finds that she is extremely busy on Saturday mornings, so she is considering hiring a part-time assistant and even possibly moving to a larger building. Having obtained a master's degree in operations research (OR) prior to embarking upon her career, she elects to analyze the situation carefully before making a decision.

She thus keeps careful records for a succession of Saturday mornings and finds that customers seem to arrive according to a Poisson process with a mean arrival rate of 5/hr. Because of her excellent reputation (what else would you expect from someone with an M.S. in OR?), customers were always willing to wait. The data further showed that customer processing time (aggregated female and male) was exponentially distributed with an average of 10 min.

Cutt first decided to calculate the average number of customers in the shop and the average number of customers waiting for a haircut. From the data, $\lambda = 5$/hr and $\mu = \frac{1}{10}$/min $= 6$/hr. This gives a ρ of $\frac{5}{6}$. From (2.16) and (2.17) she finds $L = 5$ and $L_q = 4\frac{1}{6}$. The average number waiting when there is at least one person waiting is found from (2.18) as $L'_q = 6$. She is also interested in the percentage of time an arrival can walk right in without having to wait at all, which happens when no one is in the shop. The probability of this is $p_0 = 1 - \rho = \frac{1}{6}$. Hence approximately 16.7% of the time Cutt is idle and a customer can get into the chair without waiting. Because of the Poisson process governing arrivals and its completely random property, as discussed in Section 1.7, the percentage of customers that can go directly into service is also 16.7%. Thus 83.3% of the customers must wait prior to getting into the chair.

Cutt's waiting room has only four seats at present. She is interested in the probability that a customer, upon arrival, will not be able to find a seat and have to stand. This can be easily calculated as

$$\text{Pr\{finding no seat\}} = \text{Pr}\{N \geq 5\} = \rho^5 \doteq 0.402.$$

This tells Cutt that a little over 40% of the time a customer cannot find a seat and also that 40% of the customers will have to stand upon arrival. Cutt is also interested in learning how much time customers spend waiting, and the average system waiting time and line delay are easily computed from

Equations (2.19) and (2.20) to be $W = 1/(\mu - \lambda) = 1$ hr and $W_q = \rho/(\mu - \lambda) = \frac{5}{6}$ hr, not very pleasing outcomes.

To get even more information on the nature of customer waiting, Cutt has decided that she would like to know the precise probability that the line delay is more than 45 min. But to do this, she needs to have the probability distribution function for the waiting time in queue, which is something she has forgotten from her M.S. days. So she has decided to go through a complete derivation of the result by herself. This follows.

Waiting-Time Distributions

Let T_q denote the random variable "time spent waiting in the queue" and $W_q(t)$ represent its cumulative probability distribution. Up to now the queue discipline has had no effect on our derivations. When considering individual waiting time, however, queue discipline must be specified, and we are here assuming that it is first-come, first-served (FCFS). The waiting-time random variable has an interesting property in that it is part discrete and part continuous. Waiting time is, for the most part, a continuous random variable, except that there is a nonzero probability that the delay will be zero, that is, a customer entering service immediately upon arrival. Hence we have

$$W_q(0) = \Pr\{T_q \le 0\} = \Pr\{T_q = 0\}$$
$$= \Pr\{\text{system empty at an arrival}\} = q_0.$$

We shall denote the conditional probability of n in the system *given* that an arrival is about to occur by q_n. These probabilities are not always the same as the p_n with which we have been working, since the p_n are unconditional probabilities of n in the system at an arbitrary point in time. To find the distribution of *virtual* waiting time (i.e., the time a fictitious customer would have to wait *were* it to arrive at an arbitrary point in time), we would use p_n. However, for Poisson input, $q_n = p_n$, as we will show in Section 2.4 when deriving q_n for the truncated $M/M/c/K$ case where $q_n \ne p_n$. Thus

$$W_q(0) = p_0 = 1 - \rho.$$

It then remains to find $W_q(t)$ for $t > 0$.

Consider $W_q(t)$, the probability of a customer waiting a time less than or equal to t for service. If there are n units in the system upon arrival, then in order for the customer to go into service at a time between 0 and t, all n units must have been served by time t. Since the service distribution is memoryless, the distribution of the time required for n completions is independent of the time of the current arrival and is the convolution of n exponential random variables, which is an Erlang type n. In addition, since the input is Poisson, the arrival points are uniformly spaced and hence the probability that an arrival finds n in the system is identical to the stationary distribution of system size. Therefore we may write that

$$W_q(t) = \Pr\{T_q \le t\} = W_q(0) + \sum_{n=1}^{\infty} \Pr\{n \text{ completions in } \le t \mid$$

$$\text{arrival found } n \text{ in system}\} \cdot p_n$$

$$= 1 - \rho + (1 - \rho) \sum_{n=1}^{\infty} \rho^n \int_0^t \frac{\mu(\mu x)^{n-1}}{(n-1)!} e^{-\mu x} \, dx$$

$$= 1 - \rho + \rho(1 - \rho) \int_0^t \mu e^{-\mu x} \sum_{n=1}^{\infty} \frac{(\mu x \rho)^{n-1}}{(n-1)!} \, dx$$

$$= 1 - \rho + \rho(1 - \rho) \int_0^t \mu e^{-\mu x(1-\rho)} \, dx$$

$$= 1 - \rho e^{-\mu(1-\rho)t} \qquad (t > 0).$$

So the distribution of the waiting time in queue is

$$\boxed{W_q(t) = 1 - \rho e^{-\mu(1-\rho)t} \qquad (t \ge 0).} \tag{2.21}$$

As a result of (2.21), Ms. Cutt is now able to calculate the probability that an arriving customer has to wait more than 45 min as $\frac{5}{6}e^{-3/4} \doteq 0.3936$.

We already know from our earlier use of Little's law a few pages back that the mean of this distribution is $W_q = p/(\mu - \lambda)$. In verification of this result, we recompute the mean as

$$W_q = \int_0^{\infty} [1 - W_q(t)] \, dt = \int_0^{\infty} \rho e^{-\mu(1-\rho)t} \, dt = \frac{\rho}{\mu - \lambda}.$$

A full plot (created by the software module) of the CDF $W_q(t)$ as experienced by Cutt's customers can be seen in Figure 2.2 ($\mu = 6$, $\lambda = 5$). Note the jump discontinuity at $t = 0$ that results from the fact that there is a nonzero probability that an arrival finds an empty system.

Also of interest to us and H. R. Cutt would be the probability distribution of the total time (including service) that a customer has to spend in an $M/M/1$ system. Denote this random variable by T, its CDF by $W(t)$, its density by $w(t)$, and its expected value by W. We have already derived W, and it can further be shown that the system waits have an ordinary *negative*

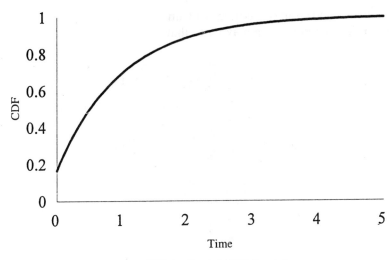

Fig. 2.2 CDF for Cutt's $M/M/1$ line delays.

exponential distribution, such that

$$\boxed{\begin{aligned} W(t) &= 1 - e^{-(\mu-\lambda)t} \qquad (t \geq 0), \\ w(t) &= (\mu - \lambda)e^{-(\mu-\lambda)t} \qquad (t > 0). \end{aligned}} \qquad (2.22)$$

The derivation of (2.22) very much follows that of the line delay distribution except that $n + 1$ service completions are required in time $\leq t$ (see Problem 2.8).

Note that the development of results in $M/M/1$ for L_q, p_n, L, L_q, W, and W_q did not depend on the order in which customers are served. Thus, all would also be valid for the general discipline $M/M/1/\infty/GD$ model. However, the CDFs, $W(t)$ and $W_q(t)$, are discipline-dependent, as can be seen in their derivations by the requirement that one's waiting time is largely determined by the amount of time it takes to serve the customers found upon arrival. We should also note that the measures of effectiveness used here are all calculated for the steady state. These results are not applicable to the initial few customers who arrive soon after opening; however, transient effects are generally "washed out" rather quickly, and the steady-state calculations suffice.

2.3   QUEUES WITH PARALLEL CHANNELS (*M/M/c*)

We now turn our attention to the multiserver *M/M/c* model in which each server has an independently and identically distributed exponential service-time distribution, with the arrival process again assumed to be Poisson. We consider the case in which there are c servers and make use of the prior theory developed, most specifically Equation (1.39), to develop this model. Clearly, since the input is Poisson and service exponential, we have a birth–death process. Hence $\lambda_n = \lambda$ for all n, and it remains to determine μ_n prior to being able to use (1.39).

If there are more than c customers in the system, all c servers must be busy with each putting out at a mean rate μ, and the mean system output rate is thus equal to $c\mu$. When there are fewer than c customers in the system, say $n < c$, only n of the c servers are busy and the system is putting out at a mean rate of $n\mu$. Hence μ_n may be written as

$$\mu_n = \begin{cases} n\mu & (1 \le n < c), \\ c\mu & (n \ge c). \end{cases} \tag{2.23}$$

Utilizing (2.23) in Equation (1.39) and the fact that $\lambda_n = \lambda$ for all n, we obtain

$$p_n = \begin{cases} \dfrac{\lambda^n}{n!\,\mu^n}\,p_0 & (1 \le n < c), \\[3mm] \dfrac{\lambda^n}{c^{n-c}c!\,\mu^n}\,p_0 & (n \ge c). \end{cases} \tag{2.24}$$

In order to find p_0, we again use the condition that probabilities must sum to 1, which gives

$$p_0 = \left(\sum_{n=0}^{c-1} \frac{\lambda^n}{n!\,\mu^n} + \sum_{n=c}^{\infty} \frac{\lambda^n}{c^{n-c}c!\,\mu^n} \right)^{-1}.$$

As in Section 1.5, we let $r = \lambda/\mu$ and $\rho = r/c = \lambda/c\mu$. Then we have

$$p_0 = \left(\sum_{n=0}^{c-1} \frac{r^n}{n!} + \sum_{n=c}^{\infty} \frac{r^n}{c^{n-c}c!} \right)^{-1}.$$

Now consider the infinite series in the above equation:

$$\sum_{n=c}^{\infty} \frac{r^n}{c^{n-c}c!} = \frac{r^c}{c!} \sum_{n=c}^{\infty} \left(\frac{r}{c}\right)^{n-c}$$

$$= \frac{r^c}{c!} \sum_{m=0}^{\infty} \left(\frac{r}{c}\right)^{m}$$

$$= \frac{r^c}{c!} \frac{1}{1 - r/c} \qquad (r/c = \rho < 1).$$

Therefore we can write

$$p_0 = \left(\sum_{n=0}^{c-1} \frac{r^n}{n!} + \frac{r^c}{c!(1-\rho)}\right)^{-1} \qquad (r/c = \rho < 1). \tag{2.25}$$

Note that the condition for existence of a steady-state solution here is $\lambda/(c\mu) < 1$; that is, the mean arrival rate must be less than the mean maximum potential service rate of the system, which is intuitively what we would expect. Also note that when $c = 1$, (2.25) reduces to the $M/M/1/\infty$ equation.

We can now derive measures of effectiveness for the $M/M/c/\infty$ model utilizing the steady-state probabilities given by Equations (2.24) and (2.25) in a manner similar to that used for the $M/M/1/\infty$ model in Section 2.2.3. We first consider the expected queue size L_q, as it is computationally easier to determine than L, since we have only to deal with p_n for $n \geq c$. Thus

$$L_q = \sum_{n=c+1}^{\infty} (n-c)p_n = \sum_{n=c+1}^{\infty} (n-c)\frac{r^n}{c^{n-c}c!}p_0$$

$$= \frac{r^c p_0}{c!} \sum_{m=1}^{\infty} m\rho^m = \frac{r^c \rho p_0}{c!} \sum_{m=1}^{\infty} m\rho^{m-1}$$

$$= \frac{r^c \rho p_0}{c!} \frac{d}{d\rho} \sum_{m=1}^{\infty} \rho^m = \frac{r^c \rho p_0}{c!} \frac{d}{d\rho}\left(\frac{1}{1-\rho} - 1\right)$$

$$= \frac{r^c \rho p_0}{c!(1-\rho)^2}.$$

Thus

$$L_q = \left(\frac{r^c \rho}{c!(1 - \rho)^2} \right) p_0. \tag{2.26}$$

To find L now, we employ Little's formula to get W_q, then use W_q to find $W = W_q + 1/\mu$, and finally employ Little's formula again to calculate $L = \lambda W$. Thus we get

$$W_q = \frac{L_q}{\lambda} = \left(\frac{r^c}{c!(c\mu)(1 - \rho)^2} \right) p_0, \tag{2.27}$$

$$W = \frac{1}{\mu} + \left(\frac{r^c}{c!(c\mu)(1 - \rho)^2} \right) p_0, \tag{2.28}$$

and

$$L = r + \left(\frac{r^c \rho}{c!(1 - \rho)^2} \right) p_0. \tag{2.29}$$

The final result for L could have been obtained directly from L_q by using $L = L_q + r$, which we showed in Section 1.5 to be valid for any $G/G/c$ system.

Although the probability distributions of the waiting times, $W(t)$ and $W_q(t)$, are not needed to get W and W_q, it is useful to obtain them whenever possible so that questions concerning probabilities of waits greater than specified amounts can be answered. Therefore we next turn our attention to the derivation of these distributions, and proceed in a manner similar to that of Section 2.2.4. Letting T_q represent the random variable "time spent waiting in queue" and $W_q(t)$ its CDF, we have

$$W_q(0) = \Pr\{T_q = 0\} = \Pr\{\leq c - 1 \text{ in system}\}$$
$$= \sum_{n=0}^{c-1} p_n = p_0 \sum_{n=0}^{c-1} \frac{r^n}{n!}.$$

Now to evaluate $\Sigma \, r^n/n!$, recall that it appears in the expression for p_0 as given in (2.25), so that

$$\sum_{n=0}^{c-1} \frac{r^n}{n!} = \frac{1}{p_0} - \frac{r^c}{c!(1-\rho)},$$

thus giving

$$W_q(0) = p_0\left(\frac{1}{p_0} - \frac{r^c}{c!(1-\rho)}\right)$$

$$= 1 - \frac{r^c p_0}{c!(1-\rho)}. \tag{2.30}$$

For $T_q > 0$ and assuming FCFS,

$$W_q(t) = \text{Pr}\{T_q \leq t\} = W_q(0) + \sum_{n=c}^{\infty} \text{Pr}\{n - c + 1 \text{ completions in } \leq t \,|$$
$$\text{arrival found } n \text{ in system}\} \cdot p_n.$$

Now when $n \geq c$, the system output is Poisson with mean rate $c\mu$, so that the time between successive completions is exponential with mean $1/(c\mu)$, and the distribution of the time for the $n - c + 1$ completions is Erlang type $n - c + 1$. Thus we can write

$$W_q(t) = W_q(0) + p_0 \sum_{n=c}^{\infty} \frac{r^n}{c^{n-c}c!} \int_0^t \frac{c\mu(c\mu x)^{n-c}}{(n-c)!} e^{-c\mu x} \, dx$$

$$= W_q(0) + \frac{r^c p_0}{(c-1)!} \int_0^t \mu e^{-c\mu x} \sum_{n=c}^{\infty} \frac{(\mu r x)^{n-c}}{(n-c)!} \, dx$$

$$= W_q(0) + \frac{r^c p_0}{(c-1)!} \int_0^t \mu e^{-\mu x(c-r)} \, dx$$

$$= W_q(0) + \frac{r^c p_0}{c!(1-\rho)}(1 - e^{-(c\mu-\lambda)t}).$$

Putting this result together with (2.30), we find that

$$\boxed{W_q(t) = 1 - \frac{r^c p_0}{c!(1-\rho)} e^{-(c\mu-\lambda)t}.} \tag{2.31}$$

From (2.31), we note that

$$\Pr\{T_q > t\} = 1 - W_q(t) = \frac{r^c p_0}{c!(1 - \rho)} e^{-(c\mu - \lambda)t},$$

so that the conditional probability $\Pr\{T_q > t \mid T_q > 0\} = e^{-(c\mu - \lambda)t}$.

The reader should note that letting $c = 1$ reduces (2.31) to the equation for $W_q(t)$ of the $M/M/1$ model given in (2.21). Similar statements would be true for the other $M/M/c$ measures of effectiveness.

We leave as an exercise (Problem 2.17) to show that

$$W_q = E[T_q] = \int_0^\infty [1 - W_q(t)] \, dt = \left(\frac{r^c}{c!(c\mu)(1 - \rho)^2} \right) p_0,$$

as given by (2.27). We also see that there is the necessary jump discontinuity in $W_q(t)$, since there is always a nonzero probability that an arrival has no delay.

To find the formula for the CDF of the system waiting time, we first split the situation into two separate possibilities, namely, those customers having no line wait [probability $W_q(0)$] and those whose system wait is a line delay plus a service time [probability $1 - W_q(0)$]. The first of these two classes of customers has a CDF which is identical to the exponential service-time distribution, with mean $1/\mu$; the second has a CDF found as the convolution of the service-time distribution with a second exponential distribution having mean $1/(c\mu - \lambda)$, the latter representing the CDF of the line waiting time, given that $T_q > 0$ (see earlier in this section). This convolution can be also written as the difference of the two exponential functions (see Problem 2.19),

$$\Pr\{T \le t\} = \frac{c(1 - \rho)}{c(1 - \rho) - 1} (1 - e^{-\mu t}) - \frac{1}{c(1 - \rho) - 1} (1 - e^{-(c\mu - \lambda)t}).$$

Thus the overall CDF of $M/M/c$ system waits may be written as

$$W(t) = W_q(0)[1 - e^{-\mu t}] + [1 - W_q(0)]$$
$$\times \left(\frac{c(1 - \rho)}{c(1 - \rho) - 1} (1 - e^{-\mu t}) - \frac{1}{c(1 - \rho) - 1} (1 - e^{-(c\mu - \lambda)t}) \right)$$
$$= \frac{c(1 - \rho) - W_q(0)}{c(1 - \rho) - 1} (1 - e^{-\mu t}) - \frac{1 - W_q(0)}{c(1 - \rho) - 1} (1 - e^{-(c\mu - \lambda)t}).$$

We now illustrate these developments with an example.

Example 2.2
City Hospital's eye clinic offers free vision tests every Wednesday evening. There are three ophthalmologists on duty. A test takes, on the average, 20 min, and the actual time is found to be approximately exponentially distributed around this average. Clients arrive according to a Poisson process with a mean of 6/hr, and patients are taken on a first-come, first-served basis. The hospital planners are interested in knowing: (1) what is the average number of people waiting; (2) the average amount of time a patient spends at the clinic; and (3) the average percentage idle time of each of the doctors. Thus we wish to calculate L_q, W, and the percentage idle time of a server.

We begin by calculating p_0, since this factor appears in all the formulas derived for the measures of effectiveness. We have that $c = 3$, $\lambda = 6/\text{hr}$, and $\mu = 1/(20 \text{ min}) = 3/\text{hr}$. Thus $r = \lambda/\mu = 2$, $\rho = \frac{2}{3}$, and, from (2.25),

$$p_0 = \left(1 + 2 + \frac{2^2}{2!} + \frac{2^3}{3!(1 - \frac{2}{3})}\right)^{-1} = \frac{1}{9}.$$

From Equation (2.26), we next find that

$$L_q = \left(\frac{(2^3)(\frac{2}{3})}{3!(1 - \frac{2}{3})^2}\right)\left(\frac{1}{9}\right) = \frac{8}{9},$$

and from Equations (2.26) and (2.28) that

$$W = \frac{1}{\mu} + \frac{L_q}{\lambda} = \frac{1}{3} + \frac{\frac{8}{9}}{6} = \frac{13}{27} \text{ hr} \doteq 28.9 \text{ min}.$$

Next, we have already shown (see Table 1.2) that the long-term average fraction of idle time for any server in an $M/M/c$ is equal to $1 - \rho$. For this problem, therefore, each physician is idle $\frac{1}{3}$ of the time, since the traffic intensity is $\rho = \frac{2}{3}$. Given the three servers on duty, two of them will be busy at any time (on average), since $r = 2$. Furthermore, the fraction of time that there is at least one idle doctor can be computed here as $p_0 + p_1 + p_2 = \text{Pr}\{W_q = 0\} = \frac{5}{9}$.

2.4 QUEUES WITH PARALLEL CHANNELS AND TRUNCATION (*M/M/c/K*)

We now take up the parallel-server birth–death model *M/M/c/K*, in which there is a limit K placed on the number allowed in the system at any time.

The approach here is identical to that of the infinite-capacity *M/M/c* except that the arrival rate λ_n must now be 0 whenever $n \geq K$. It then follows from (2.24) that the steady-state system-size probabilities are given by

$$p_n = \begin{cases} \dfrac{\lambda^n}{n! \mu^n} p_0 & (1 \leq n < c), \\[2ex] \dfrac{\lambda^n}{c^{n-c} c! \mu^n} p_0 & (c \leq n \leq K). \end{cases} \tag{2.32}$$

The usual boundary condition that the probabilities must sum to 1 will yield p_0. Again, the computation is nearly identical to that for the *M/M/c*, except that now both series in the computation are finite and thus there is going to be no requirement that the traffic intensity ρ be less than 1. So

$$p_0 = \left(\sum_{n=0}^{c-1} \frac{\lambda^n}{n! \mu^n} + \sum_{n=c}^{K} \frac{\lambda^n}{c^{n-c} c! \mu^n} \right)^{-1}.$$

To simplify, consider the second summation above, with $r = \lambda/\mu$ and $\rho = r/c$:

$$\sum_{n=c}^{K} \frac{r^n}{c^{n-c} c!} = \frac{r^c}{c!} \sum_{n=c}^{K} \rho^{n-c}$$

$$= \begin{cases} \dfrac{r^c}{c!} \dfrac{1 - \rho^{K-c+1}}{1 - \rho} & (\rho \neq 1), \\[2ex] \dfrac{r^c}{c!} (K - c + 1) & (\rho = 1). \end{cases}$$

Thus

$$p_0 = \begin{cases} \left(\displaystyle\sum_{n=0}^{c-1} \frac{r^n}{n!} + \frac{r^c}{c!} \frac{1 - \rho^{K-c+1}}{1 - \rho} \right)^{-1} & (\rho \neq 1), \\[3ex] \left(\displaystyle\sum_{n=0}^{c-1} \frac{r^n}{n!} + \frac{r^c}{c!} (K - c + 1) \right)^{-1} & (\rho = 1). \end{cases} \tag{2.33}$$

We leave as an exercise (see Problem 2.28) to show that taking the limit as $K \to \infty$ in Equations (2.32) and (2.33) and restricting $\lambda/c\mu < 1$ yield the results obtained for the $M/M/c/\infty$ model given by Equations (2.24) and (2.25). Also it is noted that letting $c = 1$ in (2.32) and (2.33) yields the results for the $M/M/1/K$ model.

We next proceed to find the expected queue length as follows ($\rho \neq 1$):

$$
\begin{aligned}
L_q &= \sum_{n=c+1}^{K} (n-c)p_n \\
&= \frac{p_0 r^c}{c^{n-c}c!} \sum_{n=c+1}^{K} (n-c)r^{n-c} \\
&= \frac{p_0 r^c \rho}{c!} \sum_{n=c+1}^{K} (n-c)\rho^{n-c-1} = \frac{p_0 r^c \rho}{c!} \sum_{i=1}^{K-c} i\rho^{i-1} \\
&= \frac{p_0 r^c \rho}{c!} \frac{d}{d\rho}\left(\frac{1-\rho^{K-c+1}}{1-\rho}\right),
\end{aligned}
$$

or

$$
L_q = \frac{p_0 r^c \rho}{c!(1-\rho)^2}[1 - \rho^{K-c+1} - (1-\rho)(K-c+1)\rho^{K-c}]. \qquad (2.34)
$$

For $\rho = 1$, it is necessary to employ L'Hôpital's rule twice.

To obtain the expected system size, recall from our work with the unrestricted $M/M/c$ model that $L = L_q + r$. However, for the finite-waiting-space case, we need to adjust this result (and Little's formula, as well), since a fraction p_K of the arrivals do not join the system, because they have come when there is no waiting space left. Thus the actual rate of arrivals to join the system must be adjusted accordingly. Since Poisson arrivals see time averages (the PASTA property), it follows that the effective arrival rate seen by the servers is $\lambda(1 - p_K)$. We henceforth denote any such adjusted input rate as λ_{eff}. The relationship between L and L_q must therefore be reframed for this model to be $L = L_q + \lambda_{\text{eff}}/\mu = L_q + \lambda(1 - p_K)/\mu = L_q + r(1 - p_K)$. We know that the quantity $r(1 - p_K)$ must be less than c, since the average number of customers in service must be less than the total number of available servers. This suggests the definition of something called $\rho_{\text{eff}} = \lambda_{\text{eff}}/c\mu$, which would thus have to be less than 1 for any $M/M/c$ model even though no such restriction exists on the value of $\rho = \lambda/c\mu$.

Expected values for waiting times can be readily obtained by use of Little's formula as

$$
\begin{aligned}
W &= \frac{L}{\lambda_{\text{eff}}} = \frac{L}{\lambda(1 - p_K)}, \\
W_q &= W - \frac{1}{\mu} = \frac{L_q}{\lambda_{\text{eff}}}.
\end{aligned}
\tag{2.35}
$$

For $M/M/1/K$, all of the above measures of effectiveness reduce to considerably simpler expressions, with key results of

$$
p_0 =
\begin{cases}
\dfrac{1 - \rho}{1 - \rho^{K+1}} & (\rho \neq 1), \\[2ex]
\dfrac{1}{K + 1} & (\rho = 1),
\end{cases}
\tag{2.36}
$$

$$
p_n =
\begin{cases}
\dfrac{(1 - \rho)\rho^n}{1 - \rho^{K+1}} & (\rho \neq 1), \\[2ex]
\dfrac{1}{K + 1} & (\rho = 1),
\end{cases}
\tag{2.37}
$$

and

$$
L_q =
\begin{cases}
\dfrac{\rho}{1 - \rho} - \dfrac{\rho(K\rho^K + 1)}{1 - \rho^{K+1}} & (\rho \neq 1), \\[2ex]
\dfrac{K(K - 1)}{2(K + 1)} & (\rho = 1),
\end{cases}
\tag{2.38}
$$

with $L = L_q + (1 - p_0)$. Note that this final relationship implies that $1 - p_0 = \lambda(1 - p_K)/\mu$, which when rewritten as $\mu(1 - p_0) = \lambda(1 - p_K)$ verifies that the system's effective output rate must equal its effective input rate.

The derivation of the waiting-time CDF is somewhat complicated, since the series are finite, though they can be expressed in terms of cumulative Poisson sums, as we shall show. Also, it is now necessary to derive the arrival-point probabilities $\{q_n\}$, since the input is no longer Poisson because of the size truncation at K, and $q_n \neq p_n$.

We use Bayes' theorem to determine the q_n, so that

$$q_n \equiv \Pr\{n \text{ in system} \mid \text{arrival about to occur}\}$$

$$= \frac{\Pr\{\text{arrival about to occur} \mid n \text{ in system}\} \cdot p_n}{\sum_{n=0}^{K} \Pr\{\text{arrival about to occur} \mid n \text{ in system}\} \cdot p_n}$$

$$= \lim_{\Delta t \to 0} \left\{ \frac{[\lambda \, \Delta t + o(\Delta t)]p_n}{\sum_{n=0}^{K-1} [\lambda \, \Delta t + o(\Delta t)]p_n} \right\}$$

$$= \lim_{\Delta t \to 0} \left\{ \frac{[\lambda + o(\Delta t)/\Delta t]p_n}{\sum_{n=0}^{K-1} [\lambda + o(\Delta t)/\Delta t]p_n} \right\}$$

$$= \frac{\lambda p_n}{\lambda \sum_{n=0}^{K-1} p_n}$$

$$= \frac{p_n}{1 - p_K} \qquad (n \le K - 1).$$

We note in passing that had this same analysis been performed for $M/M/c/\infty$, then the final portion of the equation above would be equal to p_n, since p_K goes to 0 when the capacity constraint is removed (i.e., K goes to ∞). Thus it follows for $M/M/c/\infty$ that $q_n = p_n$.

Finally, to get the CDF $W_q(t)$ for the line delays, we note, in a fashion similar to the derivation leading to (2.31), that

$$W_q(t) = \Pr\{T_q \le t\} = W_q(0) + \sum_{n=c}^{K-1} \Pr\{n - c + 1 \text{ completions in } \le t \mid$$
$$\text{arrival found } n \text{ in system}\} \cdot q_n,$$

since there cannot be arrivals joining the system whenever they encounter K customers. It follows that

$$W_q(t) = W_q(0) + \sum_{n=c}^{K-1} q_n \int_0^t \frac{c\mu(c\mu x)^{n-c}}{(n-c)!} e^{-c\mu x} \, dx$$

$$= W_q(0) + \sum_{n=c}^{K-1} q_n \left(1 - \int_t^\infty \frac{c\mu(c\mu x)^{n-c}}{(n-c)!} e^{-c\mu x} \, dx \right).$$

For the simplification of Equation (1.15), Section 1.7, we have shown that

$$\int_t^\infty \frac{\lambda(\lambda x)^m}{m!} e^{-\lambda x} \, dx = \sum_{i=0}^{m} \frac{(\lambda t)^i e^{-\lambda t}}{i!}.$$

Letting $m = n - c$ and $\lambda = c\mu$ gives

$$\int_{t}^{\infty} \frac{c\mu(c\mu x)^{n-c}}{(n-c)!} e^{-c\mu x} \, dx = \sum_{i=0}^{n-c} \frac{(c\mu t)^i e^{-c\mu t}}{i!}$$

and hence

$$W_q(t) = W_q(0) + \sum_{n=c}^{K-1} q_n - \sum_{n=c}^{K-1} q_n \sum_{i=0}^{n-c} \frac{(c\mu t)^i e^{-c\mu t}}{i!}$$

$$= 1 - \sum_{n=c}^{K-1} q_n \sum_{i=0}^{n-c} \frac{(c\mu t)^i e^{-c\mu t}}{i!}.$$

Example 2.3

Consider an automobile emission inspection station with three inspection stalls, each with room for only one car. It is reasonable to assume that cars wait in such a way that when a stall becomes vacant, the car at the head of the line pulls up to it. The station can accommodate at most four cars waiting (seven in the station) at one time. The arrival pattern is Poisson with a mean of one car every minute during the peak periods. The service time is exponential with mean 6 min. I. M. Fussy, the chief inspector, wishes to know the average number in the system during peak periods, the average wait (including service), and the expected number per hour that cannot enter the station because of full capacity.

Using minutes as the basic time unit, $\lambda = 1$ and $\mu = \frac{1}{6}$. Thus we have $r = 6$ and $\rho = 2$ for this *M/M/3/7* system. We first calculate p_0 from Equation (2.33) and find that

$$p_0 = \left(\sum_{n=0}^{2} \frac{6^n}{n!} + \frac{6^3}{3!} \frac{1 - 2^5}{1 - 2} \right)^{-1}$$

$$= \frac{1}{1141} \doteq 0.00088.$$

From Equation (2.34) we then get

$$L_q = \frac{p_0(6^3)(2)}{3!} [1 - 2^5 + 5(2^4)] = \frac{3528}{1141} \doteq 3.09 \text{ cars,}$$

so that

$$L = L_q + r(1 - p_K) = \frac{3528}{1141} + 6 \left(1 - \frac{6^7}{(3^4)(3!)(1141)} \right) = \frac{9606}{1141} \doteq 6.06 \text{ cars.}$$

To find the average wait during peak periods, Equation (2.35) is used and it is found that

$$W = \frac{L}{\lambda_{\text{eff}}} = \frac{L}{\lambda(1 - p_7)}$$

$$= \frac{L}{1 - p_0 6^7/(3^4 3!)} \doteq 12.3 \text{ min.}$$

The expected number of cars per hour that cannot enter the station is given by

$$60\lambda p_k = 60 p_7 = \frac{60 p_0 6^7}{3^4 3!} \doteq 30.4 \text{ cars/hr.}$$

Might this suggest an alternative setup for the inspection station?

2.5   ERLANG'S FORMULA (*M/M/c/c*)

The special case of the truncated queue $M/M/c/K$ for which $K = c$, that is, where no line is allowed to form, gives rise to a stationary distribution which is known as Erlang's first formula and can be readily obtained from Equations (2.32) and (2.33) with $K = c$ as

$$p_n = \frac{\dfrac{(\lambda/\mu)^n}{n!}}{\displaystyle\sum_{i=0}^{c} \dfrac{(\lambda/\mu)^i}{i!}} \qquad (0 \le n \le c). \tag{2.39}$$

The resultant formula for p_c is itself called *Erlang's loss formula* and corresponds to the probability of a full system at any time in the steady state, namely,

$$p_c = \frac{r^c/c!}{\sum_{i=0}^{c} r^i/i!} \qquad (r = \lambda/\mu).$$

Note that multiplying the numerator and denominator of the right-hand side of p_c by e^{-r} results in the ratio of a Poisson probability and a cumulative Poisson probability, so that one could make use of Poisson numbers in calculating p_c.

Since the input to the $M/M/c/c$ is Poisson, the probability that an arrival is lost is equal to the probability that all channels are busy. The original physical situation which motivated Erlang in 1917 to devise this model was the simple telephone network. Incoming calls arrive as a Poisson stream, service times are mutually independent, exponential random variables, and all calls that arrive and find every trunk line busy, that is, get a busy signal, are turned away. The model has, in fact, always been of great value in telecommunications design.

But the great importance of this formula lies in the very surprising fact that (2.39) is valid for *any M/G/c/c, independent* of the form of the service-time distribution. That is to say, the steady-state system probabilities are only a function of the mean service time, not of the underlying CDF. Erlang was also able to deduce the formula for the case when service times are constant. While this result was later shown to be correct, his proof was not quite valid. Later works by Vaulot (1927), Pollaczek (1932), Palm (1938), Kosten (1948–1949), and others smoothed out Erlang's 1917 proof and supplied proofs for the complete arbitrariness of the service-time distribution. A further addition to this sequence of papers was work by Takács (1969), which supplied some additional results for the problem. We shall prove the validity of Erlang's loss formula for general service in Chapter 5, Section 5.2.2.

2.6   QUEUES WITH UNLIMITED SERVICE ($M/M/\infty$)

We now treat a queueing model for which there is unlimited service, that is, an infinite number of servers available. This model is often referred to as the ample-server problem. A self-service situation is a good example of the use of such a model.

We make use of the general birth–death results with $\lambda_n = \lambda$ and $\mu_n = n\mu$, for all n, which yields

$$p_n = \frac{r^n}{n!}p_0, \qquad p_0 = \left(\sum_{n=0}^{\infty} \frac{r^n}{n!} \right)^{-1}.$$

But the infinite series in the expression for p_0 is clearly identical to the representation of e^r. Thus it follows that

$$\boxed{p_n = \frac{r^n e^{-r}}{n!} \qquad (n \geq 0),} \tag{2.40}$$

so that the steady-state probability distribution of n in the system is Poisson with parameter $r = \lambda/\mu$. Note that the value of λ/μ is not in any way restricted for the existence of a steady-state solution. It also turns out (we show this in Section 5.2.3) that Equation (2.40) is valid for *any M/G/∞* model, that is, p_n depends only on the mean service time and not at all on the form of the service-time distribution. It is not surprising that this is true here in light of a similar result we mentioned previously for $M/M/c/c$, since p_n of (2.40) could have been obtained from (2.39) by taking the limit as $c \to \infty$.

The expected system size is the mean of the Poisson distribution of (2.40) and is thus found as $L = r = \lambda/\mu$. Since we have as many servers as customers in the system, $L_q = 0 = W_q$. The average waiting time in the system obviously becomes merely the average service time, so that $W = 1/\mu$, and the waiting-time distribution function $W(t)$ is identical to the service-time distribution, namely, exponential with mean $1/\mu$.

Example 2.4

Television station KCAD in a large western metropolitan area wishes to know the average number of viewers it can expect on a Saturday evening prime-time program. It has found from past surveys that people turning on their television sets on Saturday evening during prime time can be described rather well by a Poisson distribution with a mean of 100,000/hr. There are five major TV stations in the area, and it is believed that a given person chooses among these essentially at random. Surveys have also shown that the average person tunes in for 90 min and that viewing times are approximately exponentially distributed.

Since the mean arrival rate (people tuning in KCAD during prime time on Saturday evening) is 100,000/5 = 20,000/hr and the mean service time is 90 min or 1.5 hr, it follows that the average number of viewers during prime time is $L = 20,000/\frac{2}{3} = 30,000$ people.

2.7 FINITE SOURCE QUEUES

In previous models we have assumed that the population from which arrivals come (the calling population) is infinite, since the number of arrivals in any time interval is a Poisson random variable with a denumerably infinite sample space. We now treat a problem where the calling population is finite, say of size M, and future event occurrence probabilities are functions of system state. A typical application of this model is that of machine repair, where the calling population is the machines, an arrival corresponds to a machine breakdown, and the repair crews are the servers. We assume c servers are available, that the service times are identical exponential random variables with mean $1/\mu$, and that the arrival process is described as follows. If a calling unit is not in the system at time t, the probability it will have entered by time $t + \Delta t$ is $\lambda \Delta t + o(\Delta t)$; that is, the time a calling unit spends outside the system is exponential with mean $1/\lambda$.

Because of these assumptions, we can use the birth–death theory developed previously, with the birth and death rates now given by

$$\lambda_n = \begin{cases} (M-n)\lambda & (0 \le n < M), \\ 0 & (n \ge M) \end{cases}$$

and

$$\mu_n = \begin{cases} n\mu & (0 \le n < c), \\ c\mu & (n \ge c). \end{cases}$$

Using (1.39) yields (with, as usual, $r = \lambda/\mu$)

$$p_n = \begin{cases} \dfrac{M!/(M-n)!}{n!} r^n p_0 & (1 \le n < c), \\[2ex] \dfrac{M!/(M-n)!}{c^{n-c}c!} r^n p_0 & (c \le n \le M), \end{cases}$$

or equivalently,

$$p_n = \begin{cases} \dbinom{M}{n} r^n p_0 & (1 \le n < c), \\[2ex] \dbinom{M}{n} \dfrac{n!}{c^{n-c}c!} r^n p_0 & (c \le n \le M). \end{cases} \tag{2.41}$$

The algebraic form of the $\{p_n\}$ does not allow the closed-form calculation of p_0 (also true for L_q, L, W, and W_q). Instead, we must calculate each of the coefficients multiplying p_0 in (2.41) (call them $\{a_n, n = 1, 2, 3, \ldots, M\}$) and then complete the computation as

$$p_0 = (1 + a_1 + a_2 + a_3 + \cdots + a_M)^{-1}.$$

To find the average number of customers in the system (if we are dealing with the machine breakdown problem, we are interested in machines "down" for repair), we use the definition of expected value and get

$$L = \sum_{n=1}^{M} n p_n = p_0 \sum_{n=1}^{M} n a_n.$$

But to obtain L_q and the expected waiting-time measures, W and W_q, we must first find the effective mean rate of arrivals into the system. As noted earlier in our presentation of the birth and death rates for this model, the mean arrival rate when the system is in state n is $(M-n)\lambda$. Summing up over all possible n after weighting each term by the appropriate probability

p_n, we find that

$$\lambda_{\text{eff}} = \sum_{n=0}^{M-1} (M - n)\lambda p_n = \lambda(M - L). \tag{2.42}$$

Equation (2.42) is certainly intuitive, since, on average, L are in the system and hence, on average, $M - L$ are outside and each has a mean arrival rate of λ. Now, for L_q, we know from our earlier work that

$$L_q = L - \frac{\lambda_{\text{eff}}}{\mu} = L - r(M - L). \tag{2.43}$$

It follows then from Little's formula that

$$W = \frac{L}{\lambda(M - L)} \quad \text{and} \quad W_q = \frac{L_q}{\lambda(M - L)}. \tag{2.44}$$

For the single-server version of this problem, the expression for the system-state probabilities found in (2.41) reduces to

$$p_n = \binom{M}{n} n! r^n p_0 \quad (0 \le n \le M),$$

and the rest of the analysis is identical.

We also point out here that there is an important *invariance* result for finite-source queues, similar in importance to the fact that $M/G/c/c$ steady-state probabilities are independent of the form of G. It is that (2.41) is valid for any finite-source system with exponential service, independent of the nature of the distribution of time to breakdown, as long as the lifetimes are independent with mean $1/\lambda$. The interested reader is referred to Bunday and Scraton (1980) for details of the proof. Furthermore, the $M/G/c/c$-type result also holds in that if the number of repairmen equal the number of machines, the repair distribution can be G, as long as the failure times are exponential.

Example 2.5
The Train SemiConductor Company uses five robots in the manufacture of its circuit boards. The robots break down periodically, and the company has two repair people to do service when robots fail. When one is fixed, the time until the next breakdown is thought to be exponentially distributed with a mean of 30 hr. The shop always has enough of a work backlog to ensure that all robots in operating condition will be working. The repair time for each service is thought to be exponentially distributed with a mean of 3 hr.

The shop manager wishes to know the average number of robots operational at any given time, the expected downtime of a robot that requires repair, and the expected percentage of idle time of each repairer.

To answer any of these questions, we must first calculate p_0. In this example, $M = 5$, $c = 2$, $\lambda = \frac{1}{30}$, and $\mu = \frac{1}{3}$, and thus $r = \lambda/\mu = \frac{1}{10}$. We use (2.41) and obtain the five $\{a_n\}$ multipliers as $a_1 = 5/10 = \frac{1}{2}$; $a_2 = \frac{1}{10}$; $a_3 = 15/1000 = \frac{3}{200}$; $a_4 = 15/10{,}000 = \frac{3}{2000}$; and $a_5 = 15/200{,}000 = \frac{3}{40000}$. It thus follows that

$$p_0 = \left(1 + \tfrac{1}{2} + \tfrac{1}{10} + \tfrac{3}{200} + \tfrac{3}{2000} + \tfrac{3}{40000}\right)^{-1} = \tfrac{40000}{64663} \doteq 0.619.$$

The average number of operational robots is $M - L$, where

$$L = p_0\left(1 \times \tfrac{1}{2} + 2 \times \tfrac{1}{10} + 3 \times \tfrac{3}{200} + 4 \times \tfrac{3}{2000} + 5 \times \tfrac{3}{40000}\right) = \tfrac{30055}{64663} \doteq 0.465.$$

Thus $5 - 0.465$, or 4.535, robots are in operating condition on average. The expected downtime can be found from (2.44) and is

$$W = \frac{\frac{30055}{64663}}{\frac{1}{30}\left(5 - \frac{30055}{64663}\right)} = \frac{901650}{29326} \doteq 3.075 \text{ hr.}$$

The average fraction of idle time of each server is

$$p_0 + \tfrac{1}{2}p_1 = p_0\left(1 + \tfrac{1}{2}a_1\right) = p_0\left(1 + \tfrac{1}{4}\right) = \tfrac{50000}{64663} \doteq 0.773,$$

so each repair person is idle approximately 77% of the time.

The manager, because of the long idle time, is interested in knowing the answer to the same questions if the repair force is reduced to one person. The results are

$$p_0 \doteq 0.564, \qquad L \doteq 0.640,$$
$$M - L \doteq 4.360, \qquad W \doteq 4.400 \text{ hr.}$$

Since over four robots are expected operational at any time under both situations and the downtime increase is about one hour on reducing to only one repair person, the manager might well decide to move one of them to work elsewhere.

This model can be generalized to include the use of spares. We assume now that we have Y spares on hand, so that when a machine fails, a spare is immediately substituted for it. If it happens that all spares are being used and a breakdown occurs, then the system becomes short. When a machine is repaired, it then becomes a spare (unless the system is short, in which case the repaired machine goes immediately into service). For this model, λ_n is different from before and is given by

$$
\lambda_n = \begin{cases} M\lambda & (0 \le n < Y), \\ (M - n + Y)\lambda & (Y \le n < Y + M), \\ 0 & (n \ge Y + M). \end{cases}
$$

Again considering c repairmen, we have

$$
\mu_n = \begin{cases} n\mu & (0 \le n < c), \\ c\mu & (n \ge c). \end{cases}
$$

We first assume $c \le Y$ and use Equation (1.39) once more to find that (with $r = \lambda/\mu$)

$$
p_n = \begin{cases} \dfrac{M^n}{n!} r^n p_0 & (0 \le n < c), \\[3mm] \dfrac{M^n}{c^{n-c}c!} r^n p_0 & (c \le n < Y), \\[3mm] \dfrac{M^Y M!}{(M - n + Y)! c^{n-c} c!} r^n p_0 & (Y \le n \le Y + M). \end{cases} \tag{2.45}
$$

If Y is very large, we essentially have an infinite calling population with mean arrival rate $M\lambda$. Letting Y go to infinity in (2.45) yields the $M/M/c/\infty$ results of (2.24) with $M\lambda$ for λ.

When $c > Y$, we have

$$
p_n = \begin{cases} \dfrac{M^n}{n!} r^n p_0 & (0 \le n \le Y), \\[3mm] \dfrac{M^Y M!}{(M - n + Y)! n!} r^n p_0 & (Y + 1 \le n < c), \\[3mm] \dfrac{M^Y M!}{(M - n + Y)! c^{n-c} c!} r^n p_0 & (c \le n \le Y + M). \end{cases} \tag{2.46}
$$

It should be noted that when $Y = 0$ (i.e., no spares), Equation (2.46) reduces to Equation (2.41).

Observe that the direct use of Equation (2.46) to get the coefficients $\{a_n\}$

might get particularly messy when M and Y are large. Fortunately, we can avoid such difficulty by using the recursion relating p_{n+1} to p_n, which comes quite naturally out of the birth–death formulation and is a direct consequence of local balance. The balance equation may be written in recursive form as

$$p_{n+1} = \left(\frac{\lambda_n}{\mu_{n+1}} \right) p_n,$$

so that it follows for the no-spares problem that

$$p_{n+1} = \begin{cases} \dfrac{M-n}{n+1} rp_n & (0 \le n \le c-1), \\ \dfrac{M-n}{c} rp_n & (c \le n \le M-1). \end{cases}$$

Similar recursions can be developed when $Y > 0$, and, indeed, all of the more complicated birth–death modules in the text's software use this sort of recursion to do their computations.

The empty-system probability, p_0, can be found as previously for the no-spares model by once more using the fact that the probabilities must sum to 1, so that the computation of p_0 is made up of finite sums. The same is true for L and L_q. To obtain results for W and W_q comparable to Equation (2.44), we must again obtain the *effective* mean arrival rate λ_{eff}. To get λ_{eff}, one can use Equation (2.43) directly or obtain it using logic similar to that used for Equation (2.42), namely,

$$\begin{aligned} \lambda_{\mathrm{eff}} &= \sum_{n=0}^{Y-1} M\lambda p_n + \sum_{n=Y}^{Y+M} (M-n+Y)\lambda p_n \\ &= \lambda \left(M - \sum_{n=Y}^{Y+M} (n-Y)p_n \right). \end{aligned} \tag{2.47}$$

The calculations for the spares model conclude in a similar fashion to those of the no-spares case.

As a final point to close this section, suppose that we desired the waiting-time distribution in addition to the mean waiting-time value. The standard procedure in the past has been to find the waiting time of an arriving customer conditioned on the existence of n in the system at the point of arrival, and then unconditioning with respect to the stationary distribution $\{q_n\}$, where $\{q_n\}$ are the state probabilities given an arrival occurs. As was the case for $M/M/c/K$, we have $q_n \ne p_n$, and therefore, before we can obtain the CDF of the system or line wait, we have to relate the general-time probability p_n to the probability q_n that an *arrival* finds n in the system. For the general finite-source queue (machine-repair problem without spares), the two probabilities are related as

$$q_n = \frac{(M - n)p_n}{k},$$

where k is an appropriate normalizing constant determined from summing the $\{q_n\}$ to 1. To prove this, we again use Bayes' theorem as in the $M/M/c/K$ situation of Section 2.4:

$\Pr\{n \text{ in system} \mid \text{arrival is about to occur}\}$

$$= \frac{\Pr\{n \text{ in system}\} \Pr\{\text{arrival is about to occur} \mid n \text{ in system}\}}{\Sigma_n (\Pr\{n \text{ in system}\} \Pr\{\text{arrival is about to occur} \mid n \text{ in system}\})}$$

$$= \lim_{\Delta t \to 0} \frac{p_n[(M - n)\lambda \, \Delta t + o(\Delta t)]}{\Sigma_n p_n[(M - n)\lambda \, \Delta t + o(\Delta t)]}$$

$$= \frac{(M - n)p_n}{\Sigma_n (M - n)p_n} = \frac{(M - n)p_n}{M - L}.$$

As an interesting sidelight it can be shown (see Problem 2.41) that $q_n(M)$, the arrival-point probability for the no-spares case with M machines, equals $p_n(M - 1)$, the general-time probability for the no-spares machine-repair situation with $M - 1$ machines. This is not necessarily the case when there are spares. In fact, when spares are present, $q_n(M)$ can be shown to be (see Problem 2.42)

$$q_n = \begin{cases} \dfrac{Mp_n}{M - \Sigma_{n=y}^{y+M} (n - Y)p_n} & (0 \le n \le Y - 1), \\[3mm] \dfrac{(M - n + Y)p_n}{M - \Sigma_{n=y}^{y+M} (n - Y)p_n} & (Y \le n \le Y + M - 1). \end{cases} \quad (2.48)$$

This is not equal to $p_n(M - 1)$, but rather is equal to $p_n(Y - 1)$; that is, if we reduce the population size by one by reducing spares, not operating machines, then the general-time probabilities of the reduced population equal the arrival-point probabilities of the original population.

The waiting-time distributions again turn out to be in terms of cumulative Poisson sums, as in the $M/M/c/K$ model. The analysis proceeds as follows:

$$W_q(t) = \Pr\{T_q \leq t\} = W_q(0) + \sum_{n=c}^{Y+M-1} [\Pr\{n-c+1 \text{ completions in } \leq t \mid \text{arrival found } n \text{ in system}\} \cdot q_n]$$

$$= W_q(0) + \sum_{n=c}^{Y+M-1} q_n \int_0^t \frac{c\mu(c\mu x)^{n-c}}{(n-c)!} e^{-c\mu x} dx$$

$$= W_q(0) + \sum_{n=c}^{Y+M-1} q_n \left[1 - \int_t^\infty \frac{c\mu(c\mu x)^{n-c}}{(n-c)!} e^{-c\mu x} dx \right]$$

$$= 1 - \sum_{n=c}^{Y+M-1} q_n \sum_{i=0}^{n-c} \frac{(c\mu t)^i}{i!} e^{-c\mu t}.$$

2.8 STATE-DEPENDENT SERVICE

In this section we treat Markovian queues with state-dependent service; that is, the mean service rate depends on the state of the system (number in the system). In many real situations, the server (or servers) may speed up when seeing a long line forming. On the other hand, it may happen if the server is inexperienced that he/she/it becomes flustered and the mean service rate actually decreases as the system becomes more congested. It is these types of situations that are now considered.

The first model we consider is one in which a single server has two mean rates, say *slow* and *fast*. Work is performed at the slow rate until there are k in the system, at which point there is a switch to the fast rate (e.g., the service mechanism here might be a machine with two speeds.) We still assume the service times are Markovian, but the mean rate μ_n now explicitly depends on the system state n. Furthermore, no limit on the number in the system is imposed. Thus μ_n is given as

$$\mu_n = \begin{cases} \mu_1 & (1 \leq n < k), \\ \mu & (n \geq k). \end{cases} \tag{2.49}$$

Assuming the arrival process is Poisson with parameter λ and utilizing Equation (1.39), we have

$$p_n = \begin{cases} \rho_1^n p_0 & (0 \leq n < k), \\ \rho_1^{k-1} \rho^{n-k+1} p_0 & (n \geq k), \end{cases} \tag{2.50}$$

where $\rho_1 = \lambda/\mu_1$ and $\rho = \lambda/\mu < 1$. Because the probabilities must sum to 1, it follows that

$$p_0 = \left(\sum_{n=0}^{k-1} \rho_1^n + \sum_{n=k}^{\infty} \rho_1^{k-1} \rho^{n-k+1} \right)^{-1},$$

so that

$$p_0 = \begin{cases} \left(\dfrac{1 - \rho_1^k}{1 - \rho_1} + \dfrac{\rho \rho_1^{k-1}}{1 - \rho} \right)^{-1} & (\rho_1 \neq 1, \quad \rho < 1), \\[4mm] \left(k + \dfrac{\rho}{1 - \rho} \right)^{-1} & (\rho_1 = 1, \quad \rho < 1). \end{cases} \tag{2.51}$$

The reader should note that if $\mu_1 = \mu$, Equations (2.50) and (2.51) reduce of those of $M/M/1$.

To find the expected system size, we proceed as in Section 2.2.4 (assuming $\rho_1 \neq 1$):

$$\begin{aligned} L = \sum_{n=0}^{\infty} n p_n &= p_0 \left(\sum_{n=0}^{k-1} n \rho_1^n + \sum_{n=k}^{\infty} n \rho_1^{k-1} \rho^{n-k+1} \right) \\ &= p_0 \left[\rho_1 \sum_{n=0}^{k-1} n \rho_1^{n-1} + \rho_1 \left(\frac{\rho_1}{\rho} \right)^{k-2} \sum_{n=k}^{\infty} n \rho^{n-1} \right] \\ &= p_0 \left[\rho_1 \frac{d}{d\rho_1} \sum_{n=0}^{k-1} \rho_1^n + \rho_1 \left(\frac{\rho_1}{\rho} \right)^{k-2} \frac{d}{d\rho} \sum_{n=k}^{\infty} \rho^n \right] \\ &= p_0 \left[\rho_1 \frac{d}{d\rho_1} \left(\frac{1 - \rho_1^k}{1 - \rho_1} \right) + \rho_1 \left(\frac{\rho_1}{\rho} \right)^{k-2} \frac{d}{d\rho} \left(\frac{1}{1 - \rho} - \frac{1 - \rho^k}{1 - \rho} \right) \right]. \end{aligned}$$

So, finally,

$$L = p_0 \left(\frac{\rho_1 [1 + (k-1)\rho_1^k - k\rho_1^{k-1}]}{(1 - \rho_1)^2} + \frac{\rho \rho_1^{k-1} [k - (k-1)\rho]}{(1 - \rho)^2} \right). \tag{2.52}$$

We can find L_q using the last two relationships of Table 1.2 as

$$L_q = L - (1 - p_0),$$

and W and W_q from Little's formulas as

$$W = L/\lambda \quad \text{and} \quad W_q = L_q/\lambda.$$

Note that the relation $W = W_q + 1/\mu$ cannot be used here, since μ is not constant but depends on the system-state switch point k. However, by combining the above equations, we see that

$$W = W_q + \frac{1 - p_0}{\lambda},$$

which implies that the expected service time is $(1 - p_0)/\lambda$.

Example 2.6

Sonia Schine and John B. Goode have invented and applied for a patent on a machine which polishes automobiles. They have formed a partnership called the Goode-Schine Garage, and have rented an old building in which they have set up their machine. Since this is a part-time job for both partners, the garage is open on Saturdays only. Customers are taken on a first-come, first-served basis, and since their garage is in a low-density population and traffic area, there is virtually no limit on the number of customers who can wait. The car-polishing machine can run at two speeds. At the low speed, it takes 40 min, on the average, to polish a car. On the high speed, it takes only 20 min on the average. Once a switch is made, the actual times can be assumed to follow an exponential distribution.

It is estimated that customers will arrive according to a Poisson process with a mean interarrival time of 30 min. Ms. Schine has had a course in queueing theory and decides to calculate the effect of two policies: switching to high speed if there are any customers waiting (i.e., two or more in the system) versus switching to high speed only when more than one customer is waiting (three or more in the system). The machine speeds can be switched at any time, even while the machine is in operation. It is desired to know the average waiting time under the two policies.

It is therefore necessary to calculate W for the case when $k = 2$ and then for $k = 3$. We must first calculate p_0 from Equation (2.51), then L from Equation (2.52), and finally W from Little's formula. Before doing these computations, we first calculate ρ_1 and ρ to be

$$\rho_1 = \frac{\lambda}{\mu_1} = \frac{\frac{1}{30}}{\frac{1}{40}} = \frac{4}{3} \quad \text{and} \quad \rho = \frac{\lambda}{\mu} = \frac{\frac{1}{30}}{\frac{1}{20}} = \frac{2}{3}.$$

For case 1, $k = 2$ and

$$p_0 = \tfrac{1}{5} = 0.2, \qquad L = \tfrac{12}{5} = 2.4 \text{ cars},$$

$$W = L/\lambda = 2.4/\tfrac{1}{30} = 72 \text{ min} = 1 \text{ hr } 12 \text{ min}.$$

For case 2, $k = 3$ and

$$p_0 = \tfrac{3}{23} \doteq 0.13, \qquad L = 204/69 \doteq 2.96 \text{ cars},$$
$$W = L/\tfrac{1}{30} \doteq 89 \text{ min} = 1 \text{ hr } 29 \text{ min}.$$

Schine feels that the average wait of 17 more minutes for switching speeds at three rather than two might not have an adverse effect on their clientele. However, it costs more to run the machine at the higher speed. In fact, it is estimated that it costs \$15 per operating hour to operate the machine at low speed and \$24 per operating hour to operate at high speed. Thus the expected cost of operation when switching at k is given by

$$C(k) = 15 \sum_{n=1}^{k-1} p_n + 24 \sum_{n=k}^{\infty} p_n$$
$$= 15 \sum_{n=1}^{k-1} \rho_1^n p_0 + 24 \left(1 - \sum_{n=0}^{k-1} \rho_1^n p_0 \right).$$

For case 1 we have

$$C(2) = 15(\tfrac{4}{3})(\tfrac{1}{5}) + 24[1 - \tfrac{1}{5} - (\tfrac{4}{3})(\tfrac{1}{5})] \doteq \$16.80/\text{hr},$$

while for case 2 the average operating cost per hour is

$$C(3) = 15(\tfrac{3}{23})[\tfrac{4}{3} + \tfrac{16}{9}] + 24[1 - \tfrac{3}{23} - (\tfrac{4}{3})(\tfrac{3}{23}) - (\tfrac{16}{9})(\tfrac{3}{23})] \doteq \$17.22/\text{hr}.$$

Thus it is cheaper to switch at $k = 2$ even though the hourly cost per operating hour is higher, since switching at $k = 2$ yields a higher idle-time probability p_0, which more than makes up for the higher high-speed operating cost. In addition, this provides better customer service in that W is reduced by 17 min. If, however, the high-speed operating cost were even higher, it might turn out that switching at $k = 3$ could be more economical (see Problem 2.47). It should be noted that the costs of \$16.80/hr and \$17.22/hr, respectively, are not true costs per operating hour, but rather average costs per hour including hours or portions of hours when the machine is idle. It should further be noted that to obtain a true optimal operating policy, the cost of customer wait should also be included. This topic will be taken up in greater detail in Chapter 6, Section 6.5.1.

The model above can be generalized to a c-server system with a rate switch at $k > c$ (see Problem 2.49). One can also generalize the model by having two rate switches (or even more if desired) at k_1 and k_2 ($k_2 < k_1$) say (see Problem 2.50), and could also derive results for a multiserver, multirate, state-dependent model of this type.

Yet another possible type of model with state-dependent service is one where the mean service rate changes whenever the system size changes. We again assume a single-server, Markovian state-dependent-service model, possibly with μ_n given by

$$\mu_n = n^\alpha \mu.$$

Again from the general steady-state birth–death solution of (1.39), we have

$$p_n = \frac{\lambda^n}{n^\alpha (n-1)^\alpha (n-2)^\alpha \cdots (1)^\alpha \mu^n} p_0$$

$$= \frac{r^n}{(n!)^\alpha} p_0 \qquad (r = \lambda/\mu),$$

where

$$p_0 = \left(\sum_{n=0}^{\infty} \frac{r^n}{(n!)^\alpha} \right)^{-1}.$$

The infinite series for p_0 converges for any r as long as $\alpha > 0$, but it is not obtainable in closed form unless $\alpha = 1$. (In fact, when $\alpha = 1$, $\Sigma\, r^n/n! = e^r$, which reduces the model to the ample-server case presented in Section 2.6.) Thus to evaluate p_0 (and thus all the other mean measures of effectiveness) in the general case, numerical methods must be used. See Gross and Harris (1985) for a discussion of possible procedures for numerically calculating p_0.

2.9 QUEUES WITH IMPATIENCE

The intent of this section is to discuss the effects of customer impatience upon the development of waiting lines of the $M/M/c$ type. These concepts may be easily extended to other Markovian models in a reasonably straight-forward fashion and will not be explicitly pursued. However, some examples of impatience are discussed later for the $M/G/1$ queue.

Customers are said to be impatient if they tend to join the queue only when a short wait is expected and tend to remain in line if the wait has been sufficiently small. The impatience that results from an excessive wait is just as important in the total queueing process as the arrivals and departures. When this impatience becomes sufficiently strong and customers leave before being served, the manager of the enterprise involved must take action to reduce the congestion to levels that customers can tolerate. The models subsequently developed find practical application in this attempt of manage-ment to provide adequate service for its customers with tolerable waiting.

Impatience generally takes three forms. The first is balking, the reluctance of a customer to join a queue upon arrival; the second reneging, the reluctance to remain in line after joining and waiting; and the third jockeying between lines when each of a number of parallel lines has its own queue.

2.9.1 *M/M/1* Balking

In real practice, it often happens that arrivals become discouraged when the queue is long and do not wish to wait. One such model is the $M/M/c/K$; that is, if people see K ahead of them in the system, they do not join. Generally, unless K is the result of a physical restriction such as no more places to park or room to wait, people will not act quite like that voluntarily. Rarely do all customers have exactly the same discouragement limit all the time.

Another approach to balking is to employ a series of monotonically decreasing functions of the system size multiplying the average rate λ. Let b_n be this function, so that $\lambda_n = b_n \lambda$ and

$$0 \le b_{n+1} \le b_n \le 1 \qquad (n > 0, \quad b_0 \equiv 1).$$

For example, using (1.39) when $c = 1$ gives

$$p_n = p_0 \prod_{i=1}^{n} \frac{\lambda_{i-1}}{\mu_i}$$

$$= p_0 \left(\frac{\lambda}{\mu}\right)^n \prod_{i=1}^{n} b_{i-1}.$$

Possible examples that may be useful for the discouragement function b_n are $1/(n + 1)$, $1/(n^2 + 1)$, and $e^{-\alpha n}$. People are not always discouraged because of queue size, but may attempt to estimate how long they would have to wait. If the queue is moving quickly, then the person may join a long one. On the other hand, if the queue is slow-moving, a customer may become discouraged even if the line is short. Now if n people are in the system, an estimate for the average waiting time might be n/μ, if the customer had an idea of μ. We usually do, so a plausible balking function might thus be $b_n = e^{-\alpha n/\mu}$. Also note that the $M/M/1/K$ model is a special case of balking where $b_i = 1$ for $0 \le i \le K - 1$ and 0 otherwise.

2.9.2 *M/M/1* Reneging

Customers who tend to be impatient may not always be discouraged by excessive queue size, but may instead join the queue to see how long their wait may become, all the time retaining the prerogative to renege if their estimate of their total wait is intolerable. We now consider a single-channel birth–death model where both reneging and the simple balking of the previous section exist, which gives rise to a reneging function $r(n)$ defined by

$$r(n) = \lim_{\Delta t \to 0} \frac{\Pr\{\text{unit reneges during } \Delta t \text{ when there are } n \text{ customers present}\}}{\Delta t},$$

$$r(0) = r(1) \equiv 0.$$

This new process is still birth–death, but the death rate must now be adjusted to $\mu_n = \mu + r(n)$. Thus it follows from (1.39) that

$$p_n = p_0 \prod_{i=1}^{n} \frac{\lambda_{i-1}}{\mu_i}$$

$$= p_0 \lambda^n \prod_{i=1}^{n} \frac{b_{i-1}}{\mu + r(i)} \qquad (n \geq 1),$$

where

$$p_0 = \left(1 + \sum_{n=1}^{\infty} \lambda^n \prod_{i=1}^{n} \frac{b_{i-1}}{\mu + r(i)} \right)^{-1}.$$

A good possibility for the reneging function $r(n)$ is $e^{\alpha n/\mu}$, $n \geq 2$. A waiting customer would probably estimate the average system waiting time as n/μ if $n - 1$ customers were in front of him, assuming an estimate for μ were available. Again, the probability of a renege would be estimated by a function of the form $e^{\alpha n/\mu}$.

As was mentioned in the introduction to this section, there is yet an additional form of impatience called jockeying, that is, moving back and forth among the several subqueues before each of several multiple channels. Though this phenomenon is quite interesting and clearly applicable to numerous real-life situations, jockeying is analytically difficult to pursue very far, especially when we have more than two channels, since the probability functions become too complicated and the general concepts become hazy. Even though partial results can be obtained for the two-channel case, no particular insight into the multichannel cases is gained from it. If, however, the reader is specifically interested in this subject, we refer to Koenigsberg (1966).

2.10 TRANSIENT BEHAVIOR

In this section we consider the transient behavior of three specific queueing systems, namely, $M/M/1/1$ (no one allowed to wait), $M/M/1/\infty$, and $M/M/\infty$ (ample service). This discussion is restricted to these three models, since the mathematics becomes extremely complicated with the slightest relaxation of Poisson–exponential assumptions, and it is our feeling that the exhibition of some fairly simple results is sufficient for our purposes. Even these three transient derivations vary greatly in difficulty. The $M/M/1/1$ solution can be found fairly easily, but the problem becomes much more complicated when the restriction on waiting room is relaxed, or multiple servers are considered.

2.10.1   Transient Behavior of $M/M/1/1$

The derivation of the transient probabilities $\{p_n(t)\}$ that at an arbitrary time t there are n customers in a single-channel system with Poisson input, exponential service, and no waiting room is a straightforward procedure, since $p_n(t) = 0$ for all $n > 1$. It begins in the usual fashion from the birth–death differential equations as given by Equation (1.30), with $\lambda_0 = \lambda$, $\lambda_n = 0$, $n > 0$, and $\mu_1 = \mu$:

$$\frac{dp_1(t)}{dt} = -\mu p_1(t) + \lambda p_0(t),$$
$$\frac{dp_0(t)}{dt} = -\lambda p_0(t) + \mu p_1(t). \tag{2.53}$$

These differential–difference equations can be solved easily in view of the fact that it is always true that

$$p_0(t) + p_1(t) = 1.$$

Hence Equation (2.53) is equivalent to

$$\frac{dp_1(t)}{dt} \equiv p_1'(t) = -\mu p_1(t) + \lambda[1 - p_1(t)].$$

So

$$p_1'(t) + (\lambda + \mu)p_1(t) = \lambda.$$

This is just an ordinary first-order linear differential equation with constant coefficients, whose solution can clearly be seen from the discussion of Section 1.7 to be

$$p_1(t) = Ce^{-(\lambda+\mu)t} + \frac{\lambda}{\lambda + \mu}.$$

To determine C, we use the boundary value of $p_1(t)$ at $t = 0$, which is $p_1(0)$. Thus

$$C = p_1(0) - \frac{\lambda}{\lambda + \mu},$$

and consequently

$$p_1(t) = \frac{\lambda}{\lambda + \mu}(1 - e^{-(\lambda+\mu)t}) + p_1(0)e^{-(\lambda+\mu)t},$$

$$\tag{2.54}$$

$$p_0(t) = \frac{\mu}{\lambda + \mu}(1 - e^{-(\lambda+\mu)t}) + p_0(0)e^{-(\lambda+\mu)t},$$

since $p_0(t) = 1 - p_1(t)$ for all t.

The stationary solution can be found directly from (2.53) in the usual way by letting the derivatives equal zero and then, with the use of the fact that $p_0 + p_1 = 1$, solving for p_0 and p_1 ($M/M/1/K$ with $K = 1$). Also, the limiting (steady-state, equilibrium) solution can be found as the limit of the transient solution of (2.54) as t goes to ∞, and we find that

$$p_1 = \frac{\rho}{\rho + 1} \quad \text{and} \quad p_0 = \frac{1}{\rho + 1}.$$

Existence of the limiting distribution is always assured, independent of the value of $\rho = \lambda/\mu$, and thus it is identical to the stationary distribution (to see this, put $K = 1$ in the p_n expression for the $M/M/1/K$ of Section 2.4).

To get an even better feel for the behavior of this queueing system for small values of time, let us graph $p_1(t)$ from Equation (2.54). First rewrite (2.54) in the form

$$p_1(t) = a + be^{-ct},$$

where, of course,

$$a = \frac{\lambda}{\lambda + \mu} = \frac{\rho}{\rho + 1},$$

$$b = -\frac{\lambda}{\lambda + \mu} + p_1(0),$$

$$c = \lambda + \mu.$$

Therefore we would clearly have Figure 2.3 in the event, for example, that b was greater than 0. We see that $p_1(t)$ is asymptotic to $a = \rho/(\rho + 1) = p_1$.

In addition, it can be observed that if the initial probability $p_1(0)$ is assumed to be the stationary probability p_1, then b becomes 0 and $p_1(t)$ becomes equal to this equilibrium value of p_1. In other words, the queueing process can be translated into the steady state at any time by making the

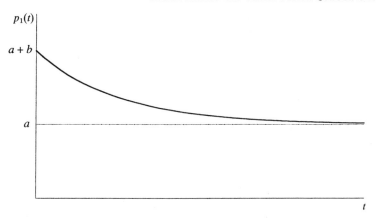

Fig. 2.3 Illustration of transient solution $p_1(t)$.

assumption that the process is already in equilibrium. This property is, in fact, true for any ergodic queueing system, independent of any assumptions about its parameters.

2.10.2   Transient Behavior of $M/M/1/\infty$

The transient derivation for $M/M/1/\infty$ is quite a complicated procedure, so presentation of it is in outline form only. A more complete picture of the details can be found in Gross and Harris (1985) and Saaty (1961). The solution of this problem postdated that of the basic Erlang work by nearly half a century, with the first published solution due to Ledermann and Reuter (1954), in which they used spectral analysis for the general birth–death process. In the same year, an additional paper appeared on the solution of this problem by Bailey (1954), and later one by Champernowne (1956). Bailey's approach to the time-dependent problem was via generating functions for the partial differential equation, and Champernowne's was via complex combinatorial methods. It is Bailey's approach that has been the most popular over the years, and this is basically the one we take. Remember that the key thing that makes this problem more difficult than may seem at first is that we are dealing with an *infinite* system of linear differential equations.

To begin, let it be assumed that the initial system size at time 0 is i. That is, if $N(t)$ denotes the number in the system at time t, then $N(0) = i$. The differential–difference equations governing the system size are given in Equation (1.30) as

$$p'_n(t) = -(\lambda + \mu)p_n(t) + \lambda p_{n-1}(t) + \mu p_{n+1}(t) \qquad (n > 0), \tag{2.55}$$

$$p'_0(t) = -\lambda p_0(t) + \mu p_1(t).$$

It turns out that we solve these time-dependent equations using a combination of probability generating functions, partial differential equations, and Laplace transforms. Therefore define

$$P(z, t) = \sum_{n=0}^{\infty} p_n(t)z^n \qquad (z \text{ complex}),$$

such that the summation is convergent in and on the unit circle (i.e., for $|z| \le 1$), with its Laplace transform defined as

$$\bar{P}(z, s) = \int_0^{\infty} e^{-st} P(z, t)\, dt \qquad (\text{Re } s > 0).$$

After the generating function is formed from (2.55), yielding a partial differential equation, it is found when the Laplace transform is taken that

$$\bar{P}(z, s) = \frac{z^{i+1} - \mu(1 - z)\bar{p}_0(s)}{(\lambda + \mu + s)z - \mu - \lambda z^2}, \qquad (2.56)$$

where $\bar{p}_0(s)$ is the Laplace transform of $p_0(t)$.

Since the Laplace transform $\bar{P}(z, s)$ converges in the region $|z| \le 1$, Re $s > 0$, wherever the denominator of the right-hand side of (2.56) has zeros in that region, so must the numerator. This fact is henceforth used to evaluate $\bar{p}_0(s)$. The denominator has two zeros, since it is a quadratic in z and they are (as functions of s)

$$z_1 = \frac{\lambda + \mu + s - \sqrt{(\lambda + \mu + s)^2 - 4\lambda\mu}}{2\lambda},$$

$$z_2 = \frac{\lambda + \mu + s + \sqrt{(\lambda + \mu + s)^2 - 4\lambda\mu}}{2\lambda}, \qquad (2.57)$$

where the square root is taken so that its real part is positive. It is clear that $|z_1| < |z_2|$, $z_1 + z_2 = (\lambda + \mu + s)/\lambda$, and $z_1 z_2 = \mu/\lambda$. In order to complete the derivation, the following important and well-known theorem of complex analysis due to Rouché is employed:

Rouché's Theorem.   *If $f(z)$ and $g(z)$ are functions analytic inside and on a closed contour C and if $|g(z)| < |f(z)|$ on C, then $f(z)$ and $f(z) + g(z)$ have the same number of zeros inside C.*

A proof of this theorem may be found in any book on complex variables. For $|z| = 1$ and $\text{Re } s > 0$ we see that

$$|f(z)| \equiv |(\lambda + \mu + s)z| = |\lambda + \mu + s| > \lambda + \mu \geq |\mu + \lambda z^2| \equiv |g(z)|;$$

hence, from Rouché's theorem, $(\lambda + \mu + s)z - \mu - \lambda z^2$ has only one zero in the unit circle. This zero is obviously z_1, since $|z_1| < |z_2|$. Thus equating the numerator of the right-hand side of (2.56) to zero for $z = z_1$ gives

$$\bar{p}_0(s) = \frac{z_1^{i+1}}{\mu(1 - z_1)}.$$

When this transform for $p_0(t)$ is inserted into (2.56) and the result written in infinite series form, we find [remembering that $i = N(0)$]

$$\bar{P}(z, s) = \frac{1}{\lambda z_2} \sum_{j=0}^{i} z_1^j z^{i-j} \sum_{k=0}^{\infty} \left(\frac{z}{z_2}\right)^k + \frac{z_1^{i+1}}{\lambda z_2(1 - z_1)} \sum_{k=0}^{\infty} \left(\frac{z}{z_2}\right)^k \qquad (|z/z_2| < 1).$$

Now, the transform of $p_n(t)$, $\bar{p}_n(s)$, is the coefficient of z^n in the Laplace transform of the generating function $P(z, t)$, $\bar{P}(z, s)$. So the next step in the process is to find $\bar{p}_n(s)$, and then this is, in turn, inverted to get $p_n(t)$, utilizing key properties of the transforms of Bessel functions in this last step. The final result is, in fact, in terms of modified Bessel functions of the first kind, $I_n(y)$, and is

$$p_n(t) = e^{-(\lambda + \mu)t} \left[\left(\frac{\lambda}{\mu}\right)^{(n-i)/2} I_{n-i}(2\sqrt{\lambda\mu}t) + \left(\frac{\lambda}{\mu}\right)^{(n-i-1)/2} I_{n+i+1}(2\sqrt{\lambda\mu}t) \right.$$

$$\left. + \left(1 - \frac{\lambda}{\mu}\right)\left(\frac{\lambda}{\mu}\right)^n \sum_{j=n+i+2}^{\infty} \left(\frac{\lambda}{\mu}\right)^{-j/2} I_j(2\sqrt{\lambda\mu}t) \right] \qquad (2.58)$$

for all $n \geq 0$, where

$$I_n(y) = \sum_{k=0}^{\infty} \frac{(y/2)^{n+2k}}{k!(n + k)!} \qquad (n > -1)$$

is the infinite series for the modified Bessel function of the first kind.

As, of course, is necessary, we can show, using the properties of the Bessel functions, that (2.58) tends to the stationary result $p_n = (1 - \rho)\rho^n$ as $t \rightarrow \infty$ when $\rho = \lambda/\mu < 1$. When $\lambda/\mu \geq 1$, $p_n(t) \rightarrow 0$ for all n, so that only when $\lambda/\mu < 1$ do we get a valid steady-state probability distribution. This then agrees with our previous steady-state result of (2.3), as it should, since this is an ergodic system.

2.10.3   Transient Behavior of *M/M/∞*

It turns out for this model that the development of the transient solution is not too difficult a task. To begin, let it be assumed that the initial system size at time 0 is 0, so that $N(0) = 0$. The differential–difference equations governing the system size are derived from Equation (1.30) with $\lambda_n = \lambda$ and $\mu_n = n\mu$ as

$$p_n'(t) = -(\lambda + n\mu)p_n(t) + \lambda p_{n-1}(t) + (n + 1)\mu p_{n+1}(t) \qquad (n > 0),$$

$$p_0'(t) = -\lambda p_0(t) + \mu p_1(t). \tag{2.59}$$

It turns out that we can solve these time-dependent equations using a combination of probability generating functions and partial differential equations, without requiring Laplace transforms. It is found that the generating function of the probabilities $\{p_n(t)\}$ is (see, e.g., Gross and Harris, 1985)

$$P(z, t) = \sum_{n=0}^{\infty} p_n(t)z^n = \exp\left((z - 1)(1 - e^{-\mu t})\frac{\lambda}{\mu}\right). \tag{2.60}$$

To obtain the state probabilities it is necessary to expand (2.60) in a power series, $a_0 + a_1 z + a_2 z^2 + \cdots$, where the coefficients a_n then are the transient probabilities $p_n(t)$ we desire, since we are expanding a probability generating function. To do this we can use a Taylor series expansion about zero (Maclaurin series), and we find that

$$p_n(t) = \frac{1}{n!}\left((1 - e^{-\mu t})\frac{\lambda}{\mu}\right)^n \exp\left(-(1 - e^{-\mu t})\frac{\lambda}{\mu}\right) \qquad (n \geq 0).$$

It is easily seen that letting $t \to \infty$ yields the steady-state solution, which is equivalent to the Poisson distribution of Equation (2.40),

$$p_n = \frac{(\lambda/\mu)^n e^{-\lambda/\mu}}{n!}.$$

We remind the reader that, in general, analytical solutions for transient queueing situations are extremely difficult to obtain. We do treat a few special cases briefly in later chapters, cases such as *M/G/1*, *G/M/1*, and *M/G/∞*. However, since transient solutions require solving sets of differential equations, numerical methods can often be quite successfully employed. We treat this topic in some detail in Chapter 7, Section 7.3.2.

2.11   BUSY-PERIOD ANALYSES FOR $M/M/1$ AND $M/M/c$

A busy period is defined to begin with the arrival of a customer to an idle channel and to end when the channel next becomes idle. A busy cycle is the sum of a busy period and an adjacent idle period, or equivalently, the time between two successive departures leaving an empty system, or two successive arrivals to an empty system. Since the arrivals are assumed to follow a Poisson process, the distribution of the idle period is exponential with mean $1/\lambda$; hence the CDF of the busy cycle for the $M/M/1$ is the convolution of this negative exponential with the CDF of the busy period itself. Therefore the CDF of the busy period is sufficient to describe the busy cycle also and is found as follows.

The CDF of the busy period is determined by considering the original $M/M/1$ differential equations given in (2.55) with an absorbing barrier imposed at zero system size (i.e., $\lambda_0 = 0$ in the birth–death equations) and an initial size of 1 [i.e., $p_1(0) = 1$]. Then it should be clear that $p_0(t)$ will, in fact, be the required busy period CDF and $p_0'(t)$ the density. The necessary equations are

$$p_0'(t) = \mu p_1(t) \qquad \text{[because of absorbing barrier]},$$

$$p_1'(t) = -(\lambda + \mu)p_1(t) + \mu p_2(t) \qquad \text{[because of absorbing barrier]},$$

$$p_n'(t) = -(\lambda + \mu)p_n(t) + \lambda p_{n-1}(t) + \mu p_{n+1}(t) \qquad \text{[same as (2.55)]}.$$

In a fashion identical to the $M/M/1$ transient, it can be shown (see Gross and Harris, 1985) that the Laplace transform of the generating function is

$$\bar{P}(z, s) = \frac{z^2 - (\mu - \lambda z)(1 - z)(z_1/s)}{\lambda(z - z_1)(z_2 - z)}, \qquad (2.61)$$

where z_1 and z_2 have the same values as in (2.57). Now the Laplace transform of $p_0(t)$, $\bar{p}_0(s)$, is the first coefficient of the power series $\bar{P}(z, s)$ and is thus found as $\bar{P}(0, s)$. This gives

$$\bar{p}_0(s) = \frac{2\mu}{s[\lambda + \mu + s + \sqrt{(\lambda + \mu + s)^2 - 4\lambda\mu}]}.$$

From the properties of Laplace transforms and Bessel functions, it can then be shown that the busy period's density function is

$$p_0'(t) = \frac{\sqrt{\mu/\lambda}\, e^{-(\lambda+\mu)t} I_1(2\sqrt{\lambda\mu t})}{t}.$$

To get the average length of the busy period, $E[T_{bp}]$, we simply find the value of the negative of the derivative of the transform of $p'_0(t)$, $sp_0(s)$, evaluated at $s = 0$. But an attractive alternative way to find the mean length of the busy period is to use the simple steady-state ratio argument that

$$\frac{1 - p_0}{p_0} = \frac{E[T_{bp}]}{E[T_{idle}]} = \frac{E[T_{bp}]}{1/\lambda}.$$

Since $p_0 = 1 - \lambda/\mu$, it follows that the expected lengths of the busy period and busy cycle, respectively, are

$$E[T_{bp}] = \frac{1}{\mu - \lambda} \quad \text{and} \quad E[T_{bc}] = \frac{1}{\lambda} + \frac{1}{\mu - \lambda}. \tag{2.62}$$

Equation (2.62) holds for all *M/G/*1-type queues, since the exponential service property played no role in the derivation.

It is not too difficult to extend the notion of the busy period conceptually to the multichannel case. Recall that for one channel a busy period is defined to begin with the arrival of a customer at an idle channel and to end when the channel next becomes idle. In an analogous fashion, let us define an *i*-channel busy period for *M/M/c* $(0 \le i \le c)$ to begin with an arrival at the system at an instant when there are $i - 1$ in the system and to end at the very next point in time when the system size dips to $i - 1$. Let us say that the case where $i = 1$ (an arrival to an empty system) defines the system busy period. In fashion similar to that for *M/M/*1, use $T_{b,i}$ to denote the random variable "length of the *i*-channel busy period." Then the CDF of $T_{b,i}$ is determined by considering the original *M/M/c* differential–difference equations of (2.55) with an absorbing barrier imposed at a system size of $i - 1$ and an initial size of *i*. Then it should be clear that $p_{i-1}(t)$ will, in fact, be the required CDF, and its derivative the density. The necessary equations are

$p'_{i-1}(t) = i\mu p_i(t)$ [because of absorbing barrier]

$p'_i(t) = -(\lambda + i\mu)p_i(t) + (i + 1)\mu p_{i+1}(t)$ [because of absorbing barrier]

$p'_n(t) = -(\lambda + n\mu)p_n(t) + \lambda p_{n-1}(t) + (n + 1)\mu p_{n+1}(t)$ $(i < n < c)$

$p'_n(t) = -(\lambda + c\mu)p_n(t) + \lambda p_{n-1}(t) + c\mu p_{n+1}(t)$ $(n \ge c)$

Proceeding further at this point would get us bogged down in great algebraic detail. Suffice it to say that any resultant CDF would be in terms of modified Bessel functions, but with enough time and patience, $p'_{i-1}(t)$, $\bar{p}_{i-1}(s)$, and $E[T_{b,i}]$ can be gotten.

PROBLEMS

Whenever a problem is best solved by the use of this book's accompanying software, we have added a boldface C to the right of the problem number.

2.1.   (a) Prove that if $\bar{f}(s)$ is the Laplace transform of $f(t)$, then $\bar{f}(s + a)$ is the Laplace transform of $e^{-at}f(t)$.

(b) Show that the Laplace transform of a linear combination of functions is the same linear combination of the transforms of the functions: symbolically, $\mathscr{L}[\Sigma_i\, a_i f_i(t)] = \Sigma_i\, a_i \mathscr{L}[f_i(t)]$.

2.2.   Use the properties of Laplace transforms to find the functions whose Laplace transforms are the following:

(a) $(s + 1)/(s^2 + 2s + 2)$.

(b) $1/(s^2 - 3s + 2)$.

(c) $1/[s^2(s^2 + 1)]$.

(d) $e^{-s}/(s + 1)$.

2.3.   For the following generating functions (not necessarily *probability* generating functions), write the sequence they generate:

(a) $G(z) = 1/(1 - z)$.

(b) $G(z) = z/(1 - z)$.

(c) $G(z) = e^z$.

2.4.   Show that the moment generating function of the sum of independent random variables is equal to the product of their moment generating functions.

2.5.   Use the result of Problem 2.4 to show that:

(a) The sum of two independent Poisson random variables is a Poisson random variable.

(b) The sum of two independent and identical exponential random variables has a gamma or Erlang distribution.

(c) The sum of two independent but nonidentical exponential random variables has a density which is a linear combination of the two original exponential densities.

2.6C.   Use the method of operators to solve the difference equation

$$p_{n+4} - 10p_{n+3} + 35p_{n+2} - 50p_{n+1} + 24p_n = 0,$$

subject to $p_0 = 0$, $p_1 = 1$, $p_2 = 0$, and $p_3 = 1$. [*Hint:* You will first need to find the roots of a fourth-degree polynomial equation and then solve a 4×4 simultaneous linear system.]

2.7. For $M/M/1$ and $M/M/c$, find $E[T_q \mid T_q > 0]$, that is, the expected time one must wait in the queue, given that one has to wait at all.

2.8. Derive $W(t)$ and $w(t)$ (the total-waiting-time CDF and its density) as given by the equations (2.22).

2.9. Equation (1.4) is valid for any queueing system and can be used to prove Little's formula for the $G/G/1$ queue in a somewhat different manner than that used in Section 1.5.1. The approach is to plot the cumulative count of arrivals on the same graph as the cumulative count of departures. Then it can be seen that the area between these two step functions from the beginning of a busy period to the beginning of the next (a busy cycle) is the accumulated total of the waiting times of all the customers who have entered into the system during this busy cycle. Use this argument to derive an empirical version of Little's formula over a busy cycle.

2.10. What effect does doubling λ and μ have on L, L_q, W, and W_q in an $M/M/1$ model?

2.11C. A graduate research assistant "moonlights" in the food court in the student union in the evenings. He is the only one on duty at the counter during the hours he works. Arrivals to the counter seem to follow the Poisson distribution with mean of 10/hr. Each customer is served one at a time and the service time is thought to follow an exponential distribution with a mean of 4 min. Answer the following questions.

(a) What is the probability of having a queue?

(b) What is the average queue length?

(c) What is the average time a customer spends in the system?

(d) What is the probability of a customer spending more than 5 min in the queue before being waited on?

(e) The graduate assistant would like to spend his idle time grading papers. If he can grade 22 papers an hour on average when working continuously, how many papers per hour can he average while working his shift?

2.12C. A rent-a-car maintenance facility has capabilities for routine maintenance (oil change, lubrication, minor tune-up, wash, etc.) for only one car at a time. Cars arrive there according to a Poisson process at a mean rate of three per day, and service time to perform this maintenance seems to have an exponential distribution with a mean of $\frac{7}{24}$ day. It costs the company a fixed $375 a day to operate the facility. The company estimates loss in profit on a car of $25/day for every day the car is tied up in the shop. The company, by changing certain procedures and hiring faster mechanics, can decrease the mean service time to $\frac{1}{4}$ day. This also increases their operating costs. Up to what value can the operating cost increase before it is no longer economically attractive to make the change?

2.13. Before parts are assembled into a final product, they must be painted. The parts arrive at the painting center, which consists of an automatic painting machine, in a random fashion as a Poisson process at an average rate of λ/hr. It is also observed that the painting process appears to be exponential with an average time of service equal to $1/\mu$ hr. It is estimated that it costs the company about $\$C_1$/unit per hour spent tied up in the painting center. The cost of owning and operating the painting machine is strictly a function of the speed of its operation, and, in particular, one that works at an average rate of μ is charged at the rate of $\$\mu C_2$/hr, whether or not it is always in operation. How large should μ be to allow this firm to minimize the cost of the painting operation?

2.14. Find the probability that k or more are in an $M/M/c/\infty$ system for any $k \geq c$.

2.15. For the $M/M/c/\infty$ model, give an expression for p_n in terms of p_c instead of p_0, and then derive L_q and $W_q(t)$ in terms of ρ and p_c.

2.16C. (a) Our local fast-food emporium, Burger Bliss, has yet to learn a lot about queueing theory. So it does not require that all waiting customers form a single line, and instead they make every arrival randomly choose one of *three* lines formed before each server during the weekday lunch period. But they are so traditional about managing their lines that barriers have been placed between the lines to prevent jockeying. Suppose that the overall stream of incoming customers has settled in at a constant rate of 60/hr (Poisson-distributed) and that the time to complete a customer's order is well described by an exponential distribution of mean 150 seconds, independent and identically from one customer to the next. Assuming steady state, what is the average total system size?

(b) The BB has now agreed that it is preferable to have one line feeding the three servers, so the barriers have been removed. What is the expected steady-state system size now?

2.17. Verify the formula for the mean line delay W_q of the $M/M/c/\infty$ queue.

2.18C. Use the $M/M/c$ software module to calculate L, L_q, W, W_q, $\Pr\{N \geq k\}$, and $\Pr\{T_q \geq t\}$ for $\lambda = 60$/hr, $\mu = 0.75$/min, $c = 1$ and 2, $k = 2$ and 4, and $t = 0.01$ and 0.03 hr.

2.19. For $M/M/c/\infty$, derive the distribution function of the system waiting time for those customers for whom $T_q > 0$, remembering that their wait is the sum of a line delay and a service time.

2.20C. Use the $M/M/c/K$ software module to calculate L, L_q, W, W_q, p_K, $\Pr\{N \geq k\}$, and $\Pr\{T_q \geq t\}$ for $\lambda = 2$/min, $\mu = 45$/hr, $c = 2$, $K = 6$, $k = 2$ and 4, and $t = 0.01$ and 0.02 hr.

2.21C. The office of the Deputy Inspector General for Inspection and Safety administers the Air Force Accident and Incident Investigation and Reporting Program. It has established 25 investigation teams to analyze and evaluate each accident or incident to make sure it is properly reported to accident investigation boards. Each of these teams is dispatched to the locale of the accident or incident as each requirement for such support occurs. Support is only rendered those commands that have neither the facilities nor qualified personnel to conduct such services. Each accident or incident will require a team being dispatched for a random amount of time, apparently exponential with mean of 3 weeks. Requirements for such support are received by the Deputy Inspector General's office as a Poisson process with mean rate of 347/yr. At any given time, two teams are not available due to personnel leaves, sickness, and so on. Find the expected time spent by an accident or incident in and waiting for evaluation.

2.22C. An organization is presently involved in the establishment of a telecommunication center so that it may provide a more rapid outgoing message capability. Overall, the center is responsible for the transmission of outgoing messages, and receives and distributes incoming messages. The center manager at this time is primarily concerned with determining the number of transmitting personnel required at the new center. Outgoing message transmitters are responsible for making minor corrections to messages, assigning numbers when absent from original message forms, maintaining an index of codes and a 30-day file of outgoing messages, and actually transmitting the messages. It has been predetermined that this process is exponential and requires a mean time of 28 min/message. Transmission personnel will operate at the center 7 hr/day, 5 days/week. All outgoing messages will be processed in the order they are received and follow a Poisson process with a mean rate of 21 per 7-hr day. Processing on messages requiring transmission must be started within an average of 2 hr from the time they arrive at the center. Determine the minimum number of transmitting personnel to accomplish this service criterion. If the service criterion were to require the probability of any message waiting for the start of processing for more than 3 hr to be less than 0.05, how many transmitting personnel would be required?

2.23C. A small branch bank has two tellers, one for receipts and one for withdrawals. Customers arrive to each teller's cage according to a Poisson distribution with a mean of 20/hr. (The total mean arrival rate at the bank is 40/hr.) The service time of each teller is exponential with a mean of 2 min. The bank manager is considering changing the setup to allow each teller to handle both withdrawals and deposits to avoid the situations which arise from time to time when the queue is sizable in front of one teller while the other is idle. However, since the tellers would have to handle both receipts and withdrawals, their efficiency would decrease to a mean service time of 2.4 min. Compare the present system with the proposed system with respect to the total expected number of people in the bank, the expected time a customer would have to spend in the bank, the probability of a customer having to wait more than 5 min, and the average idle time of the tellers.

2.24C. The Hott Too Trott Heating and Air Conditioning Company must choose between operating two types of service shops for maintaining its trucks. It estimates that trucks will arrive at the maintenance facility according to a Poisson distribution with mean rate of one every 40 min and believes that this rate is independent of which facility is chosen. In the first type of shop, there are dual facilities operating in parallel; each facility can service a truck in 30 min on the average (the service time follows an exponential distribution). In the second type there is a single facility, but it can service a truck in 15 min on the average (service times are also exponential in this case). To help management to decide, they ask their operations research analyst, Mr. K. L. Dude, to answer the following questions:

(a) How many trucks, on the average, will be in each of the two types of facilities?

(b) How long, on the average, will a truck spend in each of the two types of facilities?

(c) Management calculates that each minute a truck must spend in the shop reduces contribution to profit by two dollars. They also know from previous experience in running dual-facility shops that the cost of operating such a facility is one dollar per minute (including labor, overhead, etc.). What would the operating cost per minute have to be for operating the second-type (single facility) shop in order for there to be no difference between the two types of shops?

2.25C. The ComPewter Company, which leases out high-end computer workstations, considers it necessary to overhaul its equipment once a year. Alternative 1 is to provide two separate maintenance stations where all work is done by hand (one machine at a time) for a total annual cost of $750,000. The maintenance time for a machine has an exponential distribution with a mean of 6 hr. Alternative 2 is to provide one maintenance station with mostly automatic equipment involving an annual cost of $1 million. In this case, the maintenance time for a machine has an exponential distribution with a mean of 3 hr. For both alternatives, the machines arrive according to a Poisson input with a mean arrival rate of one every 8 hr (since the company leases such a large number of machines, we can consider the machine population as infinite). The cost of down time per machine is $150/hr. Which alternative should the company choose? Assume that the maintenance facilities are always open and that they work $(24)(365) = 8760$ hr/yr.

2.26. Show that (a) an $M/M/1/\infty$ is always better with respect to L than an $M/M/2/\infty$ with the same ρ and (b) an $M/M/2/\infty$ is always better than two independent $M/M/1/\infty$ queues with the same service rate but each getting half of the arrivals.

2.27. For Problem 2.26(a), show that the opposite is true when considering L_q. In other words, faced with a choice between two M/M systems with identical arrival rates, one with two servers and one with a single server who can work twice as fast as each of the two servers, which is the preferable system? Give some specific examples.

2.28. Show for the $M/M/c/K$ model that taking the limit for p_n and p_0 as $K \to \infty$ and restricting $\lambda/c\mu < 1$ in Equations (2.32), (2.33), and (2.34) yield the results obtained for the $M/M/c/\infty$ model.

2.29. Show that the $M/M/c/K$ equations (2.32)–(2.34) reduce to those for $M/M/1/K$ when $c = 1$.

2.30C. For the $M/M/3/K$ model, compute L_q as K goes from 3 to "∞" for each of the following ρ values: 1.5, 1, 0.8, 0.5. Comment.

2.31C. Find the probability that a customer's wait in queue exceeds 20 min for an $M/M/1/3$ model with $\lambda = 4$/hr and $1/\mu = 15$ min.

2.32C. A small drive-it-through-yourself car wash, in which the next car cannot go through the washing procedure until the car in front is completely finished, has a capacity to hold on its grounds a maximum of 10 cars (including the one in wash). The company has found its arrivals to be Poisson with mean rate of 20 cars/hr, and its service times to be exponential with a mean of 12 min. What is the average number of cars lost to the firm every 10-hr day as a result of its capacity limitations?

2.33C. Under the assumption that customers will not wait if no seats are available, Example 2.1's Hair Salon Proprietor Cutt can rent, on Saturday, the conference room of a small computer software firm adjacent to her shop for $30.00 (cost of cleanup on a Saturday). Her shop is open on Saturdays from 8:00 A.M. to 2:00 P.M., and her marginal profit is $6.75 per customer. This office can seat an additional four people. Should Cutt rent?

2.34C. The Fowler-Heir Oil Company operates a crude-oil unloading port at its major refinery. The port has six unloading berths and four unloading crews. When all berths are full, arriving ships are diverted to an overflow facility 20 miles down river. Tankers arrive according to a Poisson process with a mean of one every 2 hr. It takes an unloading crew, on the average, 10 hr to unload a tanker, the unloading time following an exponential distribution. Tankers waiting for unloading crews are served on a first-come, first-served basis. Company management wishes to know the following:

- **(a)** On the average, how many tankers are at the port?
- **(b)** On the average, how long does a tanker spend at the port?
- **(c)** What is the average arrival rate at the overflow facility?
- **(d)** The company is considering building another berth at the main port. Assume that construction and maintenance costs would amount to X dollars per year. The company estimates that to divert a tanker to the overflow port when the main port is full costs Y dollars. What is the relation between X and Y for which it would pay for the company to build an extra berth at the main port?

2.35C. Fly-Bynite Airlines has a telephone exchange with three lines, each manned by a clerk during its busy periods. During their peak three hours per 24-hr period, many callers are unable to get into the exchange (there is no provision for callers to hold if all servers are busy). The company estimates, because of severe competition, that 60% of the callers not getting through use another airline. If the number of calls during these peak periods is roughly Poisson with a mean of 20 calls/hr and each clerk spends on an average 6 min with a caller, his service time being approximately exponentially distributed, and the average customer spends $210/trip, what is the average daily loss due to the limited service facilities? (We may assume that the number of people not getting through during off-peak hours is negligible.) If a clerk's pay and fringe benefits cost the company $24/hr and a clerk must work an 8-hr shift, what is the optimum number of clerks to employ? The three peak hours occur during the 8-hr day shift. At all other times, one clerk can handle all the traffic, and since the company never closes the exchange, exactly one clerk is used on the off shifts. Assume that the cost of adding lines to the exchange is negligible.

2.36. Show that the steady-state results obtained for the ample-server model $(M/M/\infty)$ can also be developed by taking the limit as $c \to \infty$ in the results for the $M/M/c/\infty$ model.

2.37C. The Good Writers Correspondence Academy offers a go-at-your-own-pace correspondence course in good writing. New applications are accepted at any time, and the applicant can enroll immediately. Past records indicate applications follow a Poisson distribution with a mean of 8/month. An applicant's mean completion time is found to be 10 weeks, with the distribution of completion times being exponential. On the average, how many pupils are enrolled in the school at any given time?

2.38. An application of an $M/M/\infty$ model to the field of *inventory control* is as follows. A manufacturer of a very expensive, rather infrequently demanded item uses the following inventory control procedure. She keeps a safety stock of S units on hand. The customer demand for units can be described by a Poisson process with mean λ. Every time a request for a unit is made (a customer demand), an order is placed at the factory to manufacture another (this is called a one-for-one ordering policy). The amount of time required to manufacture a unit is exponential with mean $1/\mu$. There is a carrying cost for inventory on shelf of h per unit per unit time held on shelf (representing capital tied up in inventory which could be invested and earning interest, insurance costs, spoilage, etc.) and a shortage cost of p per unit (a shortage occurs when a customer requests a unit and there is none on shelf, i.e., safety stock is depleted to zero). It is assumed that customers who request an item but find that there is none immediately available will wait until stock is replenished by orders due in (this is called backordering or backlogging); thus one can look at the charge p as a discount given to the customer because he must wait for his request to be satisfied. The problem, then, becomes one of finding the optimal value of S which minimizes total expected costs per unit time, that is, find the S which minimizes

$$E[C] = h \sum_{z=1}^{S} zp(z) + p\lambda \sum_{z=-\infty}^{0} p(z) \quad (\$/\text{unit time}),$$

where z is the steady-state on-hand inventory level ($+$ means items on shelf, $-$ means items in backorder) and $p(z)$ is the probability frequency function. Note that $\sum_{z=1}^{S} zp(z)$ is the average value of the safety stock and $\lambda \sum_{z=-\infty}^{0} p(z)$ is the expected number of backorders per unit time, since the second summation is the fraction of time there is no on-shelf safety stock and λ is the average request rate. If $p(z)$ could be determined, one could optimize $E[C]$ with respect to S.

(a) Show the relationship between Z and N, where N denotes the number of orders outstanding, that is, the number of orders currently being processed at the factory. Hence relate $p(z)$ to p_n.

(b) Show that the $\{p_n\}$ are the steady-state probabilities of an $M/M/\infty$ queue if one considers the order-processing procedure as the queueing system. State explicitly what the input and service mechanisms are.

(c) Find the optimum S for $\lambda = 8/\text{month}$, $1/\mu = 3$ days, $h = \$50/\text{unit}$ per month held, and $p = \$500$ per unit backordered.

2.39C. Farecard machines (machines which dispense tickets for riding on the subway) at the Trailsend station have a mean operating time to breakdown of 45 hr. It takes a repairman (there is only one assigned to the station) on the average 4 hr to repair. Assume both time to breakdown and time to repair follow an exponential distribution. What is the number of installed machines necessary to assure that the probability of having at least five operational is greater than 0.95?

2.40C. Use the software module for the machine repair model with spares to calculate all the usual measures of effectiveness for a problem with $M = 10$, $Y = 2$, $\lambda = 1$, $\mu = 3.5$, $c = 3$. Find the $\Pr\{N \geq k\}$ for $k = 2, 4$.

2.41. Show for the basic machine repair model (no spares) that $q_n(M)$, the failure (arrival) point probabilities for a population of size M, equal $p_n(M - 1)$, the general-time probabilities for a population of size $M - 1$. The $\{q_n\}$ are sometimes referred to as *inside observer probabilities*, while the $\{p_n\}$ are referred to as *outside observer probabilities*.

2.42. Derive $q_n(M)$ given by Equation (2.48) for a machine repair problem with spares, and show that this is *not* equal to $p_n(M - 1)$, but is equal to $p_n(Y - 1)$. The algebra is quite messy, so show it only for a numerical example $[M = 2, Y = 1, c = 1, \lambda/\mu = 1]$. While that is no proof, the statement can be shown to hold in general (see Sevick and Mitrani, 1979, or Lavenberg and Reiser, 1979).

2.43C. A coin-operated dry-cleaning store has five machines. The operating characteristics of the machines are such that any machine breaks down according to a Poisson process with mean breakdown rate of one per day. A repairman can fix a machine according to an exponential distribution with a mean repair time of one-half day. Currently three repairmen are on duty. The manager, Lew Cendirt, has the option of replacing these three repairmen with a superrepairman whose salary is equal to the total of the three regulars, but who can fix a machine in one-third the time, that is, in one-sixth day. Should he be hired?

2.44C. Suppose that each of five machines in a given shop breaks down according to a Poisson law at an average rate of one every 10 hr, and the failures are repaired one at a time by two maintenance men operating as two channels, such that each machine has an exponentially distributed servicing requirement of mean 5 hr.

  (a) What is the probability that exactly one machine will be up at any one time?

  (b) If performance of the workmen is measured by the ratio of average waiting time to average service time, what is this measure for the current situation?

  (c) What is the answer to **(a)** if an identical spare machine is put on reserve?

2.45. Find the steady-state probabilities for a machine-repair problem with M machines, Y spares, and c repairmen ($c \leq Y$) but with the following discipline: if no spares are on hand and a machine fails ($n = Y + 1$), the remaining $M - 1$ machines running are stopped until a machine is repaired; that is, if the machines are to run, there must be M running simultaneously.

2.46C. Very often in real-life modeling, even when the calling population is finite, an infinite source model is used as an approximation. To compare the two models, calculate L for Example 2.5 assuming the calling population (number of machines) is infinite. Also calculate L for an exact model when the number of machines equals 10 and 5, respectively, for $M\lambda = \frac{1}{3}$ in both cases, and compare to the calculations from an approximate infinite source model. How do you think ρ affects the approximation? [*Hint:* When using an infinite source model as an approximation to a finite-source model, λ must be set equal to $M\lambda$.]

2.47. Find the average operating costs per hour of Example 2.6 when
(a) $C_1 =$ low-speed cost $= \$25/$(operating hour); $C_2 =$ high-speed cost $= \$50/$(operating hour).
(b) $C_1 = \$25/$(operating hour); $C_2 = \$60/$(operating hour).
(c) Evaluate (b) for $k = 4$. What now is the best policy?

2.48. Assume we have a two-state, state-dependent service model as in Section 2.8 with $\rho_1 = \frac{4}{3}$ and $\rho = \frac{2}{3}$. Suppose that the customers are lawn-treating machines owned by the Green Thumb Lawn Service Company and these machines require, at random times, greasing on the company's two-speed greasing machine. Furthermore, suppose that the cost per operating hour of the greaser at the lower speed, C_1, is \$25, and at the high speed, C_2, is \$110. Also, the company estimates the cost of downtime of a lawn treater to be \$5/hr. What is the optimal switch point, k? [*Hint:* Try several values of k starting at $k = 1$, and compute the total expected cost.]

2.49. Derive the steady-state system-size probabilities for a c-server model with Poisson input and exponential state-dependent service where the mean service rate switches from μ_1 to μ when $k > c$ are in the system.

2.50. Derive the steady-state system-size probabilities for a single-server model with Poisson input and exponential state-dependent service with mean rates μ_1 $(1 \leq n < k_1)$, μ_2 $(k_1 \leq n < k_2)$, and μ $(n \geq k_2)$.

2.51. For the problem with mean service rate changing continuously with change of state, treated at the end of Section 2.8 ($\mu_n = n^\alpha \mu$), show for $\alpha \geq 1$ that the tail of the infinite series for calculating p_0 discarded by truncation at some value, say N, is bounded by the tail of the series for e^r. So, given any $\epsilon > 0$, N can be found such that the discarded tail is less than ϵ. Further show that if $p_0(N)$ is the estimate of p_0 based on N terms where N is such that the discarded tail is bounded by ϵ, then the error bounds on p_0 become

$$p_0 < p_0(N) < \frac{p_0}{1 - \epsilon} \doteq p_0(1 + \epsilon).$$

2.52. It is known for an $M/M/1$ balking situation that the stationary distribution is given by the negative binomial

$$p_n = \binom{N + n - 1}{N - 1} x^n (1 + x)^{-N-n} \qquad (n \geq 0, \quad x > 0, \quad N > 1).$$

Find L, L_q, W, W_q, and b_n.

2.53. For an $M/M/1$ balking model, it is known that $b_n = e^{-\alpha n/\mu}$. Find p_n (for all n).

2.54. Suppose that the $M/M/1$ reneging model of Section 2.9.2 has the balking function $b_n = 1/n$ for $0 \leq n \leq k$ and 0 for $n > k$, and a reneging function $r(n) = n/\mu$. Find the stationary system-size distribution.

2.55. Derive the steady-state $M/M/\infty$ solution directly from the transient.

2.56. Find the mean number in an $M/M/\infty$ system at any point in time.

2.57C. For $\rho = 0.5$, 0.9, and 1 in an $M/M/1/\infty$ model with $\lambda = 1$, plot $p_0(t)$ versus t as t goes to infinity. Comment.

2.58C. For $\rho = 0.5$ in an $M/M/1/\infty$ model with $\lambda = 1$, find L at $t = 3$.

CHAPTER 3

Advanced Markovian Queueing Models

This chapter continues the development of models that are amenable to analytic methods and is concerned especially with Markovian problems of the non-birth–death type. That is, we allow changes of more than one over infinitesimal time intervals, but insist on retaining the memoryless Markovian property. The Chapman–Kolmogorov and backward and forward equations, plus the resultant balance equations, are all still valid, and together are the essence of the approach to solution for these non-birth–death Markovian problems.

3.1 BULK INPUT ($M^{[X]}/M/1$)

In continuation of our relaxation of the simple assumptions underlying $M/M/1$, let it now be assumed, in addition to the assumption that the arrival stream forms a Poisson process, that the actual number of customers in any arriving module is a random variable X, which may take on any positive integral value less than ∞ with probability c_x. It should be clear that this new queueing problem—let it be called $M^{[X]}/M/1$, is still Markovian in the sense that future behavior is a function only of the present and not the past.

We now recall the discussions of Section 1.8, particularly those of the common generalizations of the Poisson process. If λ_x is the arrival rate of a Poisson process of batches of size X, then clearly $c_x = \lambda_x/\lambda$, where λ is the composite arrival rate of all batches and is equal to $\sum_{i=1}^{\infty} \lambda_i$. This total process, which arises from the overlap of the set of Poisson processes with rates $\{\lambda_x, x = 1, 2, \ldots\}$ as previously mentioned in Section 1.8, is a *multiple* or *compound Poisson process*.

A set of Chapman–Kolmogorov equations can be derived for this problem in the usual manner with little difficulty, and they give rise (assuming steady state—under the proper parameter settings, the processes treated in this

116

chapter do meet the conditions of Theorem 1.3) to the balance equations (see Problem 3.1)

$$0 = -(\lambda + \mu)p_n + \mu p_{n+1} + \lambda \sum_{k=1}^{n} p_{n-k}c_k \qquad (n \geq 1),$$

$$0 = -\lambda p_0 + \mu p_1.$$

$$(3.1)$$

The last term in the general equation of (3.1) comes from the fact that a total of n in the system can arise from the presence of $n - k$ in the system followed by the arrival of a batch of size k. To solve the system of equations given by (3.1), we shall use a generating-function approach. Difference-equation methods are often used instead to solve the problem when the maximum batch is small. It is also not a very difficult matter to extend the results of this section to the $M^{[X]}/M/c$ model. This would be done in the same manner as $M/M/1$ is extended to $M/M/c$.

To complete the solution, we need first to define

$$P(z) = \sum_{n=0}^{\infty} p_n z^n \qquad (|z| \leq 1)$$

and

$$C(z) = \sum_{n=1}^{\infty} c_n z^n \qquad (|z| \leq 1)$$

as the generating functions of the steady-state probabilities $\{p_n\}$ and the batch-size distribution $\{c_n\}$, respectively. If each equation of (3.1) is then multiplied by the appropriate z^n, and if the resultant equations are summed, it is found that

$$0 = -\lambda \sum_{n=0}^{\infty} p_n z^n - \mu \sum_{n=1}^{\infty} p_n z^n + \frac{\mu}{z} \sum_{n=1}^{\infty} p_n z^n + \lambda \sum_{n=1}^{\infty} \sum_{k=1}^{n} p_{n-k}c_k z^n. \quad (3.2)$$

We observe that $\sum_{k=1}^{n} p_{n-k}c_k$ is the probability function for the sum of the steady-state system size and batch size, since this is merely a convolution formula for discrete random variables. It can easily be shown that the generating function of this sum is the product of the respective generating functions (a basic property of all generating functions), namely,

$$\sum_{n=1}^{\infty} \sum_{k=1}^{n} p_{n-k}c_k z^n = \sum_{k=1}^{\infty} c_k z^k \sum_{n=k}^{\infty} p_{n-k}z^{n-k} = C(z)P(z).$$

Hence (3.2) may be rewritten as

$$0 = -\lambda P(z) - \mu[P(z) - p_0] + \frac{\mu}{z}[P(z) - p_0] + \lambda C(z)P(z),$$

and thus

$$P(z) = \frac{\mu p_0(1 - z)}{\mu(1 - z) - \lambda z[1 - C(z)]} \qquad (|z| \leq 1). \qquad (3.3)$$

To find p_0 from $P(1)$ and the average number in the system, L, from $P'(1)$, let us rewrite the generating function (3.3), with $r = \lambda/\mu$, as

$$P(z) = \frac{p_0}{1 - rz\bar{C}(z)}, \qquad \bar{C}(z) \equiv \frac{1 - C(z)}{1 - z}.$$

Now note that $\bar{C}(z) = [1 - C(z)]/[1 - z]$ is the generating function of the complementary batch-size probabilities $\Pr\{X > x\} = 1 - C_x = \bar{C}_x$, since $1/(1 - z)$ is the generating function of 1 and $C(z)/(1 - z)$ is the generating function of the cumulative probabilities C_x, which can be seen by noting that

$$\sum_{x=1}^{\infty} C_x z^x = \sum_{x=1}^{\infty}\left(\sum_{i=1}^{x} c_i\right)z^x = \sum_{i=1}^{\infty}\left(c_i z^i \sum_{x=i}^{\infty} z^{x-i}\right) = \left(\sum_{i=1}^{\infty} c_i z^i\right)\frac{1}{1 - z}.$$

It follows after one application of l'Hôpital's rule that $\bar{C}(1) = E[X]$, and after two applications that $\bar{C}'(1) = E[X(X - 1)]/2$, so that

$$1 = P(1) = \frac{p_0}{1 - r\bar{C}(1)}$$

and

$$P'(1) = \frac{r[\bar{C}(1) + \bar{C}'(1)]}{1 - r\bar{C}(1)}.$$

Therefore

$$\boxed{p_0 = 1 - rE[X] = 1 - \rho}$$

and

$$\boxed{L = \frac{r\{E[X] + E[X^2]\}}{2(1 - \rho)} = \frac{\rho + rE[X^2]}{2(1 - \rho)} \qquad (\rho = \lambda E[X]/\mu).} \qquad (3.4)$$

As expected, $\rho < 1$ is the necessary and sufficient condition for stationarity. The remaining usual measures of effectiveness may be found by using the single-server results (Table 1.2) that $L_q = L - (1 - p_0) = L - \rho$ and then Little's formulas. The individual state probabilities $\{p_n\}$ can often also be obtained by the direct inversion of the generating function of (3.3).

Two interesting useful examples of these results would be to let the batch sizes be either constant (each of size K) or geometrically distributed. For the constant case, the formula for the mean system size simplifies greatly to

$$L = \frac{\rho + K\rho}{2(1 - \rho)} = \frac{K + 1}{2} \frac{\rho}{1 - \rho} \qquad (\rho = \lambda K/\mu), \qquad (3.5)$$

which is precisely equal to the $M/M/1$ mean system size multiplied by $(K + 1)/2$. Since this is a single-server system, it follows that

$$L_q = L - \rho = \frac{2\rho^2 + (K - 1)\rho}{2(1 - \rho)}. \qquad (3.6)$$

The inversion of $P(z)$ to get the individual $\{p_n\}$ is a reasonable task when K is small (similar to Example 3.1 below).

When the batch sizes are instead distributed geometrically, that is,

$$c_x = (1 - \alpha)\alpha^{x-1} \qquad (0 < \alpha < 1),$$

it follows that $\rho = \lambda E[X]/\mu = r/(1 - \alpha)$, that

$$C(z) = (1 - \alpha) \sum_{n=1}^{\infty} \alpha^{n-1}z^n = \frac{z(1 - \alpha)}{1 - \alpha z},$$

and from (3.3) that

$$P(z) = \frac{(1 - \rho)(1 - z)}{1 - z - rz[1 - C(z)]}$$

$$= \frac{(1 - \rho)(1 - z)}{1 - z - rz[1 - z(1 - \alpha)/(1 - \alpha z)]}$$

$$= \frac{(1 - \rho)(1 - \alpha z)}{1 - z[\alpha + (1 - \alpha)\rho]}$$

$$= (1 - \rho)\left(\frac{1}{1 - z[\alpha + (1 - \alpha)\rho]} - \frac{\alpha z}{1 - z[\alpha + (1 - \alpha)\rho]}\right).$$

Therefore, utilizing the formula for the sum of a geometric series, we are able to write that

$$P(z) = (1 - \rho)\left(\sum_{n=0}^{\infty} \{[\alpha + (1 - \alpha)\rho]z\}^n - \sum_{n=0}^{\infty} \alpha[\alpha + (1 - \alpha)\rho]^n z^{n+1} \right),$$

from which we get

$$p_n = (1 - \rho)\{[\alpha + (1 - \alpha)\rho]^n - \alpha[\alpha + (1 - \alpha)\rho]^{n-1}\}$$
$$= (1 - \rho)[\alpha + (1 - \alpha)\rho]^{n-1}[(1 - \alpha)\rho] \qquad (n > 0).$$

Example 3.1
Let us consider a multistage machine-line process which produces an assembly in quantity. After the first stage many items are found to have one or more defects, which must be repaired before they enter the second stage. It is the job of one worker to make the necessary adjustments to put the assembly back into the stream of the process. The number of defects per item is registered automatically, and it exceeds two an extremely small number of times. The interarrival times for both units with one and two defective parts are found to be governed closely by exponential distributions, with parameters $\lambda_1 = 1/\text{hr}$ and $\lambda_2 = 2/\text{hr}$, respectively. There are so many different types of parts which have been found defective that an exponential distribution does actually provide a good fit for the worker's service-time distribution, with mean $1/\mu = 10 \text{ min}$.

But it is subsequently noted that the rates of defects have increased, although not continuously. It is therefore decided to put another person on the job, who will concentrate on repairing those units with two defects, while the original worker works only on singles. When to add the additional person will be decided on the basis of a cost analysis.

Now there are a number of alternative cost structures available, and it is decided by management that the expected cost of the system to the company will be based upon the average delay time of assemblies in for repair, which is directly proportional to the average number of units in the system, L. To find L under the assumption that there are only two possible batch sizes, we first note that $\lambda = \lambda_1 + \lambda_2 = 3$, $r = \frac{1}{2}$, $c_1 = \lambda_1/\lambda = \frac{1}{3}$, and $c_2 = \lambda_2/\lambda = \frac{2}{3}$. Thus the batch-size mean and second moment are respectively equal to $E[X] = \frac{1}{3} + 2 \times \frac{2}{3} = \frac{5}{3}$ and $E[X^2] = \frac{1}{3} + 2^2 \times \frac{2}{3} = 3$, and it follows that the system utilization rate is $\rho = \lambda E[X]/\mu = \frac{5}{6}$ and

$$L = \frac{\rho + rE[X^2]}{2(1 - \rho)} = \frac{\frac{5}{6} + \frac{3}{2}}{2(1 - \frac{5}{6})} = 7.$$

Although not necessary for solving the above problem, the individual state probabilities may be found by writing out the generating function from (3.3), using the fact that here $C(z) = c_1 z + c_2 z^2 = \frac{1}{3}z + \frac{2}{3}z^2$:

$$P(z) = \frac{\mu(1 - \rho)(1 - z)}{\mu(1 - z) - \lambda z(1 - z/3 - 2z^2/3)}$$

$$= \frac{1}{6 - 3z - 2z^2}.$$

The roots of the denominator of this generating function are $(-3 \pm \sqrt{57})/4$, or 1.137 and -2.637 to three-decimal-place accuracy. Because the two roots are each greater than 1 in absolute value, it follows that there is a two-term linear partial-fraction expansion of $P(z)$, which turns out to be

$$P(z) = \frac{1}{7.550}\left(\frac{1}{1.137 - z} + \frac{1}{2.637 + z}\right)$$

$$= \frac{1}{7.550}\left(\frac{1/1.137}{1 - z/1.137} + \frac{1/2.637}{1 + z/2.637}\right)$$

$$= \frac{1}{7.550}\left[\frac{1}{1.137}\sum_{n=0}^{\infty}\left(\frac{z}{1.137}\right)^n + \frac{1}{2.637}\sum_{n=0}^{\infty}\left(-\frac{z}{2.637}\right)^n\right].$$

Therefore,

$$p_n = 0.116(0.880)^n + 0.050(-0.379)^n \qquad (n \geq 0).$$

Now, if C_1 is the cost per unit time per waiting repair and C_2 the cost of a worker per unit time, then the expected cost per unit time, C, of the single-server system is $C = C_1 L + C_2$. If a second repairer now sets up a separate service channel, the additional cost of his or her time is incurred, over and above the cost of the items in the queue. In this case, we have two queues. The singlet line would be a standard $M/M/1$, but the doublets would not. However, a single Poisson stream of doublets is merely a special case of the multiple Poisson bulk-input model with $\lambda_1 = 0$. The expected number of required repairs in the system is then the sum of the expected values of the two streams. Since the first stream is a standard $M/M/1$, its expected length is

$$L_1 = \frac{\lambda_1/\mu}{1 - \lambda_1/\mu}.$$

To get the expected length of the second line, we use Equation (3.5) with

$K = 2$ and $\rho = 2\lambda_2/\mu$, so that

$$L_2 = \frac{3\lambda_2/\mu}{1 - 2\lambda_2/\mu},$$

and thus

$$L = L_1 + L_2 = \frac{\lambda_1/\mu}{1 - \lambda_1/\mu} + \frac{3\lambda_2/\mu}{1 - 2\lambda_2/\mu}.$$

Therefore the new expected cost is

$$C^* = C_1 \left(\frac{\lambda_1/\mu}{1 - \lambda_1/\mu} + \frac{3\lambda_2/\mu}{1 - 2\lambda_2/\mu} \right) + 2C_2.$$

Hence any decision is based upon the comparative magnitude of C and C^*, and an additional channel is invoked whenever $C^* < C$, or

$$C_1 \left(\frac{\lambda_1/\mu}{1 - \lambda_1/\mu} + \frac{3\lambda_2/\mu}{1 - 2\lambda_2/\mu} \right) + C_2 < C_1 L = C_1 \left(\frac{\rho + r\mathrm{E}[X^2]}{2(1 - \rho)} \right)$$

$$= C_1 \left(\frac{\lambda_1/\mu + 3\lambda_2/\mu}{1 - \lambda_1/\mu - 2\lambda_2/\mu} \right),$$

that is,

$$C_2 < C_1 \left(\frac{\lambda_1/\mu + 3\lambda_2/\mu}{1 - \lambda_1/\mu - 2\lambda_2/\mu} - \frac{\lambda_1/\mu}{1 - \lambda_1/\mu} - \frac{3\lambda_2/\mu}{1 - 2\lambda_2/\mu} \right),$$

and removed when the inequality is reversed. Using our values for the parameters gives a decision criterion of $C_2 < 19C_1/5$.

3.2 BULK SERVICE ($M/M^{[Y]}/1$)

For the main problem of this section, let it be assumed that arrivals occur at a single-channel facility as an ordinary Poisson process, that they are served FCFS, that there is no waiting-capacity constraint, and that these customers are served K at a time, except when less than K are in the system and ready for service, at which time all units are served. Further, if less than K are in service, new arrivals immediately enter service up to the limit K, and finish with the others, regardless of the time into service after service begins. The amount of time required for the service of any batch is an exponentially distributed random variable, whether or not the batch is of full size K. This model will henceforth be known by the notation $M/M^{[K]}/1$.

A slight variation on this theme which we also consider is a model in which the batch size for service must be exactly K and if the number present when the server becomes idle is less than K, it waits until K accumulates.

The basic model is, of course, a non-birth–death Markovian problem. The stochastic balance equations are given as (see Problem 3.1)

$$0 = -(\lambda + \mu)p_n + \mu p_{n+K} + \lambda p_{n-1} \qquad (n \geq 1),$$
$$0 = -\lambda p_0 + \mu p_1 + \mu p_2 + \cdots + \mu p_{K-1} + \mu p_K. \tag{3.7}$$

The first equation of (3.7) may be rewritten in operator notation as

$$[\mu D^{K+1} - (\lambda + \mu)D + \lambda]p_n = 0 \qquad (n \geq 0); \tag{3.8}$$

hence if $(r_1, \ldots, r_{K+1})$ are the roots of the operator or characteristic equation, then

$$p_n = \sum_{i=1}^{K+1} C_i r_i^n \qquad (n \geq 0).$$

Since $\sum_{n=0}^{\infty} p_n = 1$, each r_i must be less than one or $C_i = 0$ for all r_i not less than one. So let us now determine the number of roots less than one. For this an appeal is made to Rouché's theorem (see Section 2.10.2), and it is found that there is, in fact, one and only one root (say r_0) in $(0, 1)$ [see Problem 3.2]. So

$$p_n = C r_0^n \qquad (n \geq 0, \quad 0 < r_0 < 1).$$

Using the boundary condition that Σp_n must total one, we find that $C = p_0 = 1 - r_0$; hence

$$\boxed{p_n = (1 - r_0)r_0^n \qquad (n \geq 0, \quad 0 < r_0 < 1).} \tag{3.9}$$

Measures of effectiveness for this model can be obtained in the usual manner. Since the stationary solution has the same geometric form as that of the $M/M/1$ (with r_0 replacing ρ), we can immediately write that

$$\boxed{L = \frac{r_0}{1 - r_0}, \qquad L_q = L - \frac{\lambda}{\mu}}$$

and

$$W = \frac{r_0}{\lambda(1 - r_0)}, \qquad W_q = W - \frac{1}{\mu}.$$

Let us now assume that the batch size must be exactly K, and if not, the server waits until such time to start. Then the equations (3.7) must be slightly rewritten to read

$$0 = -(\lambda + \mu)p_n + \mu p_{n+K} + \lambda p_{n-1} \qquad (n \geq K),$$
$$0 = -\lambda p_n + \mu p_{n+K} + \lambda p_{n-1} \qquad (1 \leq n < K), \qquad (3.10)$$
$$0 = -\lambda p_0 + \mu p_K.$$

The first equation of this second approach to bulk service is identical to that of the first approach; hence

$$p_n = C r_0^n \qquad (n \geq K - 1, \quad 0 < r_0 < 1).$$

The obtaining of C (and p_0) is a more complicated procedure here, since the foregoing formula for p_n is valid only for $n \geq K - 1$.

From the steady-state equations,

$$p_K = \frac{\lambda}{\mu} p_0 = C r_0^K,$$

and therefore

$$C = \frac{\lambda p_0}{\mu r_0^K}.$$

To get p_0 now, we must use the $K - 1$ stationary equations given in (3.10) as

$$0 = \mu p_{n+K} - \lambda p_n + \lambda p_{n-1}.$$

Substituting the above geometric formula for p_n, $n \geq K$, into these $K - 1$ equations, it can be seen that

$$p_0 r_0^n = p_n - p_{n-1} \qquad (1 \leq n \leq K).$$

These equations can be solved by iteration starting with $n = 1$ or we can note that these are nonhomogeneous linear difference equations whose solutions are

$$p_n = C_1 + C_2 r_0^n,$$

where $C_1 = p_0 - C_2$ and $C_2 = -p_0 r_0/(1 - r_0)$, and

$$p_n = \begin{cases} \dfrac{p_0(1 - r_0^{n+1})}{1 - r_0} & (1 \le n < K), \\[2ex] \dfrac{p_0 \lambda r_0^{n-K}}{\mu} & (n \ge K). \end{cases} \tag{3.11}$$

To get p_0 we use the usual boundary condition that $\sum_{n=0}^{\infty} p_n = 1$. Hence from Equation (3.11),

$$p_0 = \left(1 + \sum_{n=1}^{K-1} \frac{1 - r_0^{n+1}}{1 - r_0} + \frac{\lambda}{\mu} \sum_{n=K}^{\infty} r_0^{n-K} \right)^{-1}$$

$$= \left(1 + \frac{K-1}{1 - r_0} - \frac{r_0^2(1 - r_0^{K-1})}{(1 - r_0)^2} + \frac{\lambda}{\mu(1 - r_0)} \right)^{-1}$$

$$= \left(\frac{\mu r_0^{K+1} - (\lambda + \mu)r_0 + \lambda + \mu K(1 - r_0)}{\mu(1 - r_0)^2} \right)^{-1}.$$

But, from the characteristic equation in (3.8), we know that

$$\mu r_0^{K+1} - (\lambda + \mu)r_0 + \lambda = 0.$$

Thus

$$p_0 = \frac{\mu(1 - r_0)^2}{\mu K(1 - r_0)} = \frac{1 - r_0}{K}. \tag{3.12}$$

The formulas for the $\{p_n\}$ could have also been obtained via the probability generating function. The generating function can be shown to be (see Problem 3.4)

$$P(z) = \frac{(1 - z^K) \sum_{n=0}^{K-1} p_n z^n}{rz^{K+1} - (r + 1)z^K + 1} \qquad (r = \lambda/\mu). \tag{3.13}$$

To make use of (3.13), it is necessary to eliminate the p_n, $n = 0, 1, \ldots, K - 1$, from the numerator of the right-hand side. To do this we again appeal to Rouché's theorem (see Section 2.10.2). The generating function $P(z)$ has the property that it must converge inside the unit circle. We notice that the denominator of $P(z)$ has $K + 1$ zeros. Applying Rouché's theorem to the denominator (see Problem 3.5) tells us that K of these lie on or within the unit circle. One zero of the denominator is $z = 1$; thus $K - 1$ lie within and must coincide with those of $\sum_{n=0}^{K-1} p_n z^n$ for $P(z)$ to converge, so that when a zero appears in the denominator it is canceled by one in the numerator. Hence this leaves one zero of the denominator (since there are a total of $K + 1$) which lies outside the unit circle. We denote this by z_0.

A major observation at this point is that the roots of the denominator are precisely the reciprocals of those of the characteristic equation (3.8). Thus $z_0 = 1/r_0$. This inverse relationship can be seen in Section 2.2 for $M/M/1$ by comparing Equations (2.6) and (2.14). In fact, it can be shown that the zeros of the characteristic equation of any system of linear difference equations with constant coefficients are the reciprocals of the poles of the system's generating function. This follows because of the way that the poles show up in a partial-fraction expansion in terms like $1/(a - z)$, which then can be rewritten as the infinite series $(1/a) \sum_{n=0}^{\infty} (z/a)^n$. We have already seen this sort of thing in Example 3.1.

Returning now to the generating function, we see that dividing the denominator by the product $(z - 1)(z - z_0)$ results in a polynomial with $K - 1$ roots inside the unit circle. These must therefore match the roots of $\sum_{n=0}^{K-1} p_n z^n$, so that the two polynomials differ by at most a multiplicative constant, and we may therefore write that

$$\sum_{n=0}^{K-1} p_n z^n = A \frac{\rho z^{K+1} - (\rho + 1)z^K + 1}{(z - 1)(z - z_0)}.$$

Substituting the right-hand side above into (3.13) yields

$$P(z) = \frac{A(1 - z^K)}{(z - 1)(z - z_0)} = \frac{A}{z_0 - z} \sum_{n=0}^{K-1} z^n.$$

Since $P(1) = 1$, it follows that

$$A = \frac{z_0 - 1}{K}$$

and thus

$$P(z) = \frac{(z_0 - 1) \sum_{n=0}^{K-1} z^n}{K(z_0 - z)}. \tag{3.14}$$

Prior to showing an example, we make a final comment on batch service queues. Chaudhry and Templeton (1983) have treated the generalized batch service policy that if the number in the queue when the server becomes free (N) is less than k, the server waits until $N = k$ to start serving a batch. If $N > K$, the server takes the first K in the queue. Once service begins, future arrivals must wait until the server is free (unlike our first model). When $k = K$, this policy reduces to our second model. To treat this policy in general requires methods more advanced than the basic CK analysis used in this chapter.

Example 3.2
The Drive-It-Through-Yourself Car Wash decides to change its operating procedure. It will install new machinery which will permit the washing of two cars at once (and one if no other cars wait) and will move to a new location which will effectively have no waiting-capacity limitation. The company expects arrivals to be Poisson with mean 20/hr, and its service times to be exponential (a function of car size) with a mean of 5 min. What average line length should it anticipate?

The model for this bulk-service problem is the first of the two presented. The given parameters are $\lambda = 20$/hr, $\mu = 1/(5 \text{ min}) = 12$/hr, and $K = 2$. The operator equation is therefore

$$12r^3 - 32r + 20 = 4(3r^3 - 8r + 5) = 0.$$

One root is, of course, $r = 1$, and division by the factor $r - 1$ leaves

$$3r^2 + 3r - 5 = 0,$$

which has roots $r = (-3 \pm \sqrt{69})/6$. We select the positive root with absolute value less than 1, namely, $r_0 = (-3 + \sqrt{69})/6 \doteq 0.884$. Therefore

$$L \doteq \frac{0.884}{0.116} \doteq 7.6 \text{ cars} \quad \text{and} \quad L_q \doteq 7.6 - \frac{20}{12} \doteq 5.9 \text{ cars}.$$

3.3   ERLANGIAN MODELS ($M/E_k/1$, $E_k/M/1$, $E_j/E_k/1$)

Up to now, all probabilistic queueing models studied have assumed Poisson input (exponential interarrival times) and exponential service times, or simple variations thereof. In many practical situations, however, the exponential assumptions may be rather limiting, especially the assumption concerning service times being distributed exponentially. In this section, we allow for a more general probability distribution for describing the input process or the service mechanism.

3.3.1   The Erlang Distribution

To begin, consider a random variable T which has the gamma probability density

$$f(t) = \frac{1}{\Gamma(\alpha)\beta^\alpha} t^{\alpha-1} e^{-t/\beta} \qquad (\alpha, \beta > 0, \quad 0 < t < \infty),$$

where $\Gamma(\alpha) = \int_0^\infty t^{\alpha-1} e^{-t} dt$ is the usual gamma function, and α and β are the parameters of the distribution. The mean $E[T]$ and variance $\text{Var}[T]$ are given as

$$E[T] = \alpha\beta,$$

$$\text{Var}[T] = \alpha\beta^2.$$

If we further consider a special class of these distributions where α and β are related by

$$\alpha = k,$$

$$\beta = \frac{1}{k\mu},$$

where k is any arbitrary positive integer and μ any arbitrary positive constant, we obtain the Erlang family of probability distributions, namely,

$$f(t) = \frac{(\mu k)^k}{(k-1)!} t^{k-1} e^{-k\mu t} \qquad (0 < t < \infty).$$

The parameters of the Erlang are thus k and μ, and the mean and variance are given by

$$E[T] = \frac{1}{\mu},$$

$$\text{Var}[T] = \frac{1}{k\mu^2}.$$

For any particular value of k, the resulting Erlang is referred to as an Erlang type k or E_k distribution. Figure 3.1 illustrates the effect of k on the Erlang family of distributions; for obvious reasons, it is often called the shape parameter.

The Erlang family of probability distributions provides much more modeling flexibility than does the exponential. In fact, the exponential is a special Erlang, namely, type 1. As k increases, the Erlang becomes more symmetrical, and as it approaches ∞, the Erlang becomes deterministic with value $1/\mu$. Thus in practical situations where observed data might not bear

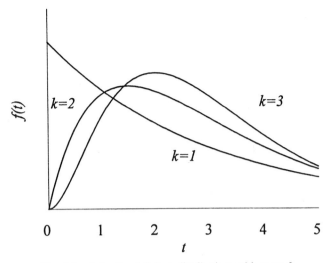

Fig. 3.1 A family of Erlang distributions with mean 3.

out the exponential-distribution assumption, the Erlang can provide greater flexibility by being better able to represent the real world.

Another reason why the Erlang is useful in queueing analyses is its relation to the exponential distribution and the latter's Markovian property. The Erlang distribution itself is of course non-Markovian (Section 1.8 showed that the exponential distribution was unique among continuous distributions in its Markovian property). However, as shown in Problem 2.5(b), one can readily see that the sum of k IID exponential random variables with mean $1/k\mu$ yields an Erlang type-k distribution. It is this relation, as we shall see a little later, that allows the analysis of queueing models with Erlangian input or service to be performed.

The relation of the Erlang to exponential distributions also allows us to describe queueing models where the service (or arrivals) may be a series of identical phases. For example, suppose, in performing a laboratory test, the technician must perform four steps, each taking the same mean time (say $1/4\mu$), with the times distributed exponentially. This is represented pictorially in Figure 3.2. The overall service function is then Erlang type 4 with mean

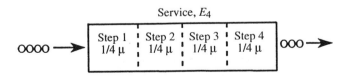

Fig. 3.2 Use of the Erlang for phased service.

$1/\mu$. If the input process were Poisson, we would have an $M/E_4/1$ queueing model.

Note what this model implies. First, all steps (or phases) of the service are independent and *identical*. Second, only one customer at a time is in the service mechanism; that is, a customer enters phase 1 of the service, then progresses through the remaining phases, and must complete the last phase before the next customer enters the first phase. This rules out assembly-line-type models where as soon as a customer finishes a phase of service another can enter it. Finally, while the phase representation for the Erlang may serve as a realistic model for some situations, the Erlang is not restricted to modeling situations where there are actually phases of service. The Erlang has a more general use in that it has greater flexibility than the exponential, and hence it is better able to fit observed data in many situations in which the exponential cannot.

3.3.2 Generalization to Phase-Type Distributions

The concept of phases can be generalized to include a much wider class of distributions than just the Erlang. To show how this can be accomplished, we first use another method for generating an Erlang distribution, illustrated by looking at a three-state continuous-parameter Markov chain with Q matrix

$$Q = \begin{pmatrix} -\mu & \mu & 0 \\ 0 & -\mu & \mu \\ 0 & 0 & 0 \end{pmatrix}.$$

Note that state 3 is an absorbing state, that is, once the process enters state 3, it never leaves. We use a starting probability vector $p(0) = (1, 0, 0)$, so that we must start in state 1. Then the CDF for the time to absorption, T, of this Markov chain will be an E_2.

Letting T_1 represent the time until the process transits from state 1 to state 2 and T_2 represent the time until the process transits from state 2 to state 3, then $T = T_1 + T_2$. Since T_i is exponential with parameter μ, we know T has an E_2 distribution. But had we not utilized this fact, we could have calculated the CDF of T by solving the forward differential equations of the Markov process for $p_3(t)$, which will be the desired CDF, since it is the probability that the absorption took place in time less than or equal to t.

To begin, we can immediately write down that the probability of being in stage 1 at time t is $p_1(t) = e^{-\mu t}$ because of the exponential distribution of phase lengths. Since state 3 is absorbing, the remaining CK equations for the chain are given by

$$p_2(t + \Delta t) = \mu \, \Delta t \, p_1(t) + (1 - \mu \, \Delta t)p_2(t) + o(\Delta t),$$

$$p_3(t + \Delta t) = p_3(t) + \mu \, \Delta t \, p_2(t) + o(\Delta t).$$

The resulting differential equation for $p_2(t)$ is

$$p_2'(t) + \mu p_2(t) = \mu p_1(t) = \mu e^{-\mu t}.$$

This is a first-order, linear differential equation, which has solution $p_2(t) = \mu t e^{-\mu t}$, found using the boundary condition $p_2(0) = 0$ and then direct substitution. Then the differential equation for $p_3(t)$ is given by

$$p_3'(t) = \mu p_2(t),$$

and thus

$$p_3'(t) = \mu^2 t e^{-\mu t},$$

which is indeed an E_2 density.

The same solution can be derived directly from the matrix–vector formulation of the defining differential equations using the first two rows and columns of Q—call this matrix $\tilde{Q}$—with $p(t) = [p_1(t), p_2(t)]$,

$$p'(t) = p(t)\tilde{Q} = p(t)\begin{pmatrix} -\mu & \mu \\ 0 & -\mu \end{pmatrix}$$

—and then using the third column of Q to write the answer once more as

$$p_3'(t) = \mu p_2(t).$$

Staying in matrix–vector form, analogous to the solution $y(t) = y(0)e^{at}$ of the scalar differential equation $y'(t) = ay(t)$, the solution to the matrix–vector system may be written as (see Problem 3.19)

$$p(t) = p(0)e^{\tilde{Q}t} \qquad [p(0) = (1, 0)]$$

$$= p(0)\begin{pmatrix} e^{-\mu t} & \mu t e^{-\mu t} \\ 0 & e^{-\mu t} \end{pmatrix},$$

so that

$$p(t) = [e^{-\mu t}, \mu t e^{-\mu t}]$$

and $p_3'(t) = \mu p_2(t)$ as before.

This procedure can also be used to obtain more general phase-type distributions, and much more detail will be provided on their use later in the text.

For now, let us consider some basic examples. For the first one, assume that the matrices Q and $\tilde{Q}$ are now instead given by

$$Q = \begin{pmatrix} -\mu_1 & \mu_1 & 0 \\ 0 & -\mu_2 & \mu_2 \\ 0 & 0 & 0 \end{pmatrix}, \qquad \tilde{Q} = \begin{pmatrix} -\mu_1 & \mu_1 \\ 0 & -\mu_2 \end{pmatrix}.$$

Then, using the same arguments as in the Erlang example will give us a distribution which is the convolution of two nonidentical exponential random variables (not an Erlang, since Erlangs require IID exponentials). We get

$$p_1(t) = e^{-\mu_1 t},$$

$$p_2(t) = \frac{\mu_1}{\mu_2 - \mu_1} e^{-\mu_1 t} - \frac{\mu_1}{\mu_2 - \mu_1} e^{-\mu_2 t},$$

$$p_3'(t) = \mu_2 p_2(t).$$

If we now instead change the matrices Q and $\tilde{Q}$ to

$$Q = \begin{pmatrix} -\mu_1 & 0 & \mu_1 \\ 0 & -\mu_2 & \mu_2 \\ 0 & 0 & 0 \end{pmatrix}, \qquad \tilde{Q} = \begin{pmatrix} -\mu_1 & 0 \\ 0 & -\mu_2 \end{pmatrix}$$

and use $p(0) = (q, 1 - q)$ as the initial state vector, we can generate a distribution which is a mixture of exponentials (hyperexponential, H_2). This time we get

$$p_1(t) = q e^{-\mu_1 t},$$

$$p_2(t) = (1 - q) e^{-\mu_2 t},$$

$$p_3'(t) = \mu_1 p_1(t) + \mu_2 p_2(t).$$

Thus we see that this device provides a method for obtaining considerably more general distributions than merely Erlangs. We shall have much more material on these matters and their application in queueing in later sections.

3.3.3   Erlang Service Model ($M/E_k/1$)

We now consider a model in which the service time has an Erlang type-k distribution. Even though the service may not actually consist of k phases, it is convenient in analyzing this model to consider the Erlang as being made up of k exponential phases, each with mean $1/k\mu$. Let $p_{n,i}(t)$ represent the probability of n in the system and the customer in service being in phase i ($i = 1, 2, \ldots, k$), where we now number the phases backward; that is, k is the first phase of service and 1 is the last (a customer leaving phase 1 actually leaves the system). We can write the following steady-state balance equations (see Problem 3.8):

$$0 = -(\lambda + k\mu)p_{n,i} + k\mu p_{n,i+1} + \lambda p_{n-1,i} \qquad (n \geq 2, \ 1 \leq i \leq k - 1),$$

$$0 = -(\lambda + k\mu)p_{n,k} + k\mu p_{n+1,1} + \lambda p_{n-1,k} \qquad (n \geq 2),$$

$$0 = -(\lambda + k\mu)p_{1,i} + k\mu p_{1,i+1} \qquad (1 \leq i < k - 1), \qquad (3.15)$$

$$0 = -(\lambda + k\mu)p_{1,k} + k\mu p_{2,1} + \lambda p_0,$$

$$0 = -\lambda p_0 + k\mu p_{1,1}.$$

These equations are not particularly easy to handle because of their bivariate nature. However, it is not too difficult to obtain the expected measures of effectiveness (L, L_q, W, W_q) and the state probabilities from (3.15) by working directly on the point process which instead counts phases of service in the system [with the bivariate state pair (n, i) transformed to $(n - 1)k + i$] and then converts back to customers. This can be interpreted as the number of phases in the system requiring service, since $n - 1$ customers are waiting ($n \geq 1$), each requiring k phases of service, and the customer in service requires i phases more. This is essentially equivalent to modeling the Erlang service problem as a constant bulk-input model (the $M^{[K]}/M/1$ of Section 3.1), where each input unit is considered to bring in $K = k$ phases and the (phase) service rate μ is to be replaced by $k\mu$. (The phase approach to the Erlang service problem and the analogous bulk-input model are not identical, since completed intermediate phases of service cannot leave the system Erlang queue, unlike what happens to an individual completed customer in the bulk model.)

The first key result of this connection is that the average line delay for an $M/E_k/1$ customer can be found from the average system size of the bulk-input model given in Equation (3.5) multiplied by $1/k\mu$ [since the time to serve L phases is $L \cdot (1/k\mu)$] with the service rate μ of the bulk model itself replaced by the phase service rate $k\mu$. This yields

$$\boxed{W_q = \frac{k + 1}{2k} \frac{\rho}{\mu(1 - \rho)} \qquad (\rho = \lambda/\mu).} \qquad (3.16)$$

It follows that

$$L_q = \lambda W_q = \frac{k+1}{2k} \frac{\rho^2}{1-\rho}, \tag{3.17}$$

and $L = L_q + \rho$, $W = L/\lambda = W_q + 1/\mu$.

Now, to compute the steady-state probabilities themselves, we can immediately get the empty probability, since we know for all single-channel, one-at-a-time-service queues that $p_0 = 1 - \rho$ (see Table 1.2). Next, we shall convert Equation (3.15) to a (batch-input) system based on a single variable counting phases in the system using the aforementioned transformation $(n, i) = (n-1)k + i$.

Making the above transformation in (3.15) yields

$$0 = -(\lambda + k\mu)p_{(n-1)k+i} + k\mu p_{(n-1)k+i+1}$$
$$+ \lambda p_{(n-2)k+i} \qquad (n \geq 1, \quad 1 \leq i \leq k), \tag{3.18}$$
$$0 = -\lambda p_0 + k\mu p_1,$$

where any p turning out to have a negative subscript is assumed to be zero. It can be seen quite readily that by writing out the top equation in (3.18) sequentially starting at $n = 1$, $i = 1$, (3.18) can be simplified to

$$0 = -(\lambda + k\mu)p_n + k\mu p_{n+1} + \lambda p_{n-k} \qquad (n \geq 1),$$
$$0 = -\lambda p_0 + k\mu p_1, \tag{3.19}$$

which is precisely what (3.1) would give for a constant batch size k and service rate $k\mu$. Now if we let $p_n^{(P)}$ represent the probability of n in the phase or bulk-input system defined by (3.19), then it follows that the probability of n in the Erlang service system, p_n, is given by

$$p_n = \sum_{j=(n-1)k+1}^{nk} p_j^{(P)} \qquad (n \geq 1). \tag{3.20}$$

The waiting-time CDF can also be obtained for this problem, but the method of solution is quite different from those presented so far. The results also follow nicely from the general theory of $G/G/1$ queues as presented later in Chapter 6, Section 6.1.

While we have here utilized the relation between the $M^{[k]}/M/1$ and the $M/E_k/1$, it is important to note that a similar partnership holds between the $M/M^{[k]}/1$ queue and the $E_k/M/1$, so that the previous bulk results of Section 3.2 can be useful in deriving results about the Erlang arrival model to be treated in the following section. Prior to considering an Erlang arrival model, we illustrate Erlang service models by the following three examples.

Example 3.3
The Grabeur-Money Savings and Loan has a drive-up window. During the busy periods for drive-up service, customers arrive according to a Poisson distribution with a mean of 16/hr. From observations on the teller's performance, the mean service time is estimated to be 2.5 min, with a standard deviation of $\frac{5}{4}$ min. It is thought that the Erlang would be a reasonable assumption for the distribution of the teller's service time. Also, since the building (and drive-up window) is located in a large shopping center, there is virtually no limit on the number of vehicles that can wait. The company officials wish to know, on the average, how long a customer must wait until reaching the window for service, and how many vehicles are waiting for service.

The appropriate model, of course, is an $M/E_k/1$ model. To determine k, we first note that $1/\mu = 2.5$ min and that $\sigma^2 = 1/(k\mu^2) = \frac{25}{16}$, which yields $k = 4$. Thus we have an $M/E_4/1$ model with $\rho = \frac{2}{3}$, so from (3.17) we have

$$L_q = \frac{5}{8}\frac{\frac{4}{9}}{1-\frac{2}{3}} = \frac{5}{6}, \qquad W_q = \frac{60}{16}\frac{5}{6} = \frac{25}{8}\text{ min.}$$

Example 3.4
A small heating-oil distributor, the Phil R. Upp Company, has only one truck. The capacity of the truck is such that after delivering to a customer, it must return to be refilled. Customers call in for deliveries on an average of once every 50 min during the height of the heating season. The distribution of time between calls has been found to be approximately exponential. It has also been observed that it takes on average 20 min for the truck to get to a customer and 20 min for the truck to return. Both the time for driving out and the time for returning have approximately exponential distributions. The time it takes to unload and load the truck has been included in the driving times. W. A. R. Mup, company general manager, is considering the possibility of purchasing an additional truck and would like to know, under the current situation, how long on average a customer has to wait from the time he places a call until the truck arrives. All customers are served on a first-come, first-served basis.

The service time in this problem corresponds to the time a truck is tied up, and is made up of two identical exponential stages, one consisting of loading and traveling out to the customer, and one consisting of unloading and traveling back to the terminal. Hence we have an $M/E_2/1$ model, where $\lambda = \frac{6}{5}$/hr and $\mu = \frac{3}{2}$/hr, so that $\rho = \frac{4}{5}$. The average time that a customer must wait from the time the call is placed until service starts (the truck begins loading for his delivery) is W_q and is given by (3.17) as

$$W_q = \frac{3}{4} \frac{\frac{16}{25}}{\frac{6}{5}(1 - \frac{4}{5})} = 2 \text{ hr.}$$

The average time it takes the truck to arrive once it is dispatched (loading for a given customer commences) is one-half an average service time, that is, 20 min. Thus the average wait from the time a customer calls in until the truck arrives and begins unloading is 2 hr 20 min.

Example 3.5
A manufacturer of a special electronic guidance-system component has a quality control checkpoint at the end of the production line to assure that the component is properly calibrated. If it fails the test, the component is sent to a special repair center, where it is readjusted. There are two specialists at the center, and each can adjust a component in an average of 5 min, their repair time being exponentially distributed. The average number of rejects from the quality control point per hour is 18, and the sequence of rejections appears to be well described by a Poisson process. The company can lease one machine which can adjust the component to the same degree of accuracy as the repair staff in exactly $2\frac{2}{3}$ min, that is, with no variation in repair time. The machine leasing costs are roughly equivalent to the salary and fringe-benefit costs for the staff. (If the repairers are replaced, they can be used elsewhere in the company and there will be no morale or labor problems.) Should the company lease the machine?

We wish to compare the expected waiting time W and system size L under each alternative. Alternative 1, keeping the staff of two, is an $M/M/2$ model, while alternative 2, leasing the machine, is an $M/D/1$ model. The calculations for the alternatives are as follows.

$M/M/2$:

$$\lambda = 18/\text{hr}, \ \mu = 0.2/\text{min} = 12/\text{hr}, \qquad W_q = \tfrac{3}{28} \text{ hr} \doteq 6.4 \text{ min,}$$

$$W \doteq 6.4 + 5 = 11.4 \text{ min} \quad \text{and} \quad L = \lambda W \doteq 3.42.$$

$M/D/1$: To obtain the results for this model, we use $\lim_{k \to \infty} M/E_k/1$;

$$\lambda = 18/\text{hr}, \quad \mu = \tfrac{3}{8} \text{ min} = 22.5/\text{hr},$$

$$W_q = \lim_{k \to \infty} \left(\frac{k+1}{2k} \frac{\rho^2}{\lambda(1-\rho)} \right) = \frac{\rho^2}{2\lambda(1-\rho)} = \frac{4}{45} \text{ hr} = \frac{16}{3} \text{ min,}$$

$$W \doteq \tfrac{16}{3} + \tfrac{8}{3} = 8 \text{ min} \quad \text{and} \quad L = \lambda W = 2.4.$$

Thus alternative 2 (the machine) is shown to be preferable.

3.3.4 Erlang Arrival Model ($E_k/M/1$)

As mentioned in the previous section, we can utilize the results of the second bulk-service model of Section 3.2 to develop results for the Erlang input model. We assume that the interarrival times are Erlang type k distributed, with a mean of $1/\lambda$. We can look, therefore, at an arrival having passed through k phases, each with a mean time of $1/k\lambda$, prior to actually entering the system. Here we number the phases frontwards from 0 to $k - 1$. Again, we remind the reader that this is a device convenient for analysis that does not have to correspond to the actual arrival mechanism—the only assumption on interarrival times is that they follow an Erlang type-k distribution with mean $1/\lambda$.

We define the state variable as the number of arrival phases in the system, so that we desire to find the probability of n arrival phases in the system in the steady state, which we denote by $p_n^{(P)}$. It is an easy matter once we have this to obtain the probability of n customers in the system by utilizing a relation similar to (3.20), that is,

$$p_n = \sum_{j=nk}^{nk+k-1} p_j^{(P)}. \tag{3.21}$$

We can get $p_n^{(P)}$ from $p_n^{(B)}$, the steady-state probability of n in the bulk-service model given by (3.11) but with λ replaced by $k\lambda$ (why?).

We know from (3.11) and (3.12) that

$$p_j^{(P)} = \frac{k\lambda p_0^{(P)}}{\mu} r_0^{j-k} \qquad (j \geq k)$$

$$= \rho(1 - r_0)r_0^{j-k} \qquad (\rho = \lambda/\mu).$$

So, for $n > 0$,

$$\begin{aligned}
p_n &= \sum_{j=nk}^{nk+k-1} p_j^{(P)} \\
&= \rho(1 - r_0)(r_0^{nk-k} + r_0^{nk-k+1} + \cdots + r_0^{nk-1}) \\
&= \rho(1 - r_0)r_0^{nk-k}(1 + r_0 + \cdots + r_0^{k-1}) \\
&= \rho(1 - r_0^k)(r_0^k)^{n-1}.
\end{aligned} \tag{3.22}$$

We see that this is a geometric distribution (as in $M/M/1$), but with r_0^k as the multiplier instead of ρ. In Chapter 5, Section 5.3, and in Chapter 6, Section 6.3, we show that the general-time steady-state system-size probabilities for all $G/M/1$ queues are also geometric.

It follows from (3.22) that

$$L = \rho(1 - r_0^k) \sum_{n=1}^{\infty} n(r_0^k)^{n-1}$$

$$= \rho(1 - r_0^k) \frac{1}{(1 - r_0^k)^2} = \frac{\rho}{1 - r_0^k},$$

from which we can get $L = L_q + \rho$, $W = L/\lambda$, and $W_q = W - 1/\mu$.

We can also derive the waiting-time distribution for this model as follows (recall that this is a composite distribution, because there is always a nonzero probability of no wait for service to commence). Denoting the probability that an arrival into the system finds n customers already there by q_n, we find that

$$q_n = \Pr\{n \text{ in system} \mid \text{arrival about to occur}\}$$

$$= \Pr\{n \text{ in system} \mid \text{arrival in phase } k - 1\}$$

$$= \frac{\Pr\{n \text{ in system and arrival in phase } k - 1\}}{\Pr\{\text{arrival in phase } k-1\}}$$

$$= \frac{p_{nk+k-1}^{(P)}}{1/k},$$

since it is equally likely that an arrival is in any one of the k phases with probability $1/k$. Thus we have

$$q_n = k p_{nk+k-1}^{(P)}$$

$$= \frac{k\rho(1 - r_0)}{r_0} r_0^{kn}$$

$$= (1 - r_0^k) r_0^{kn}, \tag{3.23}$$

where the final step follows from the characteristic equation (3.8) with λ replaced by $k\lambda$. Now if there are n customers in the system upon arrival, the conditional waiting time is the time it takes to serve these n people, which is the convolution of n exponentials, each with mean $1/\mu$. This yields an Erlang type-n distribution, and the unconditional line-delay distribution function can thus be written as

$$W_q(t) = q_0 + \sum_{n=1}^{\infty} q_n \int_0^t \frac{\mu(\mu x)^{n-1}}{(n-1)!} e^{-\mu x} \, dx$$

$$= q_0 + \sum_{n=1}^{\infty} (1 - r_0^k) r_0^{kn} \int_0^t \frac{\mu(\mu x)^{n-1}}{(n-1)!} e^{-\mu x} \, dx$$

$$= q_0 + r_0^k \int_0^t (1 - r_0^k)\mu e^{-\mu x} \sum_{n=1}^{\infty} \frac{(\mu x r_0^k)^{n-1}}{(n-1)!} \, dx$$

$$= q_0 + r_0^k \int_0^t (1 - r_0^k)\mu e^{-\mu(1-r_0^k)x} \, dx$$

$$= q_0 + r_0^k [1 - e^{-\mu(1-r_0^k)t}].$$

The probability of no wait for service upon arrival is given by (3.23) as

$$q_0 = 1 - r_0^k,$$

and thus

$$\boxed{W_q(t) = 1 - r_0^k e^{-\mu(1-r_0^k)t} \qquad (t \geq 0).} \qquad (3.24)$$

Example 3.6
Arrivals coming to a single-server queueing system are found to have an Erlang type-2 distribution with mean interarrival time of 30 min. The mean service time is 25 min, and service times are exponentially distributed. Find the steady-state system-size probabilities and the expected-value measures of effectiveness for this system.

The inputs required to answer these questions are $\lambda = 1/(30 \text{ min}) = 2$/hr, $\mu = 1/(25 \text{ min}) = \frac{12}{5}$/hr, and $k = 2$. We must first find the root r_0 for the operator equation (3.8) with λ replaced by $k\lambda$, namely,

$$\mu r_0^{k+1} - (k\lambda + \mu)r_0 + k\lambda = 0,$$

or

$$\frac{12}{5} r_0^3 - \frac{32}{5} r_0 + 4 = \frac{1}{5}(12r_0^3 - 32r_0 + 20) = 0.$$

Upon simplifying we have

$$3r_0^3 - 8r_0 + 5 = 0,$$

which happens to be the same operator equation as found in Example 3.2 having positive root with absolute value less than 1 approximately equal to 0.884. Thus, from (3.22),

$$p_n = \rho(1 - r_0^k)(r_0^k)^{n-1}$$
$$\doteq 0.233(0.781)^n \qquad (n \geq 1)$$

and

$$p_0 = 1 - \rho = 1 - \tfrac{10}{12} = \tfrac{1}{6}.$$

The mean system size is found as

$$L = \frac{\rho}{1 - r_0^k} \doteq \frac{\tfrac{5}{6}}{1 - 0.781} \doteq 3.81,$$

and $W = L/\lambda \doteq 3.81/2 \doteq 1.91$ hr, while $W_q = 3.81/2 - 5/12 \doteq 1.49$ hr and $L_q \doteq 3.81 - 5/6 \doteq 2.98$.

3.3.5 $E_j/E_k/1$ Models

The complete analysis of the $E_j/E_k/1$ queue is more complicated than either the $M/E_k/1$ or $E_k/M/1$, and we leave its details to later sections in Chapter 6. However, there are some things that can be said at this point that nicely connect the results of the immediately prior sections with those for the $E_j/E_k/1$ queue. The most important of these is that the solution to this latter problem is to be found in terms of roots to a characteristic polynomial, which in a sense has a form combining the characteristic equations of the $M/E_k/1$ and $E_j/M/1$, which are, respectively, derived as

$$k\mu D^{k+1} - (\lambda + k\mu)D^k + \lambda = 0 \qquad \text{[for } M/E_k/1, \text{ from (3.19)],}$$

$$\mu D^{j+1} - (j\lambda + \mu)D^j + j\lambda = 0$$
$$\text{[for } E_j/M/1, \text{ from (3.8), with } j\lambda \text{ replacing } \lambda\text{].}$$

In Chapter 6, we show that the $E_j/E_k/1$ queue has a characteristic equation of

$$k\mu z^{j+k} - (j\lambda + k\mu)z^k + j\lambda = 0.$$

As in our earlier discussions, we can apply Rouché's theorem to help determine the location of the roots of this polynomial equation, and it turns out that whenever $k\lambda/j\mu < 1$, there are k roots inside the complex unit circle plus the root of 1 on the boundary of the unit circle. Furthermore, by showing that the derivative of the polynomial cannot vanish inside the unit circle, it follows that all of the roots are distinct, with one always real and positive, while a second real and negative root exists whenever k is an even number. As a first illustration of this, let us consider the $M/E_2/1$ model of Example 3.4. Here, $\lambda = \tfrac{6}{5}$, $\mu = \tfrac{3}{2}$, $j = 1$, and $k = 2$. Thus the characteristic equation becomes

$$3z^3 - \tfrac{21}{5}z^2 + \tfrac{6}{5} = 0,$$

with roots 1, $(1 \pm \sqrt{11})/5$. Clearly, the latter two roots have the requisite property that their absolute values are less than 1.

For a second illustration, suppose that the interarrival distribution for the above example were instead assumed to an Erlang type 3. Then the characteristic equation becomes

$$3z^5 - \tfrac{33}{5}z^2 + \tfrac{18}{5} = \tfrac{3}{5}(5z^5 - 11z^2 + 6) = 0,$$

whose roots are found (using a mathematics package) to be 1 and (approximately) 0.914, -0.689, $-0.612 \pm 1.237i$, so that there are two real roots with absolute values less than 1 and two other roots with absolute values more than 1. Suppose, now, that we let the service-time shape parameter for this problem be 3; then the characteristic equation will be

$$\tfrac{9}{2}z^6 - \tfrac{81}{10}z^3 + \tfrac{18}{5} = \tfrac{9}{10}(5z^6 - 9z^3 + 4) = 0,$$

whose roots are found to be 1 and (approximately) 0.928, $-0.464 \pm 0.804i$, $-0.5 \pm 0.866i$, so that now there are two complex roots outside the unit circle, one real root less than 1, and two complex conjugate roots inside the unit circle.

The complete details on how these roots are manipulated to get state probabilities and a line-delay distribution will come later in the text. In the meantime, suffice it to say that the roots found in this process play a key role in the complete obtaining of measures of effectiveness for the general Erlang–Erlang queue. The multiserver problems $M/E_k/c$, $E_k/M/c$, and $E_j/E_k/c$ will be discussed later. This latter material also involves more complicated matrix-analytic methods utilizing phase-type distribution functions; this will also be presented later.

3.4 PRIORITY QUEUE DISCIPLINES

Up to this point, all the models considered have had the property that units proceed to service on a first-come, first-served basis. This is obviously not the only manner of service, and there are many alternatives, such as last-come, first-served, selection in random order, and selection by priority. In priority schemes customers with the highest priorities are selected for service ahead of those with lower priorities, independent of their time of arrival into the system. There are two further refinements possible in priority situations, namely, preemption and nonpreemption. In preemptive cases, the customer with the highest priority is allowed to enter service immediately even if another with lower priority is already present in service when the higher customer arrives. In addition, a decision has to be made whether to continue the preempted customer's service from the point of preemption when resumed or to start anew. On the other hand, a priority discipline is said to be nonpreemptive if there is no interruption and the highest-priority customer just goes to the head of the queue to wait its turn.

Obviously, a considerable portion of real-life queueing situations contains priority considerations. Priority queues are generally more difficult to model than nonpriority ones, but the priority models should not be oversimplified merely to permit solution. Full consideration of priorities is absolutely essential when considering costs of queueing systems and optimal design.

Recall some of the arguments of Chapters 1 and 2. There is nothing in the $M/M/1$ (and related models) state-probability derivation that depends upon queue discipline. Indeed, it can be shown that as long as selection for service is in no way a function of the relative size of the service time, then the $\{p_n\}$ are independent of queue discipline. The proof of Little's formula also remains unchanged, and since the average system size is unaltered, the average waiting time is likewise. But there will be changes in the waiting-time distribution.

It turns out that the waiting times are stochastically smallest under FCFS (all other things kept constant). That is, the introduction of any scheme of priorities not depending on service times is going to make all higher moments worse than under FCFS. With the background available to us at this point, it is difficult to prove this assertion in general. But we can give a basic idea of what is happening by reference to Figure 1.4, presented here with specific values as Figure 3.3.

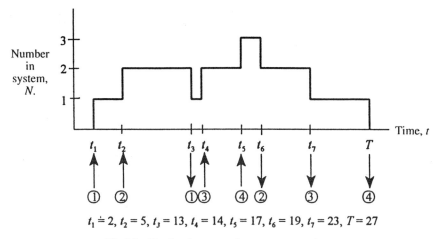

$t_1 \doteq 2,\ t_2 = 5,\ t_3 = 13,\ t_4 = 14,\ t_5 = 17,\ t_6 = 19,\ t_7 = 23,\ T = 27$

Fig 3.3 Number in a queueing system over time.

To illustrate the point, let us compare the variances computed for that sample path under FCFS and under LCFS. The busy period in the figure contains four customers, and their system waits sum to 44 (and thus average 11) under any queue discipline unrelated to service times. Under FCFS, the waiting times in system are respectively 11, 14, 9, 10, with a resultant (biased) variance of $14/4 = 3.5$. But under LCFS, they are 11, 14, 6, 13, with a variance of $37/4 = 9.25$, a substantial increase.

In further expansion of some of these ideas, we note that the remaining total service or work required for a single server at any point during an arbitrary busy period is independent of the order of service as long as the system is conservative. That is, this is true whenever no service needs are created or destroyed within the system. This means, for example, that there is no renege in the midst of service, no preemption repeat when service is not exponential, no forced idleness of the server, and so on. Nonconservative systems will in fact violate Little's formula. In the multichannel case, work conservation results when every channel has the same service-time distribution. Otherwise, preemption may lead to a change in a customer's server and thus affect the waiting times of others more than normally.

We may also tie the concept of conservation to Little's formula. First we note that the remaining total work in a conserving system must be stable across the steady state. Thus, for example, the expected gain in work present over a fixed time interval must be equal to the expected loss due to departing customers during the same period. We see then that Little's formula says that the work found by an arriving customer in the steady state under FCFS is always proportional to the general-time average queue size (i.e., $W = L\lambda$). Furthermore, in work-conserving priority situations, the formula holds for the individual priority classes as well.

3.4.1 Single Exponential Channel with Priorities

Let us begin by assuming that customers arrive as a Poisson process to a single exponential channel and that upon arrival to the system each unit will be designated to be a member of one of two priority classes (see Morse, 1958). The usual convention is to number the priority classes so that the smaller the number, the higher the priority. Let it further be assumed that the (Poisson) arrivals of the first or higher priority have mean rate λ_1 and that those of the second or lower priority have mean rate λ_2, such that $\lambda \equiv \lambda_1 + \lambda_2$. We also suppose that the first-priority items have the right to be served ahead of the others, but that there is no preemption.

In view of the foregoing assumptions, a system of steady-state balance equations may be established for $p_{mnr} \equiv \Pr\{$in steady state, m units of priority 1 and n units of priority 2 are in the system, and a unit of priority $r = 1$ or 2 is in service$\}$. These then lead to the following difference equations (see

Problem 3.21) in the event that $\rho = \lambda/\mu < 1$:

$$0 = -(\lambda + \mu)p_{mn2} + \lambda_1 p_{m-1,n,2} + \lambda_2 p_{m,n-1,2} \qquad (m > 0, \quad n > 1),$$

$$0 = -(\lambda + \mu)p_{mn1} + \lambda_1 p_{m-1,n,1} + \lambda_2 p_{m,n-1,1} + \mu(p_{m+1,n,1} + p_{m,n+1,2})$$
$$(m > 1, \quad n > 0),$$

$$0 = -(\lambda + \mu)p_{m12} + \lambda_1 p_{m-1,1,2},$$

$$0 = -(\lambda + \mu)p_{1n1} + \lambda_2 p_{1,n-1,1} + \mu(p_{2n1} + p_{1,n+1,2}),$$

$$0 = -(\lambda + \mu)p_{0n2} + \lambda_2 p_{0,n-1,2} + \mu(p_{1n1} + p_{0,n+1,2}), \qquad (3.25)$$

$$0 = -(\lambda + \mu)p_{m01} + \lambda_1 p_{m-1,0,1} + \mu(p_{m+1,0,1} + p_{m12}),$$

$$0 = -(\lambda + \mu)p_{012} + \lambda_2 p_0 + \mu(p_{111} + p_{022}),$$

$$0 = -(\lambda + \mu)p_{101} + \lambda_1 p_0 + \mu(p_{201} + p_{112}),$$

$$0 = -\lambda p_0 + \mu(p_{101} + p_{012}).$$

It should be clear that p_0 is still $1 - \rho$, since the ordering of service in no way affects the probability of idleness, and that

$$p_n = \sum_{m=0}^{n-1} (p_{n-m,m,1} + p_{m,n-m,2}) = (1 - \rho)\rho^n \qquad (n > 0).$$

Also, since the percentage of time the system is busy is ρ, the percentage of time it is busy with a type-r customer will be $\rho\lambda_r/\lambda$, so that

$$\sum_{m=1}^{\infty} \sum_{n=0}^{\infty} p_{mn1} = \frac{\lambda_1}{\mu} \quad \text{and} \quad \sum_{m=0}^{\infty} \sum_{n=1}^{\infty} p_{mn2} = \frac{\lambda_2}{\mu}.$$

It turns out, however, that obtaining a reasonable solution to these stationary equations is a very difficult matter, as we might have anticipated in view of the triple subscripts. The most we can do comfortably is obtain expected values via two-dimensional generating functions. So define

$$P_{m1}(z) = \sum_{n=0}^{\infty} z^n p_{mn1}, \qquad P_{m2}(z) = \sum_{n=1}^{\infty} z^n p_{mn2},$$

$$H_1(y, z) = \sum_{m=1}^{\infty} y^m P_{m1}(z) \qquad [\text{with } H_1(1, 1) = \lambda_1/\mu],$$

$$H_2(y, z) = \sum_{m=0}^{\infty} y^m P_{m2}(z) \qquad [\text{with } H_2(1, 1) = \lambda_2/\mu],$$

and

$$H(y, z) = H_1(y, z) + H_2(y, z) + p_0$$

$$= \sum_{m=1}^{\infty} \sum_{n=0}^{\infty} y^m z^n p_{mn1} + \sum_{m=0}^{\infty} \sum_{n=1}^{\infty} y^m z^n p_{mn2} + p_0$$

$$= \sum_{m=1}^{\infty} \sum_{n=1}^{\infty} y^m z^n (p_{mn1} + p_{mn2}) + \sum_{m=1}^{\infty} y^m p_{m01} + \sum_{n=1}^{\infty} z^n p_{0n2} + p_0,$$

where $H(y, z)$ then is merely the joint generating function for the two priorities, regardless of which type is in service. Note that $H(y, y) = p_0/(1 - \rho y)$ [with $H(1, 1) = 1$], since $H(y, z)$ collapses to the generating function of $M/M/1$ when z is set equal to y and thus no priority distinction is made [see (2.6)]. Hence if L_1 and L_2 are used to denote the mean number of customers present in system for each of the two priority classes, then

$$\frac{\partial H(y, z)}{\partial y}\bigg|_{y=z=1} = L_1$$

$$= L_{q_1} + \frac{\lambda_1}{\mu}$$

$$= \lambda_1 W_1,$$

and

$$\frac{\partial H(y, z)}{\partial z}\bigg|_{y=z=1} = L_2$$

$$= L_{q_2} + \frac{\lambda_2}{\mu}$$

$$= \lambda_2 W_2,$$

where L_{q_1} and L_{q_2} are the respective mean queue lengths. If we then multiply Equations (3.25) by the appropriate powers of y and z and sum accordingly, it is found that

$$\left(1 + \rho - \frac{\lambda_1 y}{\mu} - \frac{\lambda_2 z}{\mu} - \frac{1}{y}\right) H_1(y, z) = \frac{H_2(y, z)}{z} + \frac{\lambda_1 y p_0}{\mu} - P_{11}(z) - \frac{P_{02}(z)}{z}$$

$$(3.26)$$

and

$$\left(1 + \rho - \frac{\lambda_1 y}{\mu} - \frac{\lambda_2 z}{\mu}\right) H_2(y, z) = P_{11}(z) + \frac{P_{02}(z)}{z} - \left(\rho - \frac{\lambda_2 z}{\mu}\right) p_0. \qquad (3.27)$$

To determine the generating functions H_1 and H_2 fully, we need to know the values of $P_{11}(z)$, $P_{02}(z)$, and p_0. One equation relating $P_{11}(z)$, $P_{02}(z)$, and p_0 may be found by summing z^n $(n = 2, 3, \ldots)$ times the equation of (3.25) which involves p_{0n2}, and then using the final three equations of (3.25), thus obtaining

$$P_{11}(z) = \left(1 + \rho - \frac{\lambda_2 z}{\mu} - \frac{1}{z}\right)P_{02}(z) + \left(\rho - \frac{\lambda_2 z}{\mu}\right)p_0.$$

Substitution of this equation into (3.26) and (3.27) then gives H_1 and H_2 as functions of p_0 and P_{02}, and thus also $H(y, z)$ as

$$H(y, z) = H_1(y, z) + H_2(y, z) + p_0$$

$$= \frac{(1 - y)p_0}{1 - y - \rho y(1 - z - \lambda_1 y/\lambda + \lambda_1 z/\lambda)}$$

$$+ \frac{(1 + \rho - \rho z + \lambda_1 z/\mu)(z - y)P_{02}(z)}{z[1 + \rho - \lambda_1 y/\mu - \lambda_2 z/\mu][1 - y - \rho y(1 - z - \lambda_1 y/\lambda + \lambda_1 z/\lambda)]}.$$

By employing the condition that $H(1, 1) = 1$, it is found that

$$P_{02}(1) = \frac{\lambda_2 p_0/\mu}{(1 + \lambda_1/\mu)(1 - \rho)} \qquad (\rho = \lambda/\mu).$$

We next take the partial derivatives of H with respect to both y and z and then evaluate at $(1, 1)$ to find the means L_1 and L_2. In so doing, the exact functional relationship for $P_{02}(z)$ turns out not to be needed, and $P_{02}(1)$ alone suffices. Without suffering through the details of the differentiation, the final results are

$$
\begin{aligned}
L_1 &= \frac{(\lambda_1/\mu)(1 + \rho - \lambda_1/\mu)}{1 - \lambda_1/\mu}, \\[2mm]
L_{q1} &= \frac{\rho \lambda_1/\mu}{1 - \lambda_1/\mu}, \\[2mm]
W_{q1} &= \frac{\rho}{\mu - \lambda_1}, \\[2mm]
L_2 &= \frac{(\lambda_2/\mu)(1 + \rho \lambda_1/\mu - \lambda_1/\mu)}{(1 - \rho)(1 - \lambda_1/\mu)}, \\[2mm]
L_{q2} &= \frac{\rho \lambda_2/\mu}{(1 - \rho)(1 - \lambda_1/\mu)}, \\[2mm]
W_{q2} &= \frac{\rho}{(1 - \rho)(\mu - \lambda_1)}.
\end{aligned}
\qquad (3.28)
$$

Using the more sophisticated theory of multidimensional birth–death processes, Miller (1981) has shown that the actual probabilities for priority-1 customers are

$$p_{n_1} = (1 - \rho)\left(\frac{\lambda_1}{\mu}\right)^{n_1} + \frac{\lambda_2}{\lambda_1}\left(\frac{\lambda_1}{\mu}\right)^{n_1}\left(1 - \frac{(\lambda_1/\mu)^{n_1}}{(1 + \lambda_1/\mu)^{n_1+1}}\right) \qquad (n_1 \geq 0).$$

Some interesting and important observations may be made about the mean-value results. First, it should be noted that, as expected, units of priority-2 wait in queue longer than the higher-priority units, so that

$$W_{q2} = \frac{\mu}{\mu - \lambda} W_{q1}.$$

We can also see that as $\rho \to 1$, the expectations L_2, L_{q2}, $W_{q2} \to \infty$, while the corresponding means for the first priority approach finite limits. These first-priority expectations go to ∞ only when $\lambda_1/\mu \to 1$. That is, the first-priority units need not accumulate, even though a steady state does not exist. In addition, it should be noted that the expected wait in line for first-priority units in the absence of second-priority customers is smaller than their wait in the presence of the lower-priority units by a factor of λ_1/λ. In fact, the relative performance of the higher priority with respect to the lower increases as $\rho \to 1$ and $\lambda_1 \to 0$. However, had the units of higher priority had the power of preemption, then the two expected waits above would be identical. Also, the average number in queue is $L_q = L_{q1} + L_{q2} = \rho^2/(1 - \rho)$ (the same as for the nonpriority case), which implies that the unconditional average wait, $W_q = (\lambda_1/\lambda)W_{q1} + (\lambda_2/\lambda)W_{q2}$, is the same as that for the nonpriority case.

Similar equations can be obtained for the model which slightly generalizes this problem by serving the priority-1 customers at the rate μ_1 and the priority-2 customers at μ_2. These results are (details may be found in Morse, 1958)

$$L_{q1} = \frac{\lambda_1 \hat{\rho}}{\mu_1} \frac{\lambda_1/\lambda + (\lambda_2/\lambda)(\mu_1^2/\mu_2^2)}{1 - \lambda_1/\mu_1},$$

$$L_{q2} = \frac{(\lambda_2/\mu_1)\hat{\rho}}{1 - \lambda_1/\mu_1} \frac{\lambda_1/\lambda + (\lambda_2/\lambda)(\mu_1^2/\mu_2^2)}{1 - \lambda_1/\mu_1 - \lambda_2/\mu_2},$$

$$L_q = L_{q1} + L_{q2}$$

$$= \frac{\lambda_1\hat{\rho}/\mu_1 + \lambda_2\lambda/\mu_2^2}{1 - \lambda_1/\mu_1 - \lambda_2/\mu_2} \frac{1 - (\lambda_1/\mu_1)[\lambda_1/\lambda + (\lambda_2/\lambda)(\mu_1/\mu_2)]}{1 - \lambda_1/\mu_1}$$

$$(\hat{\rho} = \lambda/\mu_1).$$

(3.29)

The above may be presented in slightly different form by introducing the actual system utilization factor, say

$$\rho^* = \text{fraction of system busy time} = \frac{\lambda_1}{\mu_1} + \frac{\lambda_2}{\mu_2}.$$

Observe that $\rho^* < \hat{\rho}$ if

$$\frac{\lambda_1}{\mu_1} + \frac{\lambda_2}{\mu_2} < \frac{\lambda}{\mu_1} = \frac{\lambda_1 + \lambda_2}{\mu_1},$$

or, $\mu_2 > \mu_1$, and, similarly, $\rho^* > \hat{\rho}$ when $\mu_2 < \mu_1$. Then, for example, the third equation of (3.29) can be rewritten as

$$L_q = \frac{(\rho^*)^2}{1 - \rho^*} \frac{\lambda_1/\lambda + (\lambda_2/\lambda)(\mu_1^2/\mu_2^2)}{[\lambda_1/\lambda + (\lambda_2/\lambda)(\mu_1/\mu_2)]^2} \frac{1 - \lambda_1\rho^*/\lambda}{1 - \lambda_1\hat{\rho}/\lambda}. \qquad (3.30)$$

Miller's (1981) work, using the theory of multidimensional birth–death processes, displayed the actual probabilities for priority-1 customers as

$$p_{n_1} = (1 - \rho^*)\left(\frac{\lambda_1}{\mu_1}\right)^{n_1} + \frac{\lambda_2}{\lambda_1 + \mu_2 - \mu_1}\left[\left(\frac{\lambda_1}{\mu_1}\right)^{n_1} - \frac{\mu_1\lambda_1^{n_1}}{(\lambda_1 + \mu_2)^{n+1}}\right] \qquad (n_1 \geq 0).$$

Further insight can be gained into this latest model by comparing it, on the one hand, with the first model of this section, which assumed constant service rates, for which the results are given in Equations (3.28), and, on the other hand, with a model having no priorities but unequal service rates for customers of two major types, say, for example, adults and children. The analysis of the latter model is not a terribly difficult matter (see Problem 3.22), and the expected line lengths are given as

$$L_{q_1} = \frac{\lambda_1\hat{\rho}}{\mu_1} \frac{1 - (1 - \mu_1/\mu_2)(\lambda_2/\mu_2)}{1 - \lambda_1/\mu_1 - \lambda_2/\mu_2},$$

$$L_{q_2} = \frac{\lambda_2\hat{\rho}}{\mu_1} \frac{\mu_1^2/\mu_2^2 + (1 - \mu_1/\mu_2)(\lambda_1/\mu_2)}{1 - \lambda_1/\mu_1 - \lambda_2/\mu_2},$$

$$L_q = \frac{(\lambda_1/\mu_1)\hat{\rho} + (\lambda_2/\mu_2)\hat{\rho}(\mu_1/\mu_2)}{1 - \lambda_1/\mu_1 - \lambda_2/\mu_2} \qquad (3.31)$$

$$= \frac{(\rho^*)^2}{1 - \rho^*} \frac{\lambda_1/\lambda + (\lambda_2/\lambda)(\mu_1^2/\mu_2^2)}{[\lambda_1/\lambda + (\lambda_2/\lambda)(\mu_1/\mu_2)]^2}$$

$$(\hat{\rho} = \lambda/\mu_1, \quad \rho^* = \lambda_1/\mu_1 + \lambda_2/\mu_2).$$

As a sidelight, it is interesting to note that the L_q of (3.31) is always greater than that of the standard $M/M/1$ with mean service time equal to the weighted average of the respective means, namely, $1/\mu = (\lambda_1/\lambda)/\mu_1 + (\lambda_2/\lambda)/\mu_2$ (see Problem 3.23).

Now to compare the L_q's of (3.29) [or (3.30)], the two-priority, two-rate case, with the L_q's of (3.28), the two-priority, one-rate case, we assume that the single rate chosen for use in (3.28) lies somewhere between the μ_1 and μ_2 of the other model. One possibility is to choose $1/\mu$ equal to the weighted average, but the choice could be otherwise. When the weighted average is used, the ρ^* of (3.30) becomes equivalent to the ρ of (3.28). If the value of μ is chosen to equal the larger of μ_1 and μ_2, then it can be easily shown that all three measures, L_{q_1}, L_{q_2}, and L_q, are less in (3.28) than they are in (3.29), while this comparison is completely reversed when $\mu = \min(\mu_1, \mu_2)$ (see Problem 3.24). Therefore, as we would intuitively expect, any comparison between the L_{q_1}, L_{q_2}, or L_q of (3.28) and that of (3.29) in the event that μ lies strictly between μ_1 and μ_2 is going to depend on the values of the parameters involved, namely, μ_1, μ_2, μ, λ_1, and λ_2.

In comparing the two-priority, two-rate case of (3.29) now with the no-priority, two-rate case of (3.31), it can be shown (after much algebraic manipulation) that the imposition of priorities increases the mean number of priority-2 items waiting, while shortening the mean queue for priority-1 items (see Problem 3.25). Thus, if a queueing system is to be designed in such a way as to reduce expected waits for one particular group of customers, then that group should be given priority. This is especially true if the nonpriority units (essentially priority 2) have a higher mean service time.

A comparison now of Equations (3.30) and (3.31) will show that the L_q's (thus the W_q's also) differ by the factor

$$\frac{1 - \lambda_1 \rho^*/\lambda}{1 - \lambda_1 \hat{\rho}/\lambda}.$$

When this ratio is larger than one, there is more waiting if priorities are imposed than otherwise, while when the ratio is less than one, the imposition of priorities will reduce the mean values. The ratio will exceed one whenever $\rho^* < \hat{\rho}$. But from our previous discussion preceding Equation (3.30), this is equivalent to requiring that $\mu_2 > \mu_1$ or $1/\mu_1 > 1/\mu_2$. Hence the numerator is greater than the denominator as long as the expected service time for the priority units is greater than that for the nonpriority ones, and vice versa when $1/\mu_1 < 1/\mu_2$.

These comparative results have very important implications for the design of queueing systems and specifically give rise to an optimal design rule called "priority assignment by shortest processing time" or simply "the shortest processing time (SPT) rule" (see Schrage and Miller, 1966). That is, if the design criterion of a queue is the reduction of the total number waiting, or

Table 3.1 Comparison of Priority Model

Model	Results in Equation
(1) 2 priorities, 1 service rate	$(3.28)^a$
(2) 2 priorities, 2 service rates	(3.29)
(3) no priority, 2 service rates	(3.31)

Versus	$M/M/1$	(2)	(3)
(1)	$\lambda = \lambda_1 + \lambda_2$ $L_{q_1} < L_{q_1}\,(M/M/1)$ $L_{q_2} > L_{q_2}\,(M/M/1)$ $L_q < L_q\,(M/M/1)$	All comparisons depend on the values of the parameters involved $\mu_1, \mu_2, \mu, \lambda_1, \lambda_2$	Not applicable
(2)	$\lambda = \lambda_1 + \lambda_2$ $\dfrac{1}{\mu} = \dfrac{\lambda_1/\lambda}{\mu_1} + \dfrac{\lambda_2/\lambda}{\mu_2}$ $L_q > L_q\,(M/M/1),\ \mu_1 < \mu_2$ $L_q < L_q\,(M/M/1),\ \mu_1 > \mu_2$	—	$L_{q_1}:\ (2) < (3)$ $L_{q_2}:\ (2) > (3),$ $L_q: \begin{cases}(2) < (3),\ \mu_1 > \mu_2 \\ (2) > (3),\ \mu_1 < \mu_2\end{cases}$
(3)	$\lambda = \lambda_1 + \lambda_2$ $\dfrac{1}{\mu} = \dfrac{\lambda_1/\lambda}{\mu_1} + \dfrac{\lambda_2/\lambda}{\mu_2}$ $L_q > L_q\,(M/M/1)$	See cell (2, 3)	—

$^a\mu$ used in (3.28) chosen to lie between μ_1 and μ_2.

equivalently the overall mean delay, then if it is at all possible, it pays to give priority to that group of customers which seems to have the faster service rate. The degree of improvement is then a function of the ratio of the two mean arrival rates, λ_1/λ_2.

To summarize the comparisons between the three main models of this section, we present Table 3.1.

Example 3.7
To further our analysis let us consider the following problem: our friend the hair-salon proprietor, Ms. H. R. Cutt, has decided to explore the possibility of giving priority to customers who wish only a trim cut. Ms. Cutt estimates that the time required to trim a customer (needed one-third of the time) is still exponential, but with a mean of 5 min, and that the removal of these customers from the total population increases the mean of the nonpriority class to 12.5 min, leaving the distribution exponential. If Cutt measures her performance by the value of the mean waiting time (remember that the mean

throughput will not change as long as λ remains constant, but lower waits may increase λ in the future), will this change in policy reduce average line waits in any way?

So we have $\lambda_1 = \lambda/3 = \frac{5}{3}$/hr and $\lambda_2 = 2\lambda/3 = \frac{10}{3}$/hr, while $\mu_1 = 12$/hr and $\mu_2 = \frac{24}{5}$/hr. This gives $\hat{\rho} = \lambda/\mu_1 = (\lambda_1 + \lambda_2)/\mu_1 = \frac{5}{12}$. Then the substitution of the appropriate values of the constants into the equations (3.29) gives

$$ L_{q_1} = \tfrac{75}{248} \doteq 0.302, \qquad L_{q_2} = \tfrac{225}{62} \doteq 3.63, $$

$$ L_q \doteq 3.93, \quad \text{and} \quad W_q = \frac{L_q}{\lambda} \doteq 47 \, \text{min.} $$

The correct model to compare this with is the no-priority two-service-rate model of (3.31) rather than the original $M/M/1$ example of Section 2.2.3, which gave a wait in queue of 50 min. The calculations following (3.31) give

$$ \rho^* = \tfrac{5}{6}, \qquad L_q \doteq 4.7, \quad \text{and} \quad W_q \doteq \tfrac{4.7}{5} \doteq 0.94 \, \text{hr} \doteq 56 \, \text{min.} $$

Thus introducing a priority system (faster service rate) for the customers requiring less service not only reduces their average wait, but also the average wait of *all* customers, thus illustrating the effect of the SPT rule.

Note that using the two-service-rate model of (3.31) gives a slightly larger W_q than would be obtained ignoring the fact that customers are of two types (which would be the case if we considered the results of Section 2.2.3, with the regular $M/M/1$ model and $W_q = 50$). This agrees with our previous statement to that effect immediately following the development of Equation (3.31) and also with the results as given in Table 3.1. Also observe that this is an example where priority is indeed a function of relative service time. Thus the change in W_q should not be a surprise.

3.4.2 Nonpreemptive Markovian Systems with Many Priorities

As observed in the previous section, the determination of stationary probabilities in a nonpreemptive Markovian system is an exceedingly difficult matter, well near impossible when the number of priorities exceeds two. In light of this and the difficulty of handling multiindexed generating functions when there are more than two priority classes, an alternative approach to obtaining the mean-value measures L and W is used, namely, a direct expected-value procedure.

Suppose that items of the kth priority (the smaller the number, the higher the priority) arrive before a single channel according to a Poisson distribution with parameter λ_k ($k = 1, 2, \ldots, r$) and that these customers wait on a first-come, first-served basis within their respective priorities. Let the service distribution for the kth priority be exponential with mean $1/\mu_k$. Whatever the priority of a unit in service, it completes its service before another item is admitted.

We begin by defining

$$\rho_k = \frac{\lambda_k}{\mu_k} \quad (1 \le k \le r), \qquad \sigma_k = \sum_{i=1}^{k} \rho_i \quad (\sigma_0 \equiv 0, \quad \sigma_r \equiv \rho).$$

The system is stationary for $\sigma_r = \rho < 1$. Then suppose that a customer of priority i arrives at the system at time t_0 and enters service at time t_1. Its line wait is thus $T_q = t_1 - t_0$. At t_0 assume that there are n_1 customers of priority 1 in the line ahead of this new arrival, n_2 of priority 2, n_3 of priority 3, and so on. Let S_0 be the time required to finish the item already in service, and S_k be the total time required to serve n_k. During the new customer's waiting time T_q, n'_k items of priority $k < i$ will arrive and go to service ahead of this current arrival. If S'_k is the total service time of all the n'_k, then it can be seen that

$$T_q = \sum_{k=1}^{i-1} S'_k + \sum_{k=1}^{i} S_k + S_0.$$

If expected values are taken of both sides of the above equation, we find that

$$W_q^{(i)} \equiv E[T_q] = \sum_{k=1}^{i-1} E[S'_k] + \sum_{k=1}^{i} E[S_k] + E[S_0].$$

Since $\sigma_{i-1} < \sigma_i$ for all i, $\rho < 1$ implies that $\sigma_{i-1} < 1$ for all i.

To find $E[S_0]$, we observe that the combined service distribution is the mixed exponential, which is formed from the law of total probability as

$$B(t) = \sum_{k=1}^{r} \frac{\lambda_k}{\lambda} (1 - e^{-\mu_k t}),$$

where

$$\lambda = \sum_{k=1}^{r} \lambda_k.$$

The random variable "remaining time of service," S_0, has the value 0 if the system is idle; hence

$$E[S_0] = \Pr\{\text{system is busy}\} \cdot E[S_0|\text{system is busy}].$$

But the probability that the system is busy is

$$\lambda \cdot (\text{expected service time}) = \lambda \sum_{k=1}^{r} \frac{\lambda_k}{\lambda} \frac{1}{\mu_k} = \rho,$$

and

$$E[S_0|\text{system is busy}] = \sum_{k=1}^{r} (E[S_0|\text{system is busy with priority-}k\text{ customer}]$$

$$\times \Pr\{\text{customer has priority } k\})$$

$$= \sum_{k=1}^{r} \frac{1}{\mu_k} \frac{\rho_k}{\rho}.$$

Therefore

$$E[S_0] = \rho \sum_{k=1}^{r} \frac{1}{\mu_k} \frac{\rho_k}{\rho}$$

$$= \sum_{k=1}^{r} \frac{\rho_k}{\mu_k}. \tag{3.32}$$

Since n_k and the service times of individual customers, $S_k^{(n)}$, are independent,

$$E[S_k] = E[n_k S_k^{(n)}] = E[n_k]E[S_k^{(n)}] = \frac{E[n_k]}{\mu_k}.$$

Little's formula then gives

$$E[S_k] = \frac{\lambda_k W_q^{(k)}}{\mu_k} = \rho_k W_q^{(k)}.$$

Similarly,

$$E[S_k'] = \frac{E[n_k']}{\mu_k},$$

and then using the uniform property of the Poisson, it follows that

$$E[S_k'] = \frac{\lambda_k W_q^{(i)}}{\mu_k}.$$

Therefore

$$W_q^{(i)} = W_q^{(i)} \sum_{k=1}^{i-1} \rho_k + \sum_{k=1}^{i} \rho_k W_q^{(k)} + E[S_0],$$

or

$$W_q^{(i)} = \frac{\sum_{k=1}^{i} \rho_k W_q^{(k)} + E[S_0]}{1 - \sigma_{i-1}}. \tag{3.33}$$

The solution to Equation (3.33) was found by Cobham (1954), after whom much of this analysis follows, by induction on i, after a general pattern emerged upon iteration (see Problem 3.26). That solution is

$$W_q^{(i)} = \frac{E[S_0]}{(1 - \sigma_{i-1})(1 - \sigma_i)}. \tag{3.34}$$

Using Equation (3.32) finally gives

$$W_q^{(i)} = \frac{\sum_{k=1}^{i} \rho_k / \mu_k}{(1 - \sigma_{i-1})(1 - \sigma_i)}. \tag{3.35}$$

Note that (3.35) holds as long as $\sigma_r = \sum_{k=1}^{r} \rho_k < 1$.

We also know, therefore, from Little's formula [the case $r = 2$ does in fact check with the earlier result given by Equation (3.29)], that the total expected system size is

$$L_q = \sum_{i=1}^{r} L_q^{(i)} = \sum_{i=1}^{r} \frac{\lambda_i \sum_{k=1}^{r} \rho_k / \mu_k}{(1 - \sigma_{i-1})(1 - \sigma_i)}.$$

Results for higher moments were obtained by Kesten and Runnenburg (1957), and the interested reader is referred to that paper for the appropriate derivations and formulas.

It now would seem quite logical to ask how the value of L_q obtained for the $M/M/1$ priority model compares with that of the ordinary $M/M/1$ case with service time equal to the average over all priorities (as was done earlier in the two-priority case), namely,

$$\overline{\left(\frac{1}{\mu}\right)} = \sum_{i=1}^{r} \frac{\lambda_i}{\lambda} \frac{1}{\mu_i}.$$

It seems intuitive that the average wait of an item in a priority situation would be different from its value if the discipline were nonpriority. But what about the weighted average of the waits taken over all priorities,

$$\overline{W}_q = \sum_{i=1}^{r} \frac{\lambda_i W_q^{(i)}}{\lambda} ?$$

For after all, if $\overline{W}_q$ were the same as the nonpriority W_q, then the two L_q's would have to be equal.

But some further thought and analogy with the two-priority case should convince us that the expected-value measures are the same if, and only if, the μ_i's are identical. That is to say,

$$\overline{W}_q = \frac{\lambda}{\mu(\mu - \lambda)}$$

if, and only if, $\mu_i \equiv \mu$ for all i (see Problem 3.27). In fact, it turns out that if higher-priority units have faster service rates, then the average wait over all units (also the average system size measures) is less than for a nonpriority system with a constant service time of

$$\left(\frac{1}{\mu}\right) = \sum_{i=1}^{r} \frac{\lambda_i/\lambda}{\mu_i},$$

again illustrating the SPT rule mentioned earlier. If the opposite is true, that is, lower priorities have faster service, then the "equivalent" nonpriority model gives the lower average wait and system size (see Problem 3.28). These differences increase as saturation is approached. If priorities and service-rate rankings are mixed, then the result depends upon the pairings of priorities and service rates and upon the actual values of the average service times.

This argument is essentially the same as the one given in the two-priority case to justify the SPT rule. That is, if the overriding requirement in the design of a queueing system is the reduction of the delay for one specific set of items, then this class should be given priority. This becomes especially profitable if the urgent set of items takes less time to serve on the average. If, however, the criterion for design is simply to reduce the average wait in queue of all units or to reduce the total number waiting, then it helps to give priority to that class of units which tends to have the faster service rate if such is discernible. For a further discussion of the effect of priorities on delay, the reader is referred to Morse (1958) and Jaiswal (1968).

Some additional comments on where there are no differences between the state probabilities and measures of effectiveness for $M/M/1/\infty/FCFS$ and $M/M/1/\infty/PR$ (as well as $M/M/1/\infty/GD$): the same state probabilities and measures hold for arbitrary queue disciplines provided (1) all arrivals stay in the queue until served, (2) the mean service time of all units is the same, (3) the server completes service before it starts on the next item, and (4) the service channel always admits a waiting customer immediately upon the completion of another.

The analysis for the multiple-channel case is very similar to that of the preceding model except that it must now be assumed that service is governed by identical exponential distributions for each priority at each of c channels. Unfortunately, for multichannels we must assume no service-time distinction between priorities, or else the mathematics becomes quite intractable.

Let us define

$$\rho_k = \frac{\lambda_k}{c}\mu \quad (1 \le k \le r), \qquad \sigma_k = \sum_{i=1}^{k} \rho_i \quad (\sigma_r \equiv \rho = \lambda/c\mu).$$

Again the system is completely stationary for $\rho < 1$, and

$$W_q^{(i)} = \sum_{k=1}^{i-1} E[S_k'] + \sum_{k=1}^{i} E[S_k] + E[S_0],$$

where, as before, S_k is the time required to serve n_k items of the kth priority in the line ahead of the item, S_k' is the service time of the n_k' items of priority k which arrive during $W_q^{(i)}$, and S_0 is the amount of time remaining until the next server becomes available. The first two terms of the right-hand side of the $W_q^{(i)}$ equation are exactly the same as in the single-channel case, except that the system service rate $c\mu$ is used in place of the single-service rate μ_k throughout the argument.

To derive $E[S_0]$, consider

$$E[S_0] = \Pr\{\text{all channels busy}\} \cdot E[S_0 \mid \text{all channels busy}].$$

The probability that all channels are busy is [from Equation (2.24)]

$$\sum_{n=c}^{\infty} p_n = p_0 \sum_{n=c}^{\infty} \frac{(cp)^n}{c^{n-c}c!} = p_0 \frac{(c\rho)^c}{c!(1-\rho)},$$

and

$$E[S_0 \mid \text{all channels busy}] = 1/c\mu$$

from the memorylessness of the exponential. Thus from Equation (2.25),

$$E[S_0] = \frac{(cp)^c}{c!(1-\rho)(c\mu)}\left(\sum_{n=0}^{c-1} \frac{(c\rho)^n}{n!} + \frac{(c\rho)^c}{c!(1-\rho)}\right)^{-1}.$$

Therefore from (3.35),

$$W_q^{(i)} = \frac{E[S_0]}{(1 - \sigma_{i-1})(1 - \sigma_i)} = \frac{[c!(1 - \rho)(c\mu)\sum_{n=0}^{c-1}(c\rho)^{(n-c)}/n! + c\mu]^{-1}}{(1 - \sigma_{i-1})(1 - \sigma_i)},$$

and the expected line wait taken over all priorities is thus

$$W_q = \sum_{i=1}^{r} \frac{\lambda_i}{\lambda} W_q^{(i)}.$$

3.4.3 Preemptive Priorities

Let us now extend the Markovian model of Section 3.4.1 to permit units of the higher priority to preempt. We then have to decide whether or not the ejected items lose all service performed before ejection. But since it is to be assumed that service is exponential, such a question is irrelevant in view of memorylessness. In addition, since the customer served will always be of priority 1 when at least one unit of that priority is present, we may drop the use of the third subscript of p_{mnr} as a service–customer indicator and instead use it to introduce a new priority, henceforth called 3 or lowest. So p_{mnr} is now the steady-state probability that there are m units of priority 1 in the system with arrival rate λ_1 and service rate μ_1, n units of priority 2 in the system with arrival rate λ_2 and service rate μ_2, and r units of priority 3 with arrival rate λ_3 and service rate μ_3.

Under these assumptions, a system of difference equations may be derived for the stationary probabilities ($\lambda = \lambda_1 + \lambda_2 + \lambda_3$ and $\rho = \lambda_1/\mu_1 + \lambda_2/\mu_2 + \lambda_3/\mu_3 < 1$), namely,

$$0 = -\lambda p_{000} + \mu_1 p_{100} + \mu_2 p_{010} + \mu_3 p_{001},$$

$$0 = -(\lambda + \mu_1)p_{m00} + \lambda_1 p_{m-1,0,0} + \mu_1 p_{m+1,0,0},$$

$$0 = -(\lambda + \mu_2)p_{0n0} + \mu_1 p_{1,n,0} + \lambda_2 p_{0,n-1,0} + \mu_2 p_{0,n+1,0},$$

$$0 = -(\lambda + \mu_3)p_{00r} + \mu_1 p_{10r} + \mu_2 p_{01r} + \lambda_3 p_{0,0,r-1} + \mu_3 p_{0,0,r+1},$$

$$0 = -(\lambda + \mu_1)p_{mn0} + \lambda_1 p_{m-1,n,0} + \lambda_2 p_{m,n-1,0} + \mu_1 p_{m+1,n,0},$$

$$0 = -(\lambda + \mu_1)p_{m0r} + \lambda_1 p_{m-1,0,r} + \lambda_3 p_{m,0,r-1} + \mu_1 p_{m+1,0,r}, \qquad (3.36)$$

$$0 = -(\lambda + \mu_2)p_{0nr} + \mu_1 p_{1nr} + \lambda_2 p_{0,n-1,r} + \mu_2 p_{0,n+1,r} + \lambda_3 p_{0,n,r-1},$$

$$0 = -(\lambda + \mu_1)p_{mnr} + \lambda_1 p_{m-1,n,r} + \lambda_2 p_{m,n-1,r} + \lambda_3 p_{m,n,r-1} + \mu_1 p_{m+1,n,r}.$$

(There will be a total of 2^r equations when the number of preemptive priorities is r.)

The solution to this system of stationary equations is quite complicated; suffice it to say that the steady-state generating function can be obtained (see Problem 3.29) and, from it, the mean number of priority i waiting in the system. In the event that the service rate is the same for all items, it follows for $\rho_i \equiv \lambda_i/\mu$ that

$$E[N_q^{(i)}] = \frac{\rho_i \sum_{n=1}^{i} \rho_n}{(1 - \sum_{n=1}^{i-1} \rho_n)(1 - \sum_{n=1}^{i} \rho_n)} \qquad (\sum_{i=1}^{3} \rho_i < 1),$$

with

$$E[N^{(i)}] = E[N_q^{(i)}] + \frac{\rho_i}{1 - \sum_{n=1}^{i-1} \rho_i}.$$

In addition, the variance can be found to be

$$Var[N_q^{(i)}] = 2\rho_i^3 \sum_{n=1}^{i-1} \frac{\rho_n}{(1 - \sum_{n=1}^{i-1} \rho_n)(1 - \sum_{n=1}^{i} \rho_n)}$$
$$+ \frac{\rho_i^2 + \rho_i(1 - \sum_{n=1}^{i-1} \rho_n)(1 - \sum_{n=1}^{i} \rho_n)}{(1 - \sum_{n=1}^{i-1} \rho_n)^2(1 - \sum_{n=1}^{i} \rho_n)}.$$

Both of these results generalize to $r > 3$ priorities.

There are many variations on these themes, and the reader interested in this problem area is referred to Jaiswal (1968). But to close this section, one final model will be presented, which is an interesting extension by Phipps (1956) of Cobham's models of Section 3.4.2. Phipps allowed the number of priorities to be continuous, so that the priority of a given unit is assigned according to some measure of the length of time needed to serve that unit (this might, indeed, be the case in a situation such as the loading of programs onto a mainframe). In particular, let the unit arrive as a Poisson process with mean rate λ_t ($\int_0^\infty \lambda_t \, dt = \lambda$) and be served according to an exponential distribution with mean time $1/\mu_t$. Then the total service cumulative distribution is

$$B(t) = \int_0^\infty \frac{\lambda_t}{\lambda}(1 - e^{-\mu_t t}) \, dt$$

$$= 1 - \frac{1}{\lambda} \int_0^\infty \lambda_t e^{-\mu_t t} \, dt.$$

The expected waiting time $W_q^{(t)}$ of a customer with type t priority is given by analogy with the discrete case (3.34), as

$$W_q(t) = \frac{E[S_0]}{\left(1 - \lambda \int_0^t y \, dB(y)\right)^2}.$$

It can be shown in general (this will be taken up in Chapter 5; see Problem 5.5) that

$$E[S_0] = \frac{\lambda}{2} E[S^2]$$

for any single-channel queue with Poisson input. Thus

$$W_q^{(t)} = \frac{\frac{1}{2} \int_0^\infty \lambda_t t^2 \mu_t e^{-\mu_t t} \, dt}{\left(1 - \int_0^t \lambda_y y \mu_y e^{-\mu_y y} \, dy\right)^2}.$$

The expected number of customers in the line is thus

$$L_q = \lambda \int_0^\infty W_q^{(t)} \, dt = \left[\frac{\lambda}{2} \int_0^\infty \lambda_t t^2 \mu_t e^{-\mu_t t} \, dt\right] \left[\int_0^t \left(1 - \int_0^t \lambda_y y \mu_y e^{-\mu_y y} \, dy\right)^{-2} dt\right].$$

If μ_t is assumed to be the constant μ and λ_t to be λ, then L_q simplifies to

$$W_q^{(t)} = \frac{\lambda/\mu^2}{\left(1 - \lambda \int_0^t y \mu e^{-\mu y} \, dy\right)^2} = \frac{\lambda/\mu^2}{\{1 - (\lambda/\mu)[1 - e^{-\mu t}(1 + \mu t)]\}^2}$$

and

$$L_q = \left(\frac{\lambda}{\mu}\right)^2 \int_0^\infty \left(1 - \frac{\lambda}{\mu}[1 - e^{-\mu t}(1 + \mu t)]\right)^{-2} dt.$$

PROBLEMS

Whenever a problem is best solved by the use of this book's accompanying software, we have added a boldface **C** *to the right of the problem number.*

3.1. Use stochastic balance to obtain Equations (3.1), (3.7), and (3.10).

3.2. Show that the operator equation in (3.8), $\mu r^{K+1} - (\lambda + \mu)r + \lambda = 0$, has exactly one root in the interval $(0, 1)$, using Rouché's theorem. [*Hint*: Refer to Section 2.10.2, and set $g = r^{K+1}$ and $f = -(\lambda/\mu + 1)r + \lambda/\mu$.]

3.3C. The moonlighting graduate assistant of Problem 2.11 decides that a more correct model for his short-order counter is an $M^{[X]}/M/1$ where the batch sizes are 1, 2, or 3 with equal probability, such that five batches arrive per hour on the average (this maintains the previous total arrival rate of 10 customers/hr). The mean service time remains at 4 min. Compare the average queue length and system size of this model with that of the previous $M/M/1$ model.

3.4. Consider an $M^{(2)}/M/1$ with service rate $\mu = 3$/hr, in which all arrivals come in batches of two with frequency 1 batch/(2 hr).

    (a)C Find the stationary system-size probability distribution, L, L_q, W, and W_q.

    (b) Using the difference equations of (3.1), derive a closed-form expression for p_n.

3.5. Derive the probability generating function for the bulk-service model as given in Equation (3.13).

3.6. Apply Rouché's theorem to the denominator of Equation (3.13) to show that K zeros lie on or within the unit circle. [*Hint*: Show that K zeros lie on or within $|z| = 1 + \delta$ by defining $f(z)$ and $g(z)$ such that $f(z) + g(z)$ equals the denominator.]

3.7C. A ferry loads cars for delivery across a river and must have a full ferry load of 10 cars. Loading is instantaneous, and the round-trip time is an exponential random variable with mean of 15 min. The cars arrive at the shore as a Poisson process with a mean of 30/hr. On the average, how many autos are waiting on the shore at any instant for a ferry?

3.8. Derive (3.15) by stochastic balance.

3.9C. To show the effect of service-time variation on performance measures for the $M/E_k/1$ model, calculate and plot W_q versus k ($=1, 2, 3, 4, 5, 10$) for $\lambda = 1$, $\rho(=1/\mu) = 0.5, 0.7, 0.9$.

3.10C. To show the effect of arrival-time variation on performance measures for the $E_k/M/1$ model, calculate and plot W_q versus k ($=1, 2, 3, 4, 5, 10$) for $\lambda = 1$, $\rho(=1/\mu) = 0.5, 0.7, 0.9$.

3.11. Derive the steady-state probability that a customer is in phase i for an $M/E_k/1/1$ model, that is, an Erlang service model where no queue is allowed to form.

3.12. Consider the $M/E_k/c/c$ model.
   (a) Derive the steady-state difference equation for this model. [*Hint*: Let $p_{n;s_1,s_2,\ldots,s_k}$ represent the probability of n in the system with s_1 channels in phase 1, s_2 in phase 2, etc.]
   (b) Show that $p_n = p_0 \rho^n/n!$, $\rho = \lambda/\mu$, is a solution to the problem. [*Hint*: First show that is a solution to the equation of (a). Then show that

$$p_n = \sum_{s_1+s_2+\cdots+s_k=n} p_{n;s_1,s_2,\ldots,s_k} = \frac{A\rho^n}{n!}$$

   by utilizing the multinomial expansion $(x_1 + x_2 + \cdots + x_k)^n$, then setting $x_1 = x_2 = \cdots = x_k = 1$.]
   (c) Compare this result with the $M/M/c/c$ results of Section 2.5, Equation (2.39), and comment.

3.13. Give a complete explanation why the $M^{[k]}/M/1/\infty$ bulk-arrival model can be used to represent the $M/E_k/1/\infty$ model when customers are considered to be phases of service to be completed.

3.14. Consider a single-server model with Poisson input and an average customer arrival rate of 30/hr. Currently service is provided by a mechanism that takes *exactly* 1.5 min. Suppose the system can be served instead by a mechanism that has an exponential service-time distribution. What must be the mean service time for this mechanism to ensure (a) the same average time a customer spends in the system, or (b) the same average number of customers in the system?

3.15C. For an $M/E_3/1$ model with $\lambda = 6$ and $\mu = 8$, find the probability of more than two in the system.

3.16C. A large producer of television sets has a policy of checking all sets prior to shipping them from the factory to a warehouse. Each line (large-screen, portable, etc.) has its own expert inspector. At times, the highest-volume line (color portables) has experienced a bottleneck condition (at least in the management's opinion), and a detailed study was made of the inspection performance. Sets were found to arrive at the inspector's station according to a Poisson distribution with a mean of 5/hr. In carrying out the inspection, the inspector performs 10 separate tests, each taking, on an average, 1 min. The times for each test were found to be approximately exponentially distributed. Find the average waiting time a set experiences, the average number of sets in the system, and the average idle time of the inspector.

3.17C. The Rhodehawgg Trucking Company has a choice of hiring one of two individuals to operate its single-channel truck-washing facility. In studying their performances it was found that one man's times for completely washing a truck were approximately exponentially distributed with a mean rate of 6/day, while the other man's times were distributed according to an Erlang type 2 with a mean rate of 5/day. Which man should be hired when the arrival rate is 4/day?

3.18C. Isle-Air Airlines offers air shuttle service between San Juan and Charlotte Amalie every 2 hr. The procedure calls for no advance reservations but for passengers to come directly to the gate from which the shuttle leaves to purchase their tickets. It is found that passengers arrive according to a Poisson distribution with a mean of 18/hr. There is one agent at the gate check-in counter, and a time study provided the following 50 observations on the processing time in minutes:

4.00, 1.44, 4.44, 1.74, 1.16, 4.20, 3.59, 2.14, 3.54, 2.56, 5.53, 2.02, 3.06, 1.66, 3.23, 4.84, 7.99, 3.07, 1.24, 3.40, 5.01, 2.78, 1.62, 5.19, 5.09, 3.78, 1.52, 3.94, 1.96, 6.20, 3.67, 3.37, 1.84, 1.60, 1.31, 5.64, 0.99, 3.06, 1.24, 3.11, 4.57, 0.90, 2.78, 1.64, 2.43, 5.26, 2.11, 4.27, 3.36, 4.76.

On the average, how many are in the queue waiting for tickets, and what is the average wait in the queue? [*Hint*: Find the sample mean and variance of the observed service times, and see what distribution might fit.]

3.19. Complete the details of the matrix–vector derivation for the Erlang type-2 distribution of Section 3.3.2. [*Hint*: For this problem, it is easy to expand $e^{\tilde{Q}t}$ as an infinite series of matrices.]

3.20. A generalization of the inventory-control procedure of Problem 2.38 is as follows. Again using S as the safety-stock value, the policy is to place an order for an amount Q when the on-hand plus on-order inventory reaches a level s ($Q = S - s$). Note that the one-for-one policy of Problem 2.38 is a special case for which $Q = 1$ or equivalently $s = S - 1$. This policy is called a trigger point-order quantity policy and is sometimes also referred to as a continuous review (s, S) policy. Generally, a manufacturing setup cost of K dollars per order placed is also included, so that an additional cost term of $K\lambda/Q$ (dollar per unit time) is included in $E[C]$. For this situation, again assuming Poisson demand and exponential lead times, describe the queueing model appropriate to the order-processing procedure. Then relate the steady-state probabilities resulting from the order-process- ing queueing model to $p(z)$, the probability distribution of on-hand inventory. Finally, discuss the optimization procedure for $E[C]$, now a function of two variables (either s and Q, S and Q, or S and s) and the practicality of using this type of analysis.

3.21. Derive the stationary equations (3.25) for the single exponential channel with two priorities.

3.22. Derive Equation (3.31).

3.23. Show that L_q of (3.31) is always greater than L_q of $M/M/1/\infty$, that is, $L_q = \rho^2/(1 - \rho)$ with $\rho = \lambda_1/\mu_1 + \lambda_2/\mu_2$. [*Hint*: Use differential calculus.]

3.24. Compare the L_q's of the two priority, two-rate case with those of the two-priority, one-rate case when the μ of the latter equals min (μ_1, μ_2).

3.25. Verify the results of the cell (model (2), model (3)) in Table 3.1.

3.26. Carry out the induction that leads from Equation (3.33) to (3.34).

3.27. Show that $\overline{W}_q$ of Section 3.4.2 becomes the W_q of $M/M/1/\infty/FCFS$ when $\mu_i \equiv \mu$ for all i.

3.28C. Consider the $M/M/1/\infty/PR$ model of Section 3.4.1. Let the number of priorities be two, with input rates $\lambda_1 = 1 = \lambda_2$. If the service rate of the higher-priority items is $\mu_1 = 3$ and that of the lower $\mu_2 = 2$, show that the expected line wait W_q is lower than that of the ordinary $M/M/1/\infty$ with $1/\mu = 1/(2\mu_1) + 1/(2\mu_2)$. What happens when the slower-service items get the higher priority?

3.29. Find the generating function for the steady-state probabilities given by Equation (3.36).

3.30C. As is fairly common in their business, City Hospital's operations- research analysts have always divided cases coming into their emer- gency room into three types: high priority (i.e., a life is at a stake), medium priority (i.e., patient might be in pain but his or her life is not threatened), and ordinary or low priority. On an average slow day, there are (approximately) 10 arrivals per hour, $\frac{1}{5}$ of whom are high priority, while $\frac{1}{2}$ are low priority, with all registration matters handled by one clerk. The staff would like to learn the average time it takes to complete the registration of an arbitrary case when the paperwork for high-priority cases can typically be processed in 3.0 min, medium-priority cases in 5.0 min, and low-priority cases in 7.5 min. (Assume that all processes are fully Markovian and that there is no preemption.)

3.31C. What happens in Problem 3.30 when we assume that there is preemption and it takes 5.5 min to process an average patient?

CHAPTER 4

Networks, Series, and Cyclic Queues

In this chapter we present an introduction to the very important subject of queueing networks. This is a current area of great research and application interest with many extremely difficult problems, far beyond the level of this text. As an introduction, we present some basic concepts and results that are quite useful in their own right, especially in queueing network design. Such problems have taken on special importance of late in view of their increased applicability to modeling manufacturing facilities and computer/ communication nets. The reader interested in delving into this topic further is referred to Disney (1981, 1996), Kelly (1979), Lemoine (1977), van Dijk (1993), and Walrand (1988), for example.

Networks of queues can be described as a group of nodes (say k of them) where each node represents a service facility of some kind with, let us say, c_i servers at node i, $i = 1, 2, \ldots, k$. In the most general case, customers may arrive from outside the system at any node and may depart from the system from any node. Thus customers may enter the system at some node, traverse from node to node in the system, and depart from some node, not all customers necessarily entering and leaving at the same nodes, or taking the same path once having entered the system. Customers may return to nodes previously visited, skip some nodes entirely, and even choose to remain in the system forever.

We will mainly be concerned with queueing networks with the following characteristics:

1. Arrivals from the "outside" to node i follow a Poisson process with mean rate γ_i.
2. Service (holding) times at each channel at node i are independent and exponentially distributed with parameter μ_i (a node's service rate may be allowed to depend on its queue length).
3. The probability that a customer who has completed service at node i

165

will go next to node j (routing probability) is r_{ij} (independent of the state of the system), where $i = 1, 2, \ldots, k$, $j = 0, 1, 2, \ldots, k$, and r_{i0} indicates the probability that a customer will leave the system from node i.

Networks that have these properties are called *Jackson networks* (see Jackson, 1957, 1963). As we shall see later, their steady-state probability distributions have a most interesting and useful product-form solution.

Cases for which $\gamma_i = 0$ for all i (no customer may enter the system from the outside) and $r_{i0} = 0$ for all i (no customer may leave the system) are referred to as *closed* Jackson networks (the general case described above we will refer to as *open* Jackson networks). We have already studied a special closed system in Chapter 2, Section 2.7, namely, the machine-repair problem (finite-source queue). For that system, $i = 1, 2$, $j = 0, 1, 2$, $r_{12} = r_{21} = 1$, and all other $r_{ij} = 0$. Node 1 represents the operating (plus spare) machines, while node 2 represents the repair facility. Note that customers flow in a "circle," always from node 1 to node 2 and then back to node 1, and so on. Such closed network systems are also referred to as *cyclic* queues.

In this chapter, we first treat open networks where

$$\gamma_i = \begin{cases} \lambda & (i = 1), \\ 0 & (\text{elsewhere}) \end{cases}$$

and

$$r_{ij} = \begin{cases} 1 & (j = i + 1, 1 \le i \le k - 1), \\ 1 & (i = k, j = 0), \\ 0 & (\text{elsewhere}). \end{cases}$$

These networks are called *series* or *tandem* queues, since the nodes can be viewed as forming a series system with flow always in a single direction from node to node. Customers may enter from the outside only at node 1 and depart only from node k. We will then generalize the series to a true open network, and finally come back to the case of closed queueing networks, including the special cyclic queue—which, incidentally, is a closed queueing network in series. We shall restrict ourselves mainly to Markovian systems as described above; that is, all holding times are exponential, all exogeneous inputs are Poisson, and the routing probabilities r_{ij} are known and state-independent (Jackson networks). We shall look briefly at some departures from the standard Jackson network assumptions, including the case where r_{ij} is allowed to be state-dependent.

4.1 SERIES QUEUES

In this section we consider models in which there are a series of service stations through which each calling unit must progress prior to leaving the system. Some examples of such series queueing situations are a manufacturing or assembly-line process in which units must proceed through a series of work stations, each station performing a given task; a registration process (such as university registration) where the registrant must visit a series of desks (advisor, department chairman, cashier, etc.); and a clinic physical examination procedure where the patient goes through a series of stages (lab tests, electrocardiogram, chest x-ray, etc.). In the following subsections several types of series queueing models are analyzed. (Analysis of *feed forward* queueing networks, i.e., networks of queues for which customers are not allowed to revisit previously visited nodes, is quite similar to that of the basic series queue first treated below.)

4.1.1 Queue Output

The first series model to be considered is a sequence of queues with no restriction on the waiting room's capacity between stations. Such a situation is pictured in Figure 4.1. We further assume that the calling units arrive according to a Poisson process, mean λ, and the service time of each server at station i ($i = 1, 2, \ldots, n$) is exponential with mean $1/\mu_i$. One can readily see that since there is no restriction on waiting between stations, each station can be analyzed separately as a single stage (nonseries) queueing model.

The first station is an $M/M/c_1/\infty$ model. It is necessary to find the output distribution (distribution of times between successive departures) in order to find the input distribution (times between successive arrivals) to the next station. It turns out, rather surprisingly, that the departure-time distribution from an $M/M/c/\infty$ queue is identical to the interarrival-time distribution, namely, exponential with mean $1/\lambda$; hence all stations are independent

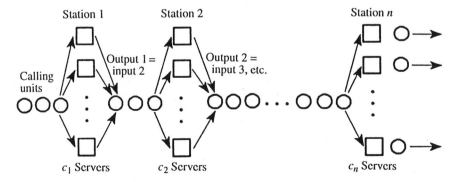

Fig. 4.1 Series queue, infinite waiting room.

$M/M/c_i/\infty$ models. Thus the results of Section 2.3 can be used on each station individually, and a complete analysis of this type of series situation is possible.

Before proceeding to the formal proof, we note that this exponential output result is a direct consequence of the *reversibility* of all birth–death Markov processes (as first mentioned in Section 2.1.1), wherein the departures from the process are the arrivals at the reversed process. To better visualize this, consider a sample path of a birth–death process over some time interval T (e.g., the one pictured in Fig. 1.4 of Section 1.5.1). If we start at T and look backward, the departures become the arrivals and the arrivals become the departures, and the probabilistic characteristics of the backward sample path appear identical to those of the forward sample path. Hence, the arrival and departure processes should be identical (note: this would not be the case for a sample path of an $M^{[2]}/M/1$ queue). Formally, in terms of the balance equations, a Markov chain is reversible if its steady-state probabilities and transition rates satisfy $\pi_i q_{ij} = \pi_j q_{ji}$.

We now proceed to verify the input–output identity with a constructive proof which utilizes a simple differential–difference argument (much like that used in the development of the birth–death process), which will show that, indeed, the interdeparture times are exponential with parameter λ. Consider an $M/M/c/\infty$ queue in steady state. Let $N(t)$ now represent the number of customers in the system at a time t after the last departure. Since we are considering steady state, we have

$$\Pr\{N(t) = n\} = p_n. \tag{4.1}$$

Furthermore, let T represent the random variable "time between successive departures" (interdeparture time), and

$$F_n(t) = \Pr\{N(t) = n \text{ and } T > t\}. \tag{4.2}$$

So $F_n(t)$ is the joint probability that there are n in the system at a time t after the last departure and that t is less than the interdeparture time; that is, another departure has not as yet occurred. The cumulative distribution of the random variable T, which will be denoted as $C(t)$, is given by

$$C(t) \equiv \Pr\{T \le t\} = 1 - \sum_{n=0}^{\infty} F_n(t), \tag{4.3}$$

since

$$\sum_{n=0}^{\infty} F_n(t) = \Pr\{T > t\}$$

is the marginal complementary cumulative distribution of T. To find $C(t)$, it is necessary to first find $F_n(t)$.

We can write the following difference equations concerning $F_n(t)$:

$$F_n(t + \Delta t) = (1 - \lambda \Delta t)(1 - c\mu \Delta t)F_n(t) + \lambda \Delta t (1 - c\mu \Delta t)F_{n-1}(t) + o(\Delta t)$$
$$(c \leq n),$$

$$F_n(t + \Delta t) = (1 - \lambda \Delta t)(1 - n\mu \Delta t)F_n(t) + \lambda \Delta t (1 - n\mu \Delta t)F_{n-1}(t) + o(\Delta t)$$
$$(1 \leq n \leq c),$$

$$F_0(t + \Delta t) = (1 - \lambda \Delta t)F_0(t) + o(\Delta t).$$

Moving $F_n(t)$ from the right side of each of the above equations, dividing by Δt, and taking the limit as $\Delta t \rightarrow 0$, we obtain the differential–difference equations as

$$\frac{dF_n(t)}{dt} = -(\lambda + c\mu)F_n(t) + \lambda F_{n-1}(t) \qquad (c \leq n),$$

$$\frac{dF_n(t)}{dt} = -(\lambda + n\mu)F_n(t) + \lambda F_{n-1}(t) \qquad (1 \leq n \leq c), \qquad (4.4)$$

$$\frac{dF_0(t)}{dt} = -\lambda F_0(t).$$

Using the boundary condition

$$F_n(0) \equiv \Pr\{N(0) = n \text{ and } T > 0\} = \Pr\{N(0) = n\} = p_n \qquad \text{[from (4.1)]}$$

and methodology similar to that of Section 1.7 where the Poisson process is derived, we find that the solution to (4.4) is (see Problem 4.1)

$$F_n(t) = p_n e^{-\lambda t}. \qquad (4.5)$$

The reader can easily verify that (4.5) is a solution to (4.4) by substitution, recalling that for $M/M/c/\infty$ models,

$$p_{n+1} = \begin{cases} \dfrac{\lambda}{(n+1)\mu} p_n & (1 \leq n \leq c), \\[2ex] \dfrac{\lambda}{c\mu} p_n & (c \leq n). \end{cases} \qquad \text{[see (2.24)].}$$

To obtain $C(t)$, the cumulative distribution of the interdeparture times, we use (4.5) in (4.3) to get

$$C(t) = 1 - \sum_{n=0}^{\infty} p_n e^{-\lambda t} = 1 - e^{-\lambda t} \sum_{n=0}^{\infty} p_n = 1 - e^{-\lambda t}, \qquad (4.6)$$

thus showing that the interdeparture times are exponential.

It is also true that the random variables $N(T)$ and T are independent and furthermore that successive interdeparture times are independent of each other (see Problem 4.2). This result was first proved by Burke (1956). So we see that the output distribution is identical to the input distribution and not at all affected by the exponential service mechanism. Intuitively one would expect the means to be identical, since we are in steady state, so that the input rate must equal the output rate; but it is not quite so intuitive that the variances and indeed the distributions are identical. Nevertheless, it is true and proves extremely useful in analyzing series queues, where the initial input rate is Poisson, the service at all stations is exponential, and there is no restriction on queue size between stations.

We now illustrate a series queueing situation of the type above with an example.

Example 4.1
Cary Meback, the president of a large Virginia supermarket chain, is experimenting with a new store design and has remodeled one of his stores as follows. Instead of the usual checkout-counter design, the store has been remodeled to include a checkout "lounge." As customers complete their shopping, they enter the lounge with their carts and, if all checkers are busy, receive a number. They then park their carts and take a seat. When a checker is free, the next number is called and the customer with that number enters the available checkout counter. The store has been enlarged so that for practical purposes, there is no limit on either the number of shoppers that can be in the food aisles or the number that can wait in the lounge, even during peak periods.

The management estimates that during peak hours customers arrive according to a Poisson process at a mean rate of 40/hr and it takes a customer, on the average, $\frac{3}{4}$ hr to fill his shopping cart, the filling times being approximately exponentially distributed. Furthermore, the checkout times are also approximately exponentially distributed with a mean of 4 min, regardless of the particular checkout counter (during peak periods each counter has a cashier and bagger, hence the low mean checkout time). Meback wishes to know the following:

1. What is the minimum number of checkout counters required in operation during peak periods?
2. If it is decided to add one more than the minimum number of counters

required in operation, what is the average waiting time in the lounge? How many people, on the average, will be in the lounge? How many people, on the average, will be in the entire supermarket?

This situation can be modeled by a two-station series queue. The first is the food portion of the supermarket. Since it is self-service and arrivals are Poissonian, we have an $M/M/\infty$ model with $\lambda = 40$ and $\mu = \frac{4}{3}$. The second station is an $M/M/c$ model, since the output of an $M/M/\infty$ queue is identical to its input. Hence the input to the checkout lounge is Poisson with a mean of 40/hr also. Since $c\mu > \lambda$ for steady-state convergence, the minimum number of checkout counters, c_m, must be greater than $\lambda/\mu = 40/15 \doteq 2.67$; hence c_m must be 3.

If it is decided to have four counters in operation, we have an $M/M/4/\infty$ model at the checkout stations with $\lambda = 40$ and $\mu = 15$. Meback desires to know W_q and L_q for the $M/M/4/\infty$ model, as well as the average total number in the supermarket, which is the sum of the L's for both models. Using Equations (2.25) and (2.27)—or the software package—for an $M/M/c/\infty$ model, we get

$$p_0 = \left[\sum_{n=0}^{3} \frac{1}{n!} \left(\frac{8}{3} \right)^n + \frac{1}{4!} \left(\frac{8}{3} \right)^4 \left(\frac{4}{4 - \frac{8}{3}} \right) \right]^{-1} \doteq 0.06$$

and

$$W_q = \frac{\left(\frac{8}{3} \right)^4 15}{3!(60 - 40)^2} (0.06) \doteq 0.019 \text{ hr} \doteq 1.14 \text{ min.}$$

To get L_q we use Little's formula and find

$$L_q = \lambda W_q \doteq 40(0.019) = 0.76,$$

so that, on the average, less than one person will be waiting in the lounge for a checker to become free.

The total number of people in the system, on the average, is the L for this $M/M/4/\infty$ model plus the L for the $M/M/\infty$ model. For the checkout station we get

$$L = \lambda W = \lambda \left(W_q + \frac{1}{\mu} \right) \doteq 40 \left(0.019 + \frac{4}{60} \right) \doteq 3.44.$$

For the supermarket proper we have from the $M/M/\infty$ model results that $L = \lambda/\mu = 40/(\frac{4}{3}) = 30$. Hence the average number of customers in the store during peak hours is 33.44 if Meback decides on four checkout counters in operation. He might do well to perform similar calculations for three checkout counters operating to see how much the congestion increases (see Problem 4.3).

For series queues, therefore, as long as there are no capacity limitations between stations and the input is Poisson, results can be rather easily obtained. Furthermore, it can be shown (see Problem 4.4) that the joint probability that there are n_1 at station 1, n_2 at station 2, . . . , and n_j at station j is merely the product $p_{n_1} p_{n_2} \cdots p_{n_j}$. This product-form type of result is quite typical of those available for Jackson networks, as we shall see in subsequent sections.

The analysis for series queues when there are limits on the capacity at a station (except for the case where the only limit is at the last station in a pure series flow situation and arriving customers who exceed the capacity are shunted out of the system—see Problem 4.5) is much more complex. This results from the blocking effect; that is, a station downstream comes up to capacity and thereby prevents any further processing at upstream stations which feed it. We treat some of these types of models in the next section.

4.1.2 Series Queues with Blocking

We consider first a simple sequential two-station, single-server-at-each-station model where no queue is allowed to form at either station. If a customer is in station 2 and service is completed at station 1, the station-1 customers must wait there until the station-2 customer is completed; that is, the system is *blocked*. Arrivals at station 1 when the system is blocked are turned away. Also, if a customer is in process at station 1, then even if station 2 is empty, arriving customers are turned away, since the system is a sequential one; that is, all customers require service at 1 and then service at 2.

We wish to find the steady-state probability p_{n_1,n_2} of n_1 in the first station and n_2 in the second station. For this model, the possible states are given in Table 4.1.

Assuming arrivals at the system (station 1) are Poisson with parameter λ

Table 4.1 Possible System States

n_1, n_2	Description
0,0	System empty
1,0	Customer in process at 1 only
0,1	Customer in process at 2 only
1,1	Customers in process at 1 and 2
b,1	Customer in process at 2 and a customer finished at 1 but waiting for 2 to become available, i.e., system is blocked

and service is exponential with parameters μ_1, and μ_2, respectively, the usual procedure leads to the steady-state equations for this multidimensional Markov chain:

$$0 = -\lambda p_{0,0} + \mu_2 p_{0,1},$$

$$0 = -\mu_1 p_{1,0} + \mu_2 p_{1,1} + \lambda p_{0,0},$$

$$0 = -(\lambda + \mu_2)p_{0,1} + \mu_1 p_{1,0} + \mu_2 p_{b,1}, \qquad (4.7)$$

$$0 = -(\mu_1 + \mu_2)p_{1,1} + \lambda p_{0,1},$$

$$0 = -\mu_2 p_{b,1} + \mu_1 p_{1,1}.$$

Using the boundary equation $\sum\sum p_{n_1,n_2} = 1$, we have six equations in five unknowns [there is some redundancy in (4.7); hence we can solve for the five steady-state probabilities]. Equation (4.7) can be used to get all probabilities in terms of $p_{0,0}$, and the boundary condition can be used to find $p_{0,0}$. If we let $\mu_1 = \mu_2$, the results are (see Problem 4.6)

$$p_{1,0} = \frac{\lambda(\lambda + 2\mu)}{2\mu^2}p_{0,0}, \qquad p_{0,1} = \frac{\lambda}{\mu}p_{0,0}, \qquad p_{1,1} = \frac{\lambda^2}{2\mu^2}p_{0,0},$$

$$p_{b,1} = \frac{\lambda^2}{2\mu^2}p_{0,0}, \qquad p_{0,0} = \frac{2\mu^2}{3\lambda^2 + 4\mu\lambda + 2\mu^2}. \qquad (4.8)$$

It is easy to see how the problem expands if one allows limits other than zero on queue length or considers more stations. For example, if one customer is allowed to wait between stations, this results in seven state probabilities for which to solve, utilizing seven equations and a boundary condition (see Problem 4.7). The complexity results from having to write a balance equation for each possible system state. Conceptually, however, these types of series queueing situations can be attacked via the methodology presented above. For large numbers of equations, as long as we have a finite set, numerical techniques for solving these simultaneous equations can also be employed.

Hunt (1956) treated a modified series model using finite-difference operators to solve a two-station sequential series queue in which no waiting is allowed between stations, but where a queue with no limit is permitted in front of the first station. He obtained the steady-state probabilities for this model, the expected system size L, and the maximum allowable ρ (call it ρ_{max}) for steady state to be assured. He also calculated ρ_{max} for some generalizations of this two-station model (infinite allowable queue in front of station 1) to three- and four-station systems with no waiting between stations, a two-station system with a capacity K allowed between stations, and a three-station system where $K = 2$ in between each of the stations. The interested reader is referred to Hunt (1956). A good general reference on networks of queues with blocking is Perros (1994).

4.2 OPEN JACKSON NETWORKS

We now treat a rather general network system which we previously described in the introduction to this chapter as the Jackson network, because of the landmark work done by Jackson (1957, 1963). To recapitulate, we consider a network of k service facilities (usually referred to as nodes). Customers can arrive from outside at any node according to a Poisson process. We will now represent the mean arrival rate to node i as γ_i (instead of the familiar λ_i) for reasons that will become clear shortly. All servers at node i work according to an exponential distribution with mean μ_i (so that all servers at a given node are identical). When customers complete service at node i, they go next to node j with probability r_{ij} (independent of the system state), $i = 1, 2, \ldots, k$. There is a probability r_{i0} that customers will leave the network at node i upon completion of service. There is no limit on queue capacity at any node; that is, we never have a blocked system or node.

Since we have a Markovian system, we can use our usual types of analyses to write the steady-state system equations. We first, however, must determine how to describe a system state. Since various numbers of customers can be at various nodes in the network, we desire the joint probability distribution for the number of customers at each node; that is, letting N_i be the random variable for the number of customers at node i in the steady state, we desire $\Pr\{N_1 = n_1, N_2 = n_2, \ldots, N_k = n_k\} \equiv p_{n_1, n_2, \ldots, n_k}$. From this joint probability distribution, we can obtain the marginal distribution for numbers of customers at a particular node by appropriately summing over the other nodes.

We shall use the method of stochastic balance to obtain the steady-state equations for this network. Rather than using the somewhat cumbersome k-component vector $(n_1, n_2, \ldots, n_k)$ for describing a state, we employ the simplified notation given in Table 4.2.

Using stochastic balance for equating flow into state $\bar{n}$ to flow out of state $\bar{n}$, and assuming that $c_i = 1$ for all i (single-server nodes) and that $n_i \geq 1$ at each node (actually, the equation that results will also hold for $n_i = 0$ if we

Table 4.2 Simplified State Descriptors

State	Simplified Notation
$n_1, n_2, \ldots, n_i, n_j, \ldots, n_k$	$\bar{n}$
$n_1, n_2, \ldots, n_i + 1, n_j, \ldots, n_k$	$\bar{n}; i^+$
$n_1, n_2, \ldots, n_i - 1, n_j, \ldots, n_k$	$\bar{n}; i^-$
$n_1, n_2, \ldots, n_i + 1, n_j - 1, \ldots, n_k$	$\bar{n}; i^+ j^-$

set terms with negative subscripts and terms containing μ_i for which the subscript $n_i = 0$ to zero), we obtain

$$\sum_{i=1}^{k} \gamma_i p_{\bar{n};i^-} + \sum_{\substack{j=1 \\ (i \neq j)}}^{k} \sum_{i=1}^{k} \mu_i r_{ij} p_{\bar{n};i^+j^-} + \sum_{i=1}^{k} \mu_i r_{i,0} p_{\bar{n};i^+} = \sum_{i=1}^{k} \mu_i (1 - r_{ii}) p_{\bar{n}} + \sum_{i=1}^{k} \gamma_i p_{\bar{n}}.$$

$$(4.9)$$

Jackson (1957, 1963) first showed that the solution to these steady-state balance equations is what has come to be generally called *product form*, where the joint probability distribution of system states can be written as

$$p_{\bar{n}} = C \rho_1^{n_1} \rho_2^{n_2} \cdots \rho_k^{n_k}.$$

We mention here that some authors consider a more restrictive definition of product form, namely that the joint distribution is made up of the product of the marginal distributions at each of the nodes. We prefer the less restrictive definition wherein the constant C itself need not separate into a product, so that a product-form solution need not be a product of true marginals.

We will present Jackson's solution and then show that it satisfies (4.9). Let λ_i be the total mean flow rate into node i (from outside and from other nodes). Then, in order to satisfy flow balance at each node, we have the traffic equations

$$\lambda_i = \gamma_i + \sum_{j=1}^{k} \lambda_j r_{ji} \qquad (4.10a)$$

or, in vector–matrix form,

$$\boldsymbol{\lambda} = \boldsymbol{\gamma} + \boldsymbol{\lambda R}. \qquad (4.10b)$$

We define ρ_i to be λ_i/μ_i for $i = 1, 2, \ldots, k$. Then, as Jackson showed, the steady-state solution to (4.9) is

$$p_{\bar{n}} \equiv p_{n_1,n_2,\ldots,n_k} = (1 - \rho_1) \rho_1^{n_1} (1 - \rho_2) \rho_2^{n_2} \ldots (1 - \rho_k) \rho_k^{n_k}. \qquad (4.11)$$

which, in this case, is a true product of marginal distributions. What this result says is that the network *acts as if* each node could be viewed as an independent $M/M/1$ queue, with parameters λ_i and μ_i, so that the joint probability distribution can be written as a product of marginal $M/M/1$'s. The reader should not be misled into believing that the network actually decomposes into individual $M/M/1$'s with the flow into each a true Poisson process with mean rate λ_i. In fact, it can be shown (see Disney, 1981) that, in general, the actual internal flow in these kinds of networks is *not* Poisson. Indeed, as long as there is any kind of *feedback* (that is, customers can return to previously visited nodes), the internal flows are not Poisson. The surprising thing is that regardless of whether internal flows are really Poisson or not, (4.11) still holds and the network behaves as if its nodes were independent $M/M/1$'s.

The global balance equation (4.9) gives rise to a simple set of local balance equations in the same spirit as the classical birth–death process, namely, $\lambda_i p_{\bar{n};i^-} = \mu_i p_{\bar{n}}$. In other words, the expected rate at which the system goes from state $\bar{n};i^-$ to $\bar{n}$ must equal the rate at which it goes in the reverse direction, $\bar{n}$ back to $\bar{n};i^-$. It thus follows by linear difference methods that $p_{\bar{n}} = \rho_i^{n_i} p_{n_1, n_2, \dots, 0, \dots, n_k}$. Therefore we can conclude that all the nodal marginal distributions are going to be *geometric* and that (4.11) is the likely form for the combined joint distribution.

To verify that (4.11) does indeed satisfy (4.9), we first show that $p_{\bar{n}} = C\rho_1^{n_1}\rho_2^{n_2}\cdots\rho_k^{n_k}$ satisfies (4.9) and then that C turns out to be $\Pi_{i=1}^{k}(1 - \rho_i)$, in order to satisfy the summability-to-one criterion. We let $\mathfrak{R}^{\bar{n}} = \rho_1^{n_1}\rho_2^{n_2}\cdots\rho_k^{n_k}$, and plugging $p_{\bar{n}} = C\mathfrak{R}^{\bar{n}}$ into (4.9) gives

$$
C\mathfrak{R}^{\bar{n}} \sum_{i=1}^{k} \frac{\gamma_i}{\rho_i} + C\mathfrak{R}^{\bar{n}} \sum_{\substack{j=1 \\ (i\neq j)}}^{k} \sum_{i=1}^{k} \mu_i r_{ij} \frac{\rho_i}{\rho_j} + C\mathfrak{R}^{\bar{n}} \sum_{i=1}^{k} \mu_i r_{i0} \rho_i
$$

$$
\overset{?}{=} C\mathfrak{R}^{\bar{n}} \sum_{i=1}^{k} \mu_i (1 - r_{ii}) + C\mathfrak{R}^{\bar{n}} \sum_{i=1}^{k} \gamma_i .
$$

Canceling out $C\mathfrak{R}^{\bar{n}}$, we have

$$
\sum_{i=1}^{k} \frac{\gamma_i \mu_i}{\lambda_i} + \sum_{\substack{j=1 \\ (i\neq j)}}^{k} \sum_{i=1}^{k} \mu_i r_{ij} \frac{\lambda_i \mu_j}{\lambda_j \mu_i} + \sum_{i=1}^{k} \mu_i r_{i0} \frac{\lambda_i}{\mu_i} \overset{?}{=} \sum_{i=1}^{k} (\mu_i - \mu_i r_{ii} + \gamma_i).
$$

From (4.10a) then,

$$
\lambda_j = \gamma_j + \sum_{\substack{i=1 \\ (i\neq j)}}^{k} r_{ij}\lambda_i + r_{jj}\lambda_j \Rightarrow \sum_{\substack{i=1 \\ (i\neq j)}}^{k} r_{ij}\lambda_i = \lambda_j - \gamma_j - r_{jj}\lambda_j,
$$

so that

$$
\sum_{i=1}^{k} \frac{\gamma_i \mu_i}{\lambda_i} + \sum_{j=1}^{k} \frac{\mu_j}{\lambda_j}(\lambda_j - \gamma_j - r_{jj}\lambda_j) + \sum_{i=1}^{k} \mu_i r_{i0} \frac{\lambda_i}{\mu_i} \overset{?}{=} \sum_{i=1}^{k} (\mu_i - \mu_i r_{ii} + \gamma_i).
$$

Changing the subscript on the second term on the left-hand side from j to i, we get

$$\sum_{i=1}^{k} \left(\frac{\gamma_i \mu_i}{\lambda_i} + \frac{\mu_i}{\lambda_i} (\lambda_i - \gamma_i - r_{ii}\lambda_i) + \lambda_i r_{i0} \right) \stackrel{?}{=} \sum_{i=1}^{k} (\mu_i - \mu_i r_{ii} + \gamma_i).$$

After moving through with the summation and canceling we have

$$\sum_{i=1}^{k} \lambda_i r_{i0} \stackrel{?}{=} \sum_{i=1}^{k} \gamma_i .$$

This is indeed true, since the left-hand side represents the total flow out of the network and the right-hand side represents the total flow in. For steady state, these must be equal.

To obtain ρ_i, we need to obtain λ_i from the traffic equations (4.10b), which are solved as $\boldsymbol{\lambda} = \boldsymbol{\gamma}(\boldsymbol{I} - \boldsymbol{R})^{-1}$. The invertibility of $\boldsymbol{I} - \boldsymbol{R}$ is assured as long as there is at least one node releasing its output to the outside and no node is totally absorbing.

Now to evaluate C, we have

$$\sum_{n_k=0}^{\infty} \cdots \sum_{n_2=0}^{\infty} \sum_{n_1=0}^{\infty} C \rho_1^{n_1} \rho_2^{n_2} \ldots \rho_k^{n_k} = 1.$$

Thus

$$C = \left(\sum_{n_k=0}^{\infty} \cdots \sum_{n_2=0}^{\infty} \sum_{n_1=0}^{\infty} \rho_1^{n_1} \rho_2^{n_2} \ldots \rho_k^{n_k} \right)^{-1} = \left(\sum_{n_k=0}^{\infty} \rho_k^{n_k} \cdots \sum_{n_2=0}^{\infty} \rho_2^{n_2} \sum_{n_1=0}^{\infty} \rho_1^{n_1} \right)^{-1}$$

$$= \left(\frac{1}{1 - \rho_k} \cdots \frac{1}{1 - \rho_2} \frac{1}{1 - \rho_1} \right)^{-1} \qquad (\rho_i < 1, \quad i = 1, 2, \ldots, k).$$

Hence

$$C = \left(\frac{1}{(1 - \rho_k) \cdots (1 - \rho_2)(1 - \rho_1)} \right)^{-1}$$

$$= \prod_{i=1}^{k} (1 - \rho_i) \qquad (\rho_i < 1, \quad i = 1, 2, \ldots, k).$$

We can obtain expected measures rather easily for individual nodes, since $L_i = \rho_i/(1 - \rho_i)$ and $W_i = L_i/\lambda_i$. This is so because of the product form of the solution for the joint probability distribution and again does not imply that the nodes are truly $M/M/1$ (which they may not be, though the system size processes are indeed independent $M/M/1$'s, since the joint probability distribution is the product of the marginals). The expected total wait within the network of any customer before its final departure would be $W = \Sigma_i L_i / \Sigma_i \gamma_i$ (Little's formula for the entire network).

The above results for Jackson networks generalize easily to c-channel nodes (see Problem 4.9). Let c_i represent the number of servers at node i, each having exponential service time with parameter μ_i. Then (4.11) becomes (again we define λ_i / μ_i as r_i to be consistent with $M/M/c$ notation, but do not confuse with the double-subscripted r_{ij}, which are routing probabilities)

$$p_{\bar{n}} \equiv p_{n_1, n_2, \ldots, n_k} = \prod_{i=1}^{k} \frac{r_i^{n_i}}{a_i(n_i)} p_{0i} \qquad (r_i \equiv \lambda_i / \mu_i), \qquad (4.12)$$

where

$$a_i(n_i) = \begin{cases} n_i! & (n_i < c_i), \\ c_i^{n_i - c_i} c_i! & (n_i \ge c_i), \end{cases} \qquad (4.13)$$

and p_{0i} is such that $\Sigma_{n_i=0}^{\infty} p_{0i} r_i^{n_i} / a_i(n_i) = 1$, which can be obtained from (2.25). Thus, again, what we have is a network that *acts as if* each node were an independent $M/M/c$.

With respect to waiting-time distributions, very little can be said. It is tempting, for example, to conclude that if we are interested in the unconditional waiting time at a particular node, since the node acts like an $M/M/c$, the waiting-time distribution should be the same as that for the $M/M/c$ model. But this is not necessarily true. A basic factor in developing $M/M/c$ waiting times was that the arrival-point steady-state probabilities $\{q_n\}$ were identical to the general-time steady-state probabilities $\{p_n\}$ because of the Poisson input. However, in a general Jackson network, as we have previously mentioned, flows are not necessarily truly Poisson, so that we cannot be sure that $q_n = p_n$, and, in fact, in general that is not true. What does appear to be the case is that the virtual waiting time (or work backlog, as it is sometimes called—see section 2.2.4) at a node that requires use of the $\{p_n\}$ is the same as that of $M/M/c$, since the $\{p_n\}$ are identical. Hence equations such as (2.31) give virtual waiting times, but unless the network has a feedforward flow (i.e., are arborescent or treelike, indicating no direct or indirect feedback) which *is* truly Poisson, nothing can be concluded about actual waiting-time distributions, though the *mean* nodal values satisfy Little's formula.

Of even more interest would be customer waiting times to traverse portions of or the entire network (often referred to as *sojourn* times). Even less can be said here, due to the complicated correlation among node waiting times. Consider, for example (Disney, 1996), a simple feedback queue where we have a single $M/M/1$ node where, with probability p, a customer after being served returns to the end of the queue for additional service and with probability $1 - p$ leaves the system. Considering the sojourn time of a specific customer to be made up of (possibly) several passes through the queue, then it is clear that at the first pass, the time spent in the system is what would normally be spent in traversing an $M/M/1$ queue, but when the customer starts the second pass (assuming a feedback customer), ahead will be a number of other customers, some of whom were originally ahead of the customer during its first pass, while others may have arrived after the customer while it was making its first pass. The number of this latter kind of customers in the system depends on how long the subject customer spent going through the queue on the first pass. Its time to get through the queue the second time, then, depends on how many customers it finds ahead of it when it rejoins the queue, which depends on how long it spent in the system on the first pass; hence the sojourn times on these two passes are dependent.

Even with feedforward networks, sojourn times can be complex. While it is true in single-server queues that the total time, $T_i^{(n)}$, that the nth customer spends in queue plus service on a single pass through node i is a sum of IID exponential random variables, and that the total system sojourn time $T^{(n)} = T_1^{(n)} + \cdots + T_k^{(n)}$ is, in the limit, the sum of IID random variables (this result is due to Reich, 1957), it is not true in multiserver queues. For example, Burke (1969) showed that in a three-station series queue with the first and third stations having a single server but the middle station having multiple servers ($c_2 \geq 2$), in the limit, $T_1^{(n)}$ and $T_2^{(n)}$ are independent, and $T_2^{(n)}$ and $T_3^{(n)}$ are independent, but $T_1^{(n)}$ and $T_3^{(n)}$ are not independent. Simon and Foley (1979) considered a three-station queueing network with single servers at each station. Customers only enter the system at station 1 and exit the system at station 3, so that

$$\lambda_i = \begin{cases} \lambda & (i = 1), \\ 0 & \text{(elsewhere)}. \end{cases}$$

However, this is not a series situation, since they allow for the possibility of bypassing, namely,

$$r_{i,j} = \begin{cases} p & (i = 1, j = 2), \\ 1 - p & (i = 1, j = 3), \\ 1 & (i = 2, j = 3; i = 3, j = 0), \\ 0 & \text{(elsewhere)}. \end{cases}$$

They also showed that, in the limit, $T_1^{(n)}$ and $T_2^{(n)}$ are independent, and $T_2^{(n)}$ and $T_3^{(n)}$ are independent, but $T_1^{(n)}$ and $T_3^{(n)}$ are not independent. The implication of these results seems to be that even in the relatively well behaved strict tandem (series) queue and feedforward types of networks, if there are multiple servers at stations other than the first or last so that customers can bypass one another, system sojourn times for successive customers are dependent, so that Reich's result holds only for single-server series systems. Thus, the major culprits in complexity of sojourn times appear to be feedback and bypassing.

One is often interested, for networks, in output processes from individual nodes, especially since they influence input processes to other nodes. We saw in Section 4.1.1 that for series situations, the output from each node is a Poisson process identical to the arrival process to the first node, so that the output from the last node (departure process out of the network) is identical to the arrival process (into the network). What now can be said concerning the more general Jackson networks? As the reader might suspect, not much is possible.

The survey by Disney (1981, 1996) summarizes key results. For single-server Jackson networks with an irreducible routing probability matrix $R = \{r_{ij}\}$, and $\rho_i < 1$, $i = 1, 2, \ldots, k$ (every entering customer eventually leaves the network), Melamed (1979) showed that the departure process for nodes from which units could leave the network are Poisson and that the collection over all nodes that yield these Poisson departure processes are mutually independent. It obviously follows that the sum total of all departures from the network must be Poisson as well.

Further, considering nodes with no feedback (there is no path a departing customer can follow that will eventually return to the same node, prior to exiting the network), the output process from this node is also Poisson. In nodes with feedback, one can think of two departing streams, one with customers who will either directly or eventually feed back, and the other with customers who will not. The nonfeedback stream is Poisson, but the feedback stream is not and, in fact, is not even a sequence of IID random variables. Disney et al. (1980) considered the single-node Poisson arrival process with direct feedback that was mentioned above and showed that the total output process, feedback plus nonfeedback streams, is also not Poisson. It is not even a sequence of IID random variables; however, as Melamed has shown, the nonfeedback process is Poisson. Disney et al. conjecture that the feedback and nonfeedback processes are dependent, but no proof either way is known. Thus feedback causes problems with Jackson network flows, just as feedback and customer bypassing cause problems with sojourn times.

In summary, as long as there is no feedback, as in series or feedforward networks, flows between nodes and to the outside are truly Poisson. Feedback destroys Poisson flows, but Jackson's solution of Equations (4.11) and (4.12) still holds.

While the results for departure processes and waiting and sojourn times are extremely complex and very little is really known other than the often counterintuitive results mentioned above, the system-size results presented earlier are quite neat, and Jackson networks have been quite useful for modeling a variety of network situations in communications, computers, and repairable-item inventory control. Of course, the Poisson–exponential assumptions must hold, as well as state-independent routing probabilities and the absence of restrictions on waiting capacities.

Example 4.2
The mid-Atlantic region of Hourfawlt Insurance Corporation has a three-node telephone system. Calls coming into the 800 number are Poisson, with a mean of 35/hr. The caller gets one of two options: press 1 for claims service and press 2 for policy service. It is estimated that the caller's listening, decision, and button-pushing time is exponential with mean of 30 seconds. Only one call at a time can be processed, so that while a call is in process, any incoming calls that arrive are put in a queue with nice background music and a nice recorded message saying how important the call is and please wait (which we assume everyone does). Approximately 55% of the calls go to claims, and the remainder to policy service. The claims processing node has three parallel servers, and it is estimated that service times are exponential with mean 6 min (mostly, basic information is taken so that appropriate forms can be mailed out). The policy service node has seven parallel servers, again with exponential service times, and here the mean service time is 20 min. All buffers in front of the nodes can be assumed to hold as many calls as come into the queues. About 2% of the customers finishing at claims then go on to policy service, and about 1% of the customers finishing at policy service go to claims. It is desired to know the average queue sizes in front of each node and the total average time a customer spends in the system.

The routing matrix for the problem (calling claims node 2 and policy service node 3) is

$$R = \begin{pmatrix} 0 & 0.55 & 0.45 \\ 0 & 0 & 0.02 \\ 0 & 0.01 & 0 \end{pmatrix},$$

and $\gamma_1 = 35$/hr, $\gamma_2 = \gamma_3 = 0$, $c_1 = 1$, $\mu_1 = 120$/hr, $c_2 = 3$, $\mu_2 = 10$/hr, $c_3 = 7$, $\mu_3 = 3$/hr.

First we must solve the traffic equations (4.10):

$$(I - R)^{-1} = \begin{pmatrix} 1 & 0.5546 & 0.4611 \\ 0 & 1.0002 & 0.02 \\ 0 & 0.01 & 1.0002 \end{pmatrix},$$

so that $\lambda = \gamma(I - R)^{-1} = (35, 19.411, 16.138)$, which yields $r_1 = 35/120 = 0.292$, $r_2 = 19.411/10 = 1.941$, $r_3 = 16.132/3 = 5.379$. Now we can use the $M/M/c$ results of Section 2.3 to obtain L_q and L for each of the nodes. These turn out to be $L_{q_1} = 0.120$, $L_{q_2} = 0.765$, $L_{q_3} = 1.402$ and $L_1 = 0.412$, $L_2 = 2.706$, $L_3 = 6.781$. The total system L is then $0.412 + 2.706 + 6.781 = 9.899$, and hence $W = 9.899/35 = 0.283$ hr or approximately 17 min.

4.2.1 Open Jackson Networks with Multiple Customer Classes

It is a rather straightforward generalization to allow customers of different types, as reflected by a different routing matrix, that is, a customer of one type has different routing probabilities than a customer of another type. The essential modification is to first solve the traffic equations separately for each customer type and then add the resulting λ's. We will use a superscript to denote customer type, so that $R^{(t)}$ is the routing probability matrix for a customer of type t ($t = 1, 2, \ldots, n$). Solving (4.10) yields a $\lambda^{(t)}$ for each customer type t. We then obtain $\lambda = \Sigma_t \, \lambda^{(t)}$. We proceed as before to obtain the L_i for each of the nodes ($i = 1, 2, \ldots, k$) using $M/M/c$ results. The average waiting time at each node (note that all customer types have the same average waiting time, since they have identical service-time distributions and wait in the same first-come, first-served queue) can be obtained by Little's formula as before. The same is true for the average system sojourn time. We can also obtain the average system size for customer type t, by simply weighting the node average total size by customer type t's relative flow rate, namely

$$L_i^{(t)} = \frac{\lambda_i^{(t)}}{\lambda_i^{(1)} + \lambda_i^{(2)} \cdots \lambda_i^{(n)}} L_i \, .$$

Example 4.3
Let us revisit Example 4.2. What was implicitly assumed in that example was that a caller going first to claims and then to policy service could go back to claims with probability 0.01. Further, a customer going first to policy service and then to claims could go back again to policy service with probability 0.02. This probably isn't very realistic; that is, for this situation, customers really would not return to a previous node. We can get around this by calling customers who first go to claims type-1 customers and customers who first go to policy service type-2 customers. Then the two routing matrices are

$$R^{(1)} = \begin{pmatrix} 0 & 1 & 0 \\ 0 & 0 & .02 \\ 0 & 0 & 0 \end{pmatrix}, \qquad R^{(2)} = \begin{pmatrix} 0 & 0 & 1 \\ 0 & 0 & 0 \\ 0 & .01 & 0 \end{pmatrix}.$$

Since 55% of the arrivals are type-1 customers, $\gamma_1^{(1)} = 19.25$, and 45% are type 2, so that $\gamma_1^{(2)} = 15.75$. Now solving the traffic equations separately gives $\lambda_1^{(1)} = 19.25$, $\lambda_2^{(1)} = 19.25$, $\lambda_3^{(1)} = 0.385$ and $\lambda_1^{(2)} = 15.75$, $\lambda_2^{(2)} = 0.1575$,

$\lambda_3^{(2)} = 15.75$. Adding to get total flows gives $\lambda = (35, 19.408, 16.135)$. Comparing this with Example 4.2, we see small differences in slightly lower flows in nodes 2 and 3 to account for the lack of recycling. The procedure now follows along as before, and using $M/M/c$ results we get $L_1 = 0.412$, $L_2 = 2.705$, and $L_3 = 6.777$, again with very slight differences from Example 4.2. The total system L is now 9.894, and the average system sojourn time is $9.894/35 = 0.283$ hr or, again, about 17 min. We can also find the average number of each type of customer at each node:

$$L_1^{(1)} = \frac{19.25}{19.25 + 15.75} L_1 = 0.227, \qquad L_2^{(1)} = \frac{19.25}{19.25 + 0.1575} L_2 = 2.683,$$

$$L_3^{(1)} = \frac{0.385}{0.385 + 15.75} L_3 = 0.162,$$

$$L_1^{(2)} = \frac{15.75}{19.25 + 15.75} L_1 = 0.185, \qquad L_2^{(2)} = \frac{0.1575}{19.25 + 0.1575} L_2 = 0.022,$$

$$L_3^{(2)} = \frac{15.75}{0.385 + 15.75} L_3 = 6.616.$$

4.3 CLOSED JACKSON NETWORKS

If we set $\gamma_i = 0$ and $r_{i0} = 0$ for all i, we have a closed Jackson network, which is equivalent to a finite-source queue of, say, N items which continuously travel inside the network. If $c_i = 1$ for all i, we can get the steady-state flow-balance equations from (4.9), the open network model, by setting $\gamma_i = r_{i0} = 0$. This yields

$$\sum_{\substack{j=1 \\ (i \neq j)}}^{k} \sum_{i=1}^{k} \mu_i r_{ij} p_{\bar{n};i^+j^-} = \sum_{i=1}^{k} \mu_i (1 - r_{ii}) p_{\bar{n}}. \qquad (4.14)$$

Since this is a special case of a general Jackson network, once more we have a product-form solution,

$$p_{\bar{n}} = C \rho_1^{n_1} \rho_2^{n_2} \cdots \rho_k^{n_k} \equiv C \mathfrak{R}^{\bar{n}}, \qquad (4.15)$$

where $\rho_i = \lambda_i/\mu_i$ must satisfy the balance equations for flow at each node i, so that the flows into and out of node i must be equal. This yields the closed network equivalent to the traffic equations of (4.10a), namely,

$$\lambda_i = \mu_i \rho_i = \sum_{j=1}^{k} \lambda_j r_{ji} = \sum_{j=1}^{k} \mu_j r_{ji} \rho_j. \qquad (4.16)$$

As for open networks, we assume that the routing matrix is irreducible and nonabsorbing. But now, one of the equations of (4.16) is redundant because the sum of the λ_i is known. Thus, we can arbitrarily set one ρ_i equal to 1 when solving $\mu_i \rho_i = \sum_{j=1}^{k} \mu_j r_{ji} \rho_j$. Problem 4.13 asks to verify that (4.15) is a solution by substituting it into (4.14).

For this case, C does not "break apart" and must be evaluated by

$$\sum_{n_1+n_2+\cdots+n_k=N} C\rho_1^{n_1}\rho_2^{n_2} \cdots \rho_k^{n_k} = 1 \Rightarrow C = \left(\sum_{n_1+n_2+\cdots+n_k=N} \rho_1^{n_1}\rho_2^{n_2} \cdots \rho_k^{n_k} \right)^{-1},$$

where the sum is taken over all possible ways the N elements can be distributed among the k nodes. The constant C is often shown as $C(N)$ to emphasize that it is a function of the total population size N. Further, the solution is often written in terms of $C^{-1}(N) \equiv G(N)$, so that

$$p_{n_1,n_2,\ldots,n_k} = \frac{1}{G(N)} \rho_1^{n_1}\rho_2^{n_2} \cdots \rho_k^{n_k},$$

where

$$G(N) = \sum_{n_1+n_2+\cdots+n_k=N} \rho_1^{n_1}\rho_2^{n_2} \cdots \rho_k^{n_k}.$$

Again, this closed network can easily be extended to c_i servers at node i (see Problem 4.14). The solution now becomes

$$p_{n_1,n_2,\ldots,n_k} = \frac{1}{G(N)} \prod_{i=1}^{k} \frac{\rho_i^{n_i}}{a_i(n_i)}, \tag{4.17}$$

where $a_i(n_i)$ is given by (4.13) and

$$G(N) = \sum_{n_1+n_2+\cdots+n_k=N} \prod_{i=1}^{k} \frac{\rho_i^{n_i}}{a_i(n_i)}. \tag{4.18}$$

Example 4.4

Two special-purpose machines are desired to be operational at all times. We call the operating node of this network node 1. The machines break down according to an exponential distribution with mean failure rate λ. Upon breakdown, a machine has a probability r_{12} that it can be repaired locally (node 2) by a single repairman who works according to an exponential distribution with parameter μ_2. With probability $1 - r_{12}$ the machine must be repaired by a specialist (node 3), who also works according to an exponential distribution, but with mean rate μ_3. Further, after completing local service at node 2, there is a probability r_{23} that a machine will also require the special service (the probability of returning to operation from node 2 is then $1 - r_{23}$). After the special service (node 3), the unit always returns to operation $(r_{31} = 1)$.

In solving this closed Jackson network, we first note that at node 1 the servers are the machines, so that $c_1 = 2$. Also, μ_1 becomes λ, that is, the mean service (or holding time) at node 1 is the mean time to failure of a machine. Thus the solution for the steady-state joint probability distribution is given by (4.17) as

$$p_{n_1,n_2,n_3} = \frac{1}{G(2)} \frac{\rho_1^{n_1}}{a_1(n_1)} \rho_2^{n_2} \rho_3^{n_3} \qquad (n_i = 0, 1, 2, \quad i = 1, 2, 3),$$

where $a_1(n_1) = 1$ for $n_1 = 0$, 1, and $a_1(2) = 2$, and we must find ρ_i from (4.16). The routing probability matrix $R = \{r_{ij}\}$ for nodes 1, 2 and 3 and is given by

$$R = \begin{pmatrix} 0 & r_{12} & 1 - r_{12} \\ 1 - r_{23} & 0 & r_{23} \\ 1 & 0 & 0 \end{pmatrix}.$$

Using the above $\{r_{ij}\}$ in (4.16) yields

$$\lambda \rho_1 = \mu_2 (1 - r_{23}) \rho_2 + \mu_3 \rho_3,$$

$$\mu_2 \rho_2 = \lambda r_{12} \rho_1,$$

$$\mu_3 \rho_3 = \lambda (1 - r_{12}) \rho_1 + \mu_2 r_{23} \rho_2.$$

One equation in the set given by (4.16) is always redundant, so we can arbitrarily set one of the $\rho_i = 1$ [the constant $G(N)$ will account for the appropriate normalizing factor]. We choose to set $\rho_2 = 1$. Thus the solutions for the $\{\rho_i\}$ are

$$\rho_2 = 1, \qquad \rho_1 = \frac{\mu_2}{r_{12}\lambda},$$

$$\rho_3 = \frac{\lambda(1 - r_{12})}{\mu_3} \frac{\mu_2}{\lambda r_{12}} + \frac{\mu_2}{\mu_3} r_{23} = \frac{(1 - r_{12})\mu_2 + r_{12}r_{23}\mu_2}{r_{12}\mu_3} = \frac{\mu_2(1 - r_{12} + r_{12}r_{23})}{r_{12}\mu_3},$$

and hence

$$p_{n_1,n_2,n_3} = \frac{1}{G(N)} \left(\frac{\mu_2}{r_{12}\lambda}\right)^{n_1} \frac{1}{a_1(n_1)} \left(\frac{\mu_2(1 - r_{12} + r_{12}r_{23})}{r_{12}\mu_3}\right)^{n_3}.$$

The normalizing constant $G(N)$ must be obtained by summing p_{n_1,n_2,n_3} over all cases for which $n_1 + n_2 + n_3 = 2$. There are actually six cases, namely, $(2, 0, 0)$, $(0, 2, 0)$, $(0, 0, 2)$, $(1, 1, 0)$, $(1, 0, 1)$, and $(0, 1, 1)$ for (n_1, n_2, n_3). To illustrate, assume $\lambda = 2$, $\mu_2 = 1$, $\mu_3 = 3$, $r_{12} = \frac{3}{4}$, and $r_{23} = \frac{1}{3}$. Then

$$p_{n_1,n_2,n_3} = \frac{1}{G(N)} \left(\frac{2}{3}\right)^{n_1} \frac{1}{a_1(n_1)} \left(\frac{2}{9}\right)^{n_3},$$

and we have

$$G(2) = \left(\frac{2}{3}\right)^2 \cdot \frac{1}{2} + 1 + \left(\frac{2}{9}\right)^2 + \frac{2}{3} + \frac{2}{3} \cdot \frac{2}{9} + \frac{2}{9} = \tfrac{187}{81} \doteq 2.3086.$$

Hence the steady-state probabilities are

$$p_{2,0,0} = \frac{\left(\frac{2}{3}\right)^2 \left(\frac{1}{2}\right)}{2.3086} = 0.0962, \qquad p_{0,2,0} = \frac{1}{2.3086} = 0.4332,$$

$$p_{0,0,2} = \frac{\left(\frac{2}{9}\right)^2}{2.3086} = 0.0214, \qquad p_{1,1,0} = \frac{\frac{2}{3}}{2.3086} = 0.2888,$$

$$p_{1,0,1} = \frac{\left(\frac{2}{3}\right)\left(\frac{2}{9}\right)}{2.3086} = 0.0642, \qquad p_{0,1,1} = \frac{\frac{2}{9}}{2.3086} = 0.0962.$$

Thus in this particular situation only 9.62% of the time are both machines operating, with at least one machine available $0.0962 + 0.2888 + 0.0642 = 44.92\%$ of the time. It may be desirable to obtain more reliable machines (lower λ) or put on more repairmen or have more machines installed, in order to have at least one available more of the time.

In the above example N was only 2 and k only 3 which made the calculation of $G(N)$ rather easy. For large N and k there are many possible ways to allocate the N customers among the k nodes, in fact, $\binom{N+k-1}{N}$ ways (see Feller, 1957). Thus it would be highly advantageous to have an efficient algorithm to calculate $G(N)$. Buzen (1973) developed one, and we now present this most useful result for closed Jackson networks.

Let $f_i(n_i) = \rho_i^{n_i}/a_i(n_i)$. Then

$$G(N) = \sum_{n_1+n_2+\cdots+n_k=N} \prod_{i=1}^{k} f_i(n_i). \tag{4.19}$$

Buzen set up an auxiliary function

$$g_m(n) = \sum_{n_1+n_2+\cdots+n_m=n} \prod_{i=1}^{m} f_i(n_i). \tag{4.20}$$

Note that $g_m(n)$ would equal $G(N)$ if $k = m$ and $N = n$; that is, it is a normalizing constant for a system with n customers and m nodes. Also, note that $G(N) = g_k(N)$. We can now set up a recursive scheme for calculating $G(N)$.

Consider $g_m(n)$ and suppose we fix $n_m = i$ first in the summation, so that we have

$$
\begin{aligned}
g_m(n) &= \sum_{i=0}^{n} \left(\sum_{n_1+n_2+\cdots+n_{m-1}+i=n} \prod_{i=1}^{m} f_i(n_i) \right) \\
&= \sum_{i=0}^{n} f_m(i) \left(\sum_{n_1+n_2+\cdots+n_{m-1}=n-i} \prod_{i=1}^{m-1} f_i(n_i) \right) \\
&= \sum_{i=0}^{n} f_m(i) g_{m-1}(n-i) \qquad (n = 0, 1, \ldots, N).
\end{aligned}
\tag{4.21}
$$

Note that from (4.21) $g_1(n) = f_1(n)$ and $g_m(0) = 1$, so that we can use (4.21) recursively to calculate $G(N) = g_k(N)$. Further, these functions aid in calculating marginal distributions as well. Suppose we desire the marginal distribution at node i, namely, $p_i(n) = \Pr\{N_i = n\}$. Let

$$
S_i = n_1 + n_2 + \cdots + n_{i-1} + n_{i+1} + \cdots + n_k.
$$

Then

$$
\begin{aligned}
p_i(n) &= \sum_{S_i=N-n} p_{n_1,\ldots,n_k} = \sum_{S_i=N-n} \frac{1}{G(N)} \prod_{i=1}^{k} f_i(N_i) \\
&= \frac{f_i(n)}{G(N)} \sum_{S_i=N-n} \prod_{\substack{j=1 \\ (j \ne i)}}^{k} f_i(n) \qquad (n = 0, 1, \ldots, N).
\end{aligned}
$$

This, however, is very cumbersome to compute. But for node k, the expression simplifies to

$$
p_k(n) = \frac{f_k(n)}{G(N)} \sum_{S_k=N-n} \prod_{i=1}^{k-1} f_i(n) = \frac{f_k(n) g_{k-1}(N-n)}{G(N)} \qquad (n = 0, 1, \ldots, N),
\tag{4.22}
$$

and to find the other marginals $p_i(n)$, $i \ne k$, Buzen (1973) suggests permuting the network to make the node i of interest $= k$. This of course, requires resolving for some of the functions $g_m(n)$. Bruell and Balbo (1980) suggest an improved algorithm for obtaining $p_i(n)$, $i \ne k$. In the next section, we present a method for the calculation of expected-value performance measures for closed Jackson networks, called *mean-value analysis*, which also can yield marginal probability distributions.

To illustrate, let us go back to Example 4.4. The factors $f_i(n_i)$ are

$$f_1(0) = 1, \quad f_1(1) = \tfrac{2}{3}, \quad f_1(2) = \tfrac{2}{9},$$

$$f_2(0) = f_2(1) = f_2(2) = 1,$$

$$f_3(0) = 1, \quad f_3(1) = \tfrac{2}{9}, \quad \text{and} \quad f_3(2) = \tfrac{4}{81}.$$

The $g_i(n)$'s are

$$G(2) = g_3(2) = \sum_{i=0}^{2} f_3(i)g_2(2 - i) = f_3(0)g_2(2) + f_3(1)g_2(1) + f_3(2)g_2(0)$$

and

$$g_2(2) = f_2(0)g_1(2) + f_2(1)g_1(1) + f_2(2)g_1(0),$$

$$g_2(1) = f_2(0)g_1(1) + f_2(1)g_1(0),$$

$$g_2(0) = f_2(0),$$

plus

$$g_1(0) = 1, \quad g_1(1) = f_1(1) = \tfrac{2}{3}, \quad g_1(2) = f_1(2) = \tfrac{2}{9}.$$

Thus we calculate from the bottom up to get

$$g_0(0) = 1,$$

$$g_2(1) = 1 \times \tfrac{2}{3} + 1 \times 1 = \tfrac{5}{3}.$$

$$g_2(2) = 1 \times \tfrac{2}{9} + 1 \times \tfrac{2}{3} + 1 \times 1 = \tfrac{17}{9},$$

$$g_3(2) = G(2) = 1 \times \tfrac{17}{9} + \tfrac{2}{9} \times \tfrac{5}{3} + \tfrac{4}{81} \times 1 = \tfrac{187}{81} \doteq 2.3086.$$

While Buzen's algorithm was not much easier for this simple example, it is quite efficient for large networks with large numbers of customers.

If now we desire the marginal distribution for the number of customers at node 3, say, we have from (4.22)

$$p_3(n) = \frac{f_3(n)g_2(2 - n)}{G(2)} = \frac{(\tfrac{2}{9})^n g_2(2 - n)}{2.3086},$$

so that

$$p_3(0) = \frac{1 \cdot g_2(2)}{2.3086} = .8182, \quad p_3(1) = \frac{\tfrac{2}{9}g_2(1)}{2.3086} = .1604,$$

$$p_3(2) = \frac{(\frac{2}{9})^2 g_2(0)}{2.3086} = .0214,$$

which is the same answer one would get by summing appropriate joint probabilities, namely,

$$p_3(0) = p_{2,0,0} + p_{0,2,0} + p_{1,1,0} = 0.0962 + .4332 + .2888 = .8182,$$

$$p_3(1) = p_{0,1,1} + p_{1,0,1} = .0962 + .0642 = .1604,$$

$$p_3(2) = p_{0,0,2} = .0214.$$

Again, in large systems, making use of the already calculated functions $g_m(n)$ is considerably more efficient than summing over the joint distribution. A good general reference on computational algorithms for closed networks is Bruell and Balbo (1980).

4.3.1 Mean-Value Analysis

The previously described methods of analyzing closed Jackson queueing networks which require computing the normalizing constant $G(N)$ are often referred to as "convolution procedures." Mean-value analysis is another approach which does not require evaluating $G(N)$. It is built on two basic principles (see Bruell and Balbo, 1980):

1. The queue length observed by an arriving customer is the same as the general-time queue length in a closed network with one less customer, that is, $q_n(N) = p_n(N-1)$.
2. Little's formula is applicable throughout the network.

The first principle allows us to write the average waiting time at a node in terms of the mean service time and average number in the system found by an arriving customer. Recall that, for the $M/M/1$ situation, it can be easily shown using Equations (2.16) and (2.19) that $W = (1 + L)/\mu$. What this implies is that the average time an arriving customer must wait is the average time to serve the queue size as seen by an arriving customer plus itself. For $M/M/c$, $q_n = p_n$, so no adjustment need be made for the fact that L is based on p_n and not q_n. For our closed network (we assume for the time being that all nodes have a single server) the equivalent equation becomes

$$W_i(N) = \frac{1 + L_i(N-1)}{\mu_i}, \tag{4.23}$$

where

$W_i(N)$ = mean waiting time at node i for a network containing N customers,

μ_i = mean service rate for the single server at node i,

$L_i(N-1)$ = mean number at node i in a network with $N-1$ customers.

The second principle, that of applying Little's formula throughout the network, allows us to write

$$L_i(N) = \lambda_i(N)W_i(N), \qquad (4.24)$$

where $\lambda_i(N)$ is the throughput (arrival rate) for node i in an N customer network. If we can find $\lambda_i(N)$, Equations (4.23) and (4.24) give us a method for recursively calculating L_i and W_i, starting with an empty network [one with no customers, for which $L_i(0) = 0$ and $W_i(1) = 1/\mu_i$] and building up to the network of interest having N customers.

To be able to compute $\lambda_i(N)$, we note that if we let $D_i(N)$ represent the average delay per customer between successive visits to node i for a network with N customers, then by the laws of conservation we have $\lambda_i(N) = N/D_i(N)$. This merely states that the number of arrivals at node i per unit time must equal the total number of customers in the system divided by the mean time it takes each customer between successive visits to node i, and is a form of Little's formula applied to the entire network, since the expected number of customers in the system is exactly N.

To get $D_i(N)$, we return to the traffic equations (4.16). Letting $v_i = \mu_i\rho_i$ (we shall see shortly why we are not using λ_i), these equations become

$$v_i = \sum_{j=1}^{k} v_j r_{ji}. \qquad (4.25)$$

Since one of the above equations is redundant, we can arbitrarily set one v_i (say v_l) equal to 1 and solve for the others. The v_i then are *relative* throughputs through node i, that is, $v_i = \lambda_i/\lambda_l$, assuming v_l is the one on which we normalize. Now we can write $D_l(N) = \sum_{i=1}^{k} v_i W_i(N)$; that is, $D_l(N)$ is a weighted average of the average delays at each node, weighted by the relative throughputs (arrival rates) of each node to node l, or equivalently weighted by the expected number of visits to each node prior to returning to the "normalized" node, node l (note that v_i can also be interpreted as the expected number of visits to node i after leaving node l prior to returning to node l). For example, if we have a two-node network with $v_1 = 1$ and $v_2 = 2$, then since the arrival rate at node 2 is twice that at node 1, the expected number of visits to node 2 after leaving node 1 prior to returning to node 1 must be two.

We can now write the mean-value analysis (MVA) algorithm for finding $L_i(N)$ and $W_i(N)$ in a k-node, single-server-per-node network with routing probability matrix $R = \{r_{ij}\}$ as follows:

(i) Solve the traffic equations (4.25), $v_i = \sum_{j=1}^{k} v_j r_{ji}$ $(i = 1, 2, \ldots, k)$, setting one of the v_j (say v_l) equal to 1.

(ii) Initialize $L_i(0) = 0$ $(i = 1, 2, \ldots, k)$.

(iii) For $n = 1$ to N, calculate

(a) $W_i(n) = \dfrac{1 + L_i(n-1)}{\mu_i}$ $(i = 1, 2, \ldots, k)$,

(b) $\lambda_l(n) = \dfrac{n}{\sum_{i=1}^{k} v_i W_i(n)}$ (assume $v_l = 1$),

(c) $\lambda_i(n) = \lambda_l(n) v_i$ $(i = 1, 2, \ldots, k, i \neq l)$,

(d) $L_i(n) = \lambda_i(n) W_i(n)$ $(i = 1, 2, \ldots, k)$.

Example 4.5

Consider Example 4.4, but now assume that there is only one machine, so that we now have single servers at each node. Solving (4.25) gives

$$(v_1, v_2, v_3) = (v_1, v_2, v_3) \begin{pmatrix} 0 & \frac{3}{4} & \frac{1}{4} \\ \frac{2}{3} & 0 & \frac{1}{3} \\ 1 & 0 & 0 \end{pmatrix},$$

or $v_1 = \frac{2}{3} v_2 + v_3$, $v_2 = \frac{3}{4} v_1$, $v_3 = \frac{1}{4} v_1 + \frac{1}{3} v_2$. Arbitrarily choosing $v_2 = 1$ $(l = 2$ here), we obtain $v_1 = \frac{4}{3}$ and $v_3 = \frac{2}{3}$.

Now $i = 1, 2, 3$ and $n = N = 1$. Applying step (iii) of the algorithm, we have for (a)

$$W_1(1) = \frac{1 + L_1(0)}{\lambda} = \frac{1}{\lambda} = \frac{1}{2} \quad \text{(note: } \mu_1 = \lambda \text{ for this example)},$$

$$W_2(1) = \frac{1 + L_2(0)}{\mu_2} = \frac{1}{\mu_2} = 1,$$

$$W_3(1) = \frac{1 + L_3(0)}{\mu_3} = \frac{1}{\mu_3} = \frac{1}{3};$$

for (b)

$$\lambda_2(1) = \frac{1}{\sum_{i=1}^{3} v_i W_i(1)} = \frac{1}{\frac{4}{3} \times \frac{1}{2} + 1 \times 1 + \frac{2}{3} \times \frac{1}{3}} = \frac{9}{17};$$

for (c)

$$\lambda_1(1) = v_i \lambda_2(1) = \tfrac{4}{3} \times \tfrac{9}{17} = \tfrac{12}{17},$$
$$\lambda_3(1) = v_3 \lambda_2(1) = \tfrac{2}{3} \times \tfrac{9}{17} = \tfrac{6}{17};$$

and for (d)

$$L_1(1) = \lambda_1(1) W_1(1) = \tfrac{12}{17} \times \tfrac{1}{2} = \tfrac{6}{17},$$
$$L_2(1) = \lambda_2(1) W_2(1) = \tfrac{9}{17} \times 1 = \tfrac{9}{17},$$
$$L_3(1) = \lambda_3(1) W_3(1) = \tfrac{6}{17} \times \tfrac{1}{3} = \tfrac{2}{17}.$$

Since we have only one machine ($N = 1$), we are finished. Had we had a spare machine, we would continue the algorithm for $n = 2$, by calculating $W_1(2) = [1 + L_1(1)]/\lambda = [1 + \tfrac{6}{17}]/2 = \tfrac{23}{24}$, etc. We can continue for any number of spares, but we can have only one operating machine, since to accommodate multiple servers, we must modify the algorithm in step (iii)(a). We will present this a little later, but first, let us check our answer using the normalizing-constant method previously given.

To do this, we must solve the traffic equations (4.16), which gives us the same answer as for Example 4.4, since we have the same matrix R, namely $\rho_1 = \tfrac{2}{3}$, $\rho_2 = 1$, and $\rho_3 = \tfrac{2}{9}$. Now we have from (4.15) that $p_{n_1,n_2,n_3} = [1/G(1)](\tfrac{2}{3})^{n_1}(\tfrac{2}{9})^{n_3}$ and $p_{100} = (\tfrac{2}{3})/G(1)$, $p_{010} = 1/G(1)$, $p_{001} = (\tfrac{2}{9})/G(1)$, so that $G(1) = (\tfrac{2}{3} + 1 + \tfrac{2}{9}) = \tfrac{17}{9}$. Thus $p_{100} = \tfrac{6}{17}$, $p_{010} = \tfrac{9}{17}$, $p_{001} = \tfrac{2}{17}$, and $L_1 = 0 \times \tfrac{11}{17} + 1 \times \tfrac{6}{17} = \tfrac{6}{17}$, $L_2 = \tfrac{9}{17}$, $L_3 = \tfrac{2}{17}$. Note here we get the steady-state probabilities and have to calculate L_i using $\sum_{n=0}^{N} n p_n$, whereas the MVA algorithm directly yields the L_i and W_i, but does not give us the steady-state probabilities.

It is, however, possible to get marginal steady-state probabilities also at each node by recursion, and in fact, we would add to the MVA algorithm a recursive relationship similar in spirit to that of $p_n = \rho p_{n-1}$ for $M/M/1$, the relation achieved from detailed (as opposed to global) stochastic balance (see Section 2.2.1),

$$p_i(n, N) = \frac{\lambda_i(N)}{\mu_i} p_i(n - 1, N - 1) \qquad (n, N \geq 1), \qquad (4.26)$$

where $p_i(n, N)$ is the marginal probability of n in an N-customer system at node i, and $p_i(0, 0) = 1$.

For our example,

$$p_1(1, 1) = \frac{\lambda_1(1)}{\lambda} p_1(0, 0) = \frac{\lambda_1(1)}{\lambda} \times 1 = \frac{\tfrac{12}{17}}{2} = \frac{6}{17},$$

$$p_2(1, 1) = \frac{\lambda_2(1)}{\mu_2} p_2(0, 0) = \frac{\tfrac{9}{17}}{1} \times 1 = \frac{9}{17},$$

$$p_3(1, 1) = \frac{\lambda_3(1)}{\mu_3} p_3(0, 0) = \frac{\frac{6}{17}}{3} \times 1 = \frac{2}{17}.$$

So

$$p_1(0, 1) = \tfrac{11}{17}, \qquad p_1(1, 1) = \tfrac{6}{17},$$
$$p_2(0, 1) = \tfrac{8}{17}, \qquad p_2(1, 1) = \tfrac{9}{17},$$
$$p_3(0, 1) = \tfrac{15}{17}, \qquad p_3(1, 1) = \tfrac{2}{17}.$$

Checking with our normalizing constant solution,

$$p_1(0, 1) = p_{010} + p_{001} = \tfrac{9}{17} + \tfrac{2}{17} = \tfrac{11}{17}, \qquad p_1(1, 1) = \tfrac{6}{17},$$
$$p_2(0, 1) = p_{100} + p_{001} = \tfrac{6}{17} + \tfrac{2}{17} = \tfrac{8}{17}, \qquad p_2(1, 1) = \tfrac{9}{17},$$
$$p_3(0, 1) = p_{100} + p_{010} = \tfrac{6}{17} + \tfrac{9}{17} = \tfrac{15}{17}, \qquad p_3(1, 1) = \tfrac{2}{17}.$$

Thus it is relatively easy to obtain the marginal steady-state probabilities at each node using MVA.

We mentioned previously that the MVA algorithm must be modified in step (iii)(a) for multiple-server cases. Here, we do not have that the workload an arriving customer (inside observer) sees is $(1/\mu)(1 + L)$, since there are multiple servers who work simultaneously on reducing the customer queue. For this situation we have

$$W_i(n) = \frac{1}{\mu_i} + \frac{1}{c_i \mu_i} \sum_{j=c_i}^{n-1} (j - c_i + 1) p_i(j, n - 1),$$

since if there are $j > c_i$ customers at node i when a customer arrives, it must wait until $j - c_i + 1$ are served at rate $c_i \mu_i$ to get into service. This may be simplified to

$$W_i(n) = \frac{1}{c_i \mu_i} \left(c_i + \sum_{j=c_i}^{n-1} j p_i(j, n - 1) - (c_i - 1) \sum_{j=c_i}^{n-1} p_i(j, n - 1) \right)$$

$$= \frac{1}{c_i \mu_i} \left[c_i + L_i(n - 1) - \sum_{j=0}^{c_i-1} j p_i(j, n - 1) - (c_i - 1) \right.$$

$$\left. \times \left(1 - \sum_{j=0}^{c_i-1} p_i(j, n - 1) \right) \right]$$

$$= \frac{1}{c_i \mu_i} \left(1 + L_i(n - 1) + \sum_{j=0}^{c_i-2} (c_i - 1 - j) p_i(j, n - 1) \right).$$

Thus for the multiserver case, even if we are interested in only the W_i and L_i, we still need to calculate the marginal probabilities $p_i(j, n - 1)$ for $j = 0$, $1, \ldots, c_i - 2$. To calculate these recursively for the multiserver case we have, in the spirit of the $M/M/c$,

$$p_i(j, n) = \frac{\lambda_i(n)}{\alpha_i(j)\mu_i} p_i(j - 1, n - 1) \qquad (i \le j \le n - 1),$$

where

$$\alpha_i(j) = \frac{a_i(j)}{a_i(j - 1)} = \begin{cases} j & (j \le c_i), \\ c_i & (j \ge c_i). \end{cases} \tag{4.27}$$

Now we can modify the single server MVA algorithm to allow for multiple servers:

(i) Solve the traffic equations (4.25) as done previously (note these are the same regardless of the number of servers at a node).
(ii) Initialize for $i = 1, 2, \ldots, k$, $L_i(0) = 0$; $p_i(0, 0) = 1$; $p_i(j, 0) = 0$, $(j \ne 0)$;
(iii) For $n = 1$ to N, calculate

(a) $W_i(n) = \dfrac{1}{c_i \mu_i} \left(1 + L_i(n - 1) + \displaystyle\sum_{j=0}^{c_i - 2} (c_i - 1 - j)p_i(j, n - 1) \right)$
    $(i = 1, 2, \ldots, k)$,
(b) $\lambda_l(n) = n / \sum_{i=1}^{k} v_i W_i(n)$ (assume $v_l = 1$),
(c) $\lambda_i(n) = \lambda_l(n)v_i$ $(i = 1, 2, \ldots, k; i \ne l)$,
(d) $L_i(n) = \lambda_i(n) W_i(n)$ $(i = 1, 2, \ldots, k)$,
(e) $p_i(j, n) = \dfrac{\lambda_i(n)}{\alpha_i(j)\mu_i} p_i(j - 1, n - 1)$
        $(j = 1, 2, \ldots, n; i = 1, 2, \ldots, k)$.

We now illustrate this on Example 4.4. Initializing, we have $L_1(0) = L_2(0) = L_3(0) = 0$, $p_1(0, 0) = p_2(0, 0) = p_3(0, 0) = 1$. The v_i's are the same as before, namely,

$$v_1 = \tfrac{4}{3}, \qquad v_2 = 1 \quad (l = 2), \qquad v_3 = \tfrac{2}{3}.$$

Now for the first iteration, steps (a) through (d) of the algorithm will produce identical results as those for the single-server case, since multiple servers are superfluous for a single-machine system. The results are, for (a),

$$W_1(1) = \tfrac{1}{2}, \qquad W_2(1) = 1, \qquad W_3(1) = \tfrac{1}{3};$$

for (b),

$$\lambda_2(1) = \tfrac{9}{17};$$

for (c),

$$\lambda_1(1) = \tfrac{12}{17}, \qquad \lambda_3(1) = \tfrac{6}{17};$$

and for (d),

$$L_1(1) = \tfrac{6}{17}, \qquad L_2(1) = \tfrac{9}{17}, \qquad L_3(1) = \tfrac{2}{17}.$$

Step (e) yields the same marginal steady-state probabilities, namely,

$$p_1(1, 1) = \tfrac{6}{17}, \qquad p_1(0, 1) = \tfrac{11}{17},$$
$$p_2(1, 1) = \tfrac{9}{17}, \qquad p_2(0, 1) = \tfrac{8}{17},$$
$$p_3(1, 1) = \tfrac{2}{17}, \qquad p_3(0, 1) = \tfrac{15}{17}.$$

The second iteration of the algorithm gives for (a)

$$W_1(2) = \frac{1}{2\lambda}[1 + L_1(1) + (2 - 1)p_1(0, 1)] = \tfrac{1}{4}[1 + \tfrac{6}{17} + \tfrac{11}{17}] = 0.5,$$

$$W_2(2) = \frac{1}{\mu_2}[1 + L_2(1) + 0] = 1[1 + \tfrac{9}{17}] = \tfrac{26}{17} = 1.530,$$

$$W_3(2) = \frac{1}{\mu_3}[1 + L_3(1) + 0] = \tfrac{1}{3}[1 + \tfrac{2}{17}] = 0.373;$$

for (b)

$$\lambda_2(2) = \frac{2}{\Sigma_{i=1}^{3} v_i W_i(2)} = \frac{2}{\tfrac{4}{3}(0.5) + 1.53 + \tfrac{2}{3}(0.373)} = 0.818;$$

for (c)

$$\lambda_1(2) = \lambda_2(2)v_1 = 0.818(\tfrac{4}{3}) = 1.091,$$
$$\lambda_3(2) = \lambda_2(2)v_3 = 0.818(\tfrac{2}{3}) = 0.545;$$

for (d)

$$L_1(2) = \lambda_1(2)W_1(2) = 0.546,$$
$$L_2(2) = \lambda_2(2)W_2(2) = 1.252,$$
$$L_3(2) = \lambda_3(2)W_3(2) = 0.203;$$

and for (e),

$$p_1(j, 2) = \frac{\lambda_1(2)}{\alpha_1(j)\lambda} p_1(j - 1, 1) \qquad (j = 1, 2),$$

which yields

$$p_1(1, 2) = \frac{1.091}{2} p_1(0, 1) = 0.353,$$

$$p_1(2, 2) = \frac{1.091}{4} p_1(1, 1) = 0.096,$$

$$p_1(0, 2) = 1 - 0.353 - 0.096 = 0.551;$$

$$p_2(j, 2) = \frac{\lambda_2(2)}{\alpha_2(j)\mu_2} p_2(j - 1, 1) \qquad (j = 1, 2),$$

which yields

$$p_2(1, 2) = 0.385, \qquad p_2(2, 2) = 0.433, \qquad p_2(0, 2) = 0.182;$$

and

$$p_3(j, 2) = \frac{\lambda_3(2)}{\alpha_3(j)\mu_3} p_3(j - 1, 1) \qquad (j = 1, 2),$$

which yields

$$p_3(1, 2) = 0.161, \qquad p_3(2, 2) = 0.021, \qquad p_3(0, 2) = 0.818.$$

Now, checking with the previous results for Example 4.4, we see that the marginal probabilities are

$p_1(0) = p_{020} + p_{002} + p_{011} = 0.5508, \qquad p_1(1) = p_{110} + p_{101} = 0.3530,$

$p_1(2) = p_{200} = 0.0962;$

$p_2(0) = p_{200} + p_{002} + p_{101} = 0.1818, \qquad p_2(1) = p_{110} + p_{011} = 0.3850,$

$p_2(2) = p_{020} = 0.4332;$

$p_3(0) = p_{200} + p_{020} + p_{110} = 0.8182, \qquad p_3(1) = p_{101} + p_{011} = 0.1604,$

$p_3(2) = p_{002} = 0.0214.$

We mention here that we have not rigorously proved the recursive relationship on the marginal probabilities given by Equations (4.26) and (4.27), but only argued intuitively based on $M/M/1$ and $M/M/c$ recursions.

We now prove (4.26) and note that the proof for (4.27) is similar when the multiserver factor $\alpha_i(n_i)$ is included. We let N_i represent the random variable "number of customers at node i (in the steady state)," so that the marginal probability distribution $p_i(n_i; N)$ is

$$p_i(n_i; N) \equiv \Pr\{N_i = n_i \,|\, N \text{ customers in network}\}$$

$$= \sum_{n_1+n_2+\cdots+n_{i-1}+n_{i+1}+\cdots+n_k=N-n_i} \frac{1}{G(N)} \rho_1^{n_1} \rho_2^{n_2} \cdots \rho_i^{n_i} \cdots \rho_k^{n_k}.$$

The complementary marginal cumulative probability distribution is

$$\bar{P}_i(n_i; N)$$

$$\equiv \Pr\{N_i \geq n_i \,|\, N \text{ customers in network}\}$$

$$= \sum_{j=n_i}^{\infty} \sum_{n_1+n_2+\cdots+n_{i-1}+n_{i+1}+\cdots+n_k=N-j} \frac{1}{G(N)} \rho_1^{n_1} \rho_2^{n_2} \cdots \rho_i^{j} \cdots \rho_k^{n_k}$$

$$= \sum_{j=n_i}^{\infty} \rho_i^{j} \sum_{n_1+n_2+\cdots+n_{i-1}+n_{i+1}+\cdots+n_k=N-j} \frac{1}{G(N)} \rho_1^{n_1} \rho_2^{n_2} \cdots \rho_{i-1}^{n_{i-1}} \rho_{i+1}^{n_{i+1}} \cdots \rho_k^{n_k}$$

$$= \sum_{j=n_i}^{\infty} \rho_i^{j} \frac{1}{G(N)} g_{k-1}(N-j) \qquad \text{[from (4.20)]}$$

$$= \frac{\rho_i^{n_i}}{G(N)} \sum_{j=1}^{\infty} \rho_i^{j-n_i} g_{k-1}(N-j) = \frac{\rho_i^{n_i}}{G(N)} \sum_{l=0}^{\infty} \rho_i^{l} g_{k-1}(N-n_i-l)$$

$$= \frac{\rho_i^{n_i}}{G(N)} g_k(N-n_i) \qquad \text{[from (4.21)]}$$

$$= \frac{\rho_i^{n_i}}{G(N)} G(N-n_i),$$

Now

$$p_i(n_i; N) = \bar{P}_i(n_i; N) - \bar{P}_i(n_i+1; N)$$

$$= \frac{\rho_i^{n_i}}{G(N)} G(N-n_i) - \frac{\rho_i^{n_i+1}}{G(N)} G(N-n_i-1)$$

$$= \frac{\rho_i^{n_i}}{G(N)} [G(N-n_i) - \rho_i G(N-n_i-1)].$$

Thus

$$
\frac{p_i(n_i; N)}{p_i(n_i - 1; N - 1)}
$$

$$
= \frac{\rho_i^{n_i}}{G(N)} \frac{G(N-1)}{\rho_i^{n_i-1}} \frac{G(N-n_i) - \rho_i G(N-n_i-1)}{G(N-1-n_i+1) - \rho_i G(N-1-n_i+1-1)}
$$

$$
= \frac{\rho_i G(N-1)}{G(N)} \frac{G(N-n_i) - \rho_i G(N-n_i-1)}{G(N-n_i) - \rho_i G(N-n_i-1)} = \frac{\rho_i G(N-1)}{G(N)}.
$$

Hence,

$$
p_i(n_i; N) = \frac{\rho_i G(N-1)}{G(N)} p_i(n_i - 1; N - 1).
$$

But note that the throughput at node i is

$$
\lambda_i(N) = \Pr\{\text{server busy at node } i\} \cdot \mu_i = \bar{P}(1; N)\mu_i = \frac{\rho_i}{G(N)} G(N-1)\mu_i,
$$

so that

$$
\frac{\rho_i G(N-1)}{G(N)} = \frac{\lambda_i(N)}{\mu_i},
$$

and finally

$$
p_i(n_i; N) = \frac{\lambda_i(N)}{\mu_i} p_i(n_i - 1; N - 1).
$$

While all the methods [including "brute force" in calculating $G(N)$] are easy to employ for these small illustrative examples, Buzen's algorithm and MVA are computationally quite superior to "brute force," with respect to efficiency (storage and speed) and stability, for larger problems (larger k and larger N). Comparing Buzen and MVA, while they are both far superior to "brute force" for large problems, they nevertheless can still face some numerical difficulties for very large state-space problems emanating from real-world modeling. For a network made up of all single-server nodes for which we desire only mean waiting times and mean system sizes at each node, MVA is superior. However, if we have multiserver nodes or we desire marginal probability distributions at the nodes, then we must calculate the probabilities recursively for MVA, and it is not clear in these situations if MVA is really better than the Buzen procedure.

4.4 CYCLIC QUEUES

If we consider a closed network of k nodes such that

$$r_{ij} = \begin{cases} 1 & (j = i + 1, 1 \le i \le k - 1), \\ 1 & (i = k, j = 1), \\ 0 & (\text{elsewhere}), \end{cases} \tag{4.28}$$

then we have a cyclic queue. A cyclic queue thus is a sort of series queue in a "circle," where the output of the last node feeds back to the first node. Since this is a special case of a closed queueing network, the results of the previous section apply. Hence, for single servers at each node, Equations (4.15) and (4.16) apply, that is,

$$p_{n_1,n_2,\ldots,n_k} = C\rho_1^{n_1}\rho_2^{n_2}\cdots\rho_k^{n_k}, \tag{4.29}$$

with $\mu_i\rho_i = \sum_{j=1}^{k} \mu_j r_{ji}\rho_j$. Using (4.28) in the traffic equation results in

$$\mu_i\rho_i = \begin{cases} \mu_{i-1}\rho_{i-1} & (i = 2, 3, \ldots, k), \\ \mu_k\rho_k & (i = 1). \end{cases}$$

Thus we have

$$\rho_i = \begin{cases} (\mu_{i-1}/\mu_i)\rho_{i-1} & (i = 2, 3, \ldots, k), \\ (\mu_k/\mu_1)\rho_k & (i = 1). \end{cases} \tag{4.30}$$

From (4.30) we see that

$$\rho_2 = \frac{\mu_1}{\mu_2}\rho_1, \quad \rho_3 = \frac{\mu_2}{\mu_3}\rho_2 = \frac{\mu_1}{\mu_3}\rho_1 \quad,\ldots, \quad \rho_{k-1} = \frac{\mu_1}{\mu_{k-1}}\rho_1, \quad \rho_k = \frac{\mu_1}{\mu_k}\rho_1.$$

Since one ρ can be set equal to one due to redundancy, we select $\rho_1 = 1$, and substituting into (4.29) we obtain

$$p_{n_1,\ldots,n_k} = \frac{1}{G(N)} \frac{\mu_1^{N-n_1}}{\mu_2^{n_2}\mu_3^{n_3}\cdots\mu_k^{n_k}}. \tag{4.31}$$

Again, $G(N)$ can be found by summing over all cases $n_1 + n_2 + \cdots + n_k = N$ or by Buzen's algorithm.

The multiple-server case can also be treated similarly, and we leave it as an exercise (see Problem 4.19). Of course, it is not necessary to develop special versions of Equations (4.15) and (4.17); these results can be used as they are, with the appropriate $\{r_{ij}\}$ given by (4.28).

We have actually treated a cyclic queue previously in the text. The machine-repair problems of Section 2.7 are really two-node cyclic queues, with node 1 representing the machine "up" node, and the number of servers at this node being the number of machines desired operational, which we denoted by M. The presence of a queue at this node means that spare machines are on hand, while an idle server represents fewer than M machines operating. The mean "service" (holding) time at this node per customer (machine) is $1/\lambda$, where λ is the machine mean failure rate. Problem 4.18 asks the reader to show that (4.31) reduces to (2.41) when $M = c = 1$.

4.5 EXTENSIONS OF JACKSON NETWORKS

Jackson networks have been extended in several ways. First, Jackson himself, in his 1963 paper, allowed state-dependent exogenous arrival processes and state-dependent internal service for open networks. The parameters of the exogenous "Poisson" arrival processes could depend on the total number of units in the network (a general birth process), while the service-time parameters at a given node could depend on the number of customers present at that node. The solution was also of product form [a more complex version of (4.17), as the general birth–death equation (1.39) is a more complex version of (2.24)]. The normalizing constant does not break apart as it did in the non-state-dependent case of Section 4.2, so that the nodes do not act like independent queues. Computation of the normalizing constant must be done similarly to that for closed networks, but is even more complicated, since the sum to one is over an infinite number of probabilities. A great deal of effort has been put forth in attempting to find efficient ways to obtain or approximate the normalizing constant. Unfortunately, no Buzen-type algorithm exists.

Another avenue of generalization of Jackson networks is to include travel times between nodes of the network. These, of course, could always be modeled as another node, but most often they are ample-server nodes (no queueing for travel).

Posner and Bernholtz (1968) treated closed Jackson networks, but allowed for ample-service travel-time "nodes" with general travel-time (holding-time) distributions. They showed that if one were interested in only marginal probability distributions of steady-state system sizes at sets of nodes exclusive of the travel-time nodes, the exact forms of the travel-time distributions did not matter—only their means. In fact, then, for any nodes in a Jackson network with ample service (number of servers equal to total number of units in the system), the forms of the service-time distributions do not explicitly enter (only the means) as long as the marginal distributions of interest do not include these nodes.

The final extension of Jackson networks we shall discuss (and probably the most significant) deals with extensions to multiple classes of customer networks, namely, multiclass Jackson networks, where, in addition to each class of customers having its own routing structure, each class also has its own mean arrival rate, and the mean service times at a node may depend on the particular customer type (class to which the customer belongs) as well.

Baskett et al. (1975) treated such multiclass Jackson networks, and obtained product-form solutions for the following three queueing disciplines: (1) processor sharing (each customer gets a share of and is served simultaneously by a single server), (2) ample service, and (3) LCFS with preemptive–resume servicing. They allowed the network to be open for some classes of customers and closed for others. Customers may switch classes after finishing at a node according to a probability distribution; that is, there is a probability $r_{is;jt}$ that a customer of class s completing service at node i next goes to node j as a class t customer. Exogenous "Poisson" input can be state-dependent (a general birth process), and service distributions can be of the phase type. Baskett et al. also considered c-server FCFS nodes, but for these, service times for *all* classes must be *IID exponential*; that is, for these nodes all customer types look alike, and service times are exponentially distributed.

Kelly's work (1975, 1976, 1979) probably represents the state of the art in the generalization of Jackson networks. In his 1975 and 1976 papers, he also considered multiple-customer classes, and set up a notational structure which allowed for unique class service times at multiserver FCFS nodes. In fact, his work is so general that it included "most" queueing disciplines (for example, all those considered by Baskett et al.; however, priority dependent on customer class is one it did not). But a price must be paid for this generality in that the description of the state space becomes much more complex. Up to now, the state space could be described by a vector consisting of the number of each class of customer at each node, but now the state space must be described by a complete customer ordering, by type, at each node. Kelly considered, as well as exponential service time, Erlang service, which further expands the state-space descriptor to include service phase. Nevertheless, he proved that the solution is still of product form.

Kelly further conjectured that many of his results can be extended to include general service-time distributions. This conjecture was based on the fact that nonnegative probability distributions can always be well approximated by finite mixtures of gamma distributions. Kelly's conjecture is proved by Barbour (1976).

While the generalizations of Baskett et al., Kelly, and Barbour are theoretically significant, obtaining computational results for these more general Jackson networks is another matter. Gross and Ince (1981) have applied Kelly's multiclass results to a closed network and obtained numerical solutions for an application in repairable-item inventory control (the machine-repair problem).

A great deal of effort has been expended in obtaining computational results for closed multiclass Jackson networks due to their use in modeling computer systems. The basic model generally considers a computer system with N terminals, one for each user logged on. While it is not strictly a closed system, since users log on and off, during busy periods one can assume all terminals are in use, so that there are always N customers (jobs) in the system. These can be at various stages in the system, such as "thinking" at the terminal, waiting in the queue to enter the central processing unit (CPU), being served by the CPU, waiting or in service at input/output stations, and so on. Multiple job classes are an important part of any such model. Bruell and Balbo (1980) provided a compendium of computing algorithms developed to treat such models.

Interest and research continue in the computational aspects of Jackson networks, particularly closed, multiclass networks, because of their immense importance for modeling a variety of systems of interest in the computer, communications, and logistics fields.

4.6 NON-JACKSON NETWORKS

The only non-Jackson networks we will consider here are those where the $\{r_{ij}\}$ are allowed to be state-dependent. In many situations, customers have flexibility when leaving a node in deciding where to go next. For example, in an open network, if a customer has two more nodes to visit prior to departing the system, which node the customer goes to next can well depend upon the relative congestion at the two nodes. Allowing this flexibility in the $\{r_{ij}\}$ destroys one of the Jackson assumptions, and hence the previous results of Section 4.2 for open networks and 4.3 for closed networks do not apply.

If, however, holding times at all nodes remain exponential and exogenous inputs (if any) remain Poisson, we can still model the network using Markov theory. A full Markov state-space analysis (a separate balance equation for each state) is required, however.

As an example, consider a closed network with three nodes and two customers. Suppose customers choose, for the node to visit next, that node with the fewest customers. If there is a tie, the customer chooses among the tied nodes with equal probability. We will further assume that customers will not directly feed back to the same node prior to visiting at least one other node.

We can develop the Q matrix for this network and solve the steady-state equations given by $\mathbf{0} = \mathbf{p}Q$, along with the boundary condition that all probabilities sum to one. Since in this example we have a finite number of customers, we have a finite state space and Q will be finite. Numerical solution techniques (see Section 7.3.1) can always be employed for finite-state-space problems.

The state space must again be described by a k (here $k = 3$) component vector (n_1, n_2, n_3). The six states for the Q matrix for this problem are respectively $(2, 0, 0)$, $(0, 2, 0)$, $(0, 0, 2)$, $(0, 1, 1)$, $(1, 1, 0)$, $(1, 0, 1)$, and the Q matrix is

$$Q = \begin{pmatrix} * & 0 & 0 & 0 & \mu_1/2 & \mu_1/2 \\ 0 & * & 0 & \mu_2/2 & \mu_2/2 & 0 \\ 0 & 0 & * & \mu_3/2 & 0 & \mu_3/2 \\ 0 & 0 & 0 & * & \mu_3 & \mu_2 \\ 0 & 0 & 0 & \mu_1 & * & \mu_2 \\ 0 & 0 & 0 & \mu_1 & \mu_3 & * \end{pmatrix},$$

where $*$ indicates the negative of the sum of the other elements in the row. Thus the steady-state equations are

$$0 = -\mu_1 p_{2,0,0},$$

$$0 = -\mu_2 p_{0,2,0},$$

$$0 = -\mu_3 p_{0,0,2},$$

$$0 = \frac{\mu_2}{2} p_{0,2,0} + \frac{\mu_3}{2} p_{0,0,2} - (\mu_3 + \mu_2) p_{0,1,1} + \mu_1 p_{1,1,0} + \mu_1 p_{1,0,1},$$

$$0 = \frac{\mu_1}{2} p_{2,0,0} + \frac{\mu_2}{2} p_{0,2,0} + \mu_3 p_{0,1,1} - (\mu_1 + \mu_2) p_{1,1,0} + \mu_3 p_{1,0,1},$$

$$0 = \frac{\mu_1}{2} p_{2,0,0} + \frac{\mu_3}{2} p_{0,0,2} + \mu_2 p_{0,1,1} + \mu_2 p_{1,1,0} - (\mu_1 + \mu_3) p_{1,0,1}.$$

We immediately see that for steady state $p_{2,0,0} = p_{0,2,0} = p_{0,0,2} = 0$, as we would expect from the facts that there are only two customers and that routing strategies are state-dependent to avoid congestion. Therefore, for this model we get the 3×3 set of equations

$$0 = -(\mu_3 + \mu_2) p_{0,1,1} + \mu_1 p_{1,1,0} + \mu_1 p_{1,0,1},$$

$$0 = \mu_2 p_{0,1,1} - (\mu_1 + \mu_2) p_{1,1,0} + \mu_3 p_{1,0,1},$$

$$0 = \mu_2 p_{0,1,1} + \mu_2 p_{1,1,0} - (\mu_1 + \mu_3) p_{1,0,1},$$

and of course $1 = p_{0,1,1} + p_{1,1,0} + p_{1,0,1}$, which can replace one of the above equations. If all the μ_i were equal, the probabilities would turn out to be equally likely, namely, the probability of finding any of the three nodes empty with one each at the other two is $\frac{1}{3}$. Even for closed networks, the number of equations can grow enormously, so that thousands or tens of thousands or even millions of equations might have to be solved. For example, a network with 50 customers and 10 nodes (not an unusually large network) could have as many as

$$\binom{N + k - 1}{N} = \binom{59}{50} \doteq 1.26 \times 10^{10}$$

equations, if all states are possible. Obviously, even for modern large-scale computers, this is a formidable task to say the least. This is why the product-form solutions of Jackson and of Gordon and Newell (1967) are so valuable.

PROBLEMS

Whenever a problem is best solved by use of this book's accompanying software, we have added a boldface **C** *to the right of the problem number.*

4.1. Find the solution (4.5) to the equations (4.4) using the methodology of Section 1.7.

4.2. Show that the number in the system just after the last departure, $N(T)$, and the time between successive departures, T, are indepen-dent random variables. [*Hint:* Find the conditional distribution of $N(T)$ given $T = t$ from (4.5), and show that it does not depend on t.] Also show that successive interdeparture times are independent.

4.3C. For Example 4.1 calculate the same performance measures using three checkout counters operating. If you were Meback's advisor, what would you recommend concerning the number of counters to have in operation?

4.4. For a two-station series queue (single server at each station) with Poisson input to the first with parameter λ, exponential service at each station with parameters μ_1, and μ_2, respectively, and no limit on number in system at either station, show that the steady-state probability that there are n_1 in the first-station system (queue and service) and n_2 in the second-station system is given by

$$p_{n_1,n_2} = p_{n_1} p_{n_2} = \rho_1^{n_1} \rho_2^{n_2} (1 - \rho_1)(1 - \rho_2).$$

[*Hint:* Find the steady-state difference equations, then show that $\rho_1^{n_1} \rho_2^{n_2} p_{0,0}$ is a solution to these equations by substitution, and then find $p_{0,0}$ by the boundary condition $\sum_{n_1=0}^{\infty} \sum_{n_2=0}^{\infty} p_{n_1,n_2} = 1$.]

4.5. Consider a three-station series queueing system (single server at each station) with Poisson input (parameter λ) and exponential service (parameters μ_1, μ_2, μ_3). There is no capacity limit on the queue in front of the first two stations, but at the third there is a limit of K allowed (including service). If K are in the third station, then any subsequent arrivals are shunted out of the system. Find the expected number in the system (all three stations) and the expected time spent in the system by a customer who completes all three stages of service.

4.6. Derive the results of (4.8) from (4.7) and the boundary condition.

4.7. Derive the steady-state difference equations for a sequential two-station, single-server system, with Poisson input (parameter λ) and exponential service (parameters μ_1, μ_2), where no queue is allowed in front of station 1 and at most one customer is allowed to wait between the stations. Blockage occurs when there is a customer waiting at station 2 and a customer completed at station 1. For the case $\mu_1 = \mu_2$, solve for the steady-state probabilities.

4.8. What are the eight system-state descriptors for a three-station series queueing model with blocking, infinite queue allowed in front of station 1, but no queues allowed at 2 or 3?

4.9. For an open Jackson network, (a) generalize the steady-state balance equation set (4.9) to allow for c_i servers at each node, and (b) show that the solution given by (4.12) satisfies (4.9). [*Hint:* Use the factor $a_i(n_i)$ in modifying (4.9).]

4.10C. Consider a seven node, open single-server Jackson network, where only nodes 2 and 4 get input from the outside (at rate 5/min). Nodes 1 and 2 have service rates of 85, nodes 3 and 4 have rates of 120, node 5 has a rate of 70, and nodes 6 and 7 have rates of 20 (all in units per minute). The routing matrix is given by

$$
\begin{pmatrix}
\frac{1}{3} & \frac{1}{4} & 0 & \frac{1}{4} & 0 & \frac{1}{6} & 0 \\
\frac{1}{3} & \frac{1}{4} & 0 & \frac{1}{3} & 0 & 0 & 0 \\
0 & 0 & \frac{1}{3} & \frac{1}{3} & \frac{1}{3} & 0 & 0 \\
\frac{1}{3} & 0 & \frac{1}{3} & 0 & \frac{1}{3} & 0 & 0 \\
0 & 0 & 0 & \frac{4}{5} & 0 & 0 & \frac{1}{6} \\
\frac{1}{6} & 0 & \frac{1}{6} & \frac{1}{6} & \frac{1}{6} & \frac{1}{6} & 0 \\
0 & \frac{1}{6} & \frac{1}{6} & \frac{1}{6} & \frac{1}{6} & 0 & \frac{1}{6}
\end{pmatrix}.
$$

Find the average system size and mean line delay at each node.

4.11C. Boobtube Boychicks, a TV repair facility that services many of the regions retail electronics stores' TV repair maintenance policies, receives sets to repair according to a Poisson process at an average rate of 9/hr. All incoming sets are first looked at by a triage specialist, who determines whether the sets go to the facility's general repair station, require special attention and go to the expert repair station, or cannot be fixed locally and must be returned to the manufacturer and are sent to shipping. About 17% of all sets are sent directly to shipping to be returned to manufacturers. Of the 83% which are kept, 57% go to general repair, while 43% are sent to the experts. All repaired sets are sent to shipping; however, 5% of the sets received at the general repair station are sent back, unrepaired, to triage for redetermination. (*Note:* it is allowed for triage to send this set back again to general repair). Because of the varied nature of the problems and varying sizes of the sets, the exponential distribution is found to be an adequate representation for triage, repair (both general and expert), and shipping times. There is a single person at triage, taking a mean time of 6 min per set. There are three general repair people, each taking, on average, 35 min per set (including the ones they can't fix and send back to triage). Barry, Barney, Billy, and Bennie Boychick (quadruplets as it turns out) are the four experts manning the expert repair station and take, on an average, 65 min to repair a set (they never fail to "fix 'em"). There are two shipping clerks, each of whom take an average time of 12.5 min to package the sets. Betty Boychick, Barry's wife and CEO of the company (who has an MS degree in OR from a major state university and teaches part-time at a local university) decides to ask her class to find the average number of sets at each node, the average time spent at each node, and the average time a set spends after it enters the receiving–triage station until it is packed and ready for delivery.

4.12C. The Won Hung Rhee Chinese carry-out restaurant serves two dishes, chow mein and spareribs. There are two separate windows, one for chow mein and one for spareribs. Customers arrive according to a Poisson process with a mean rate of 20/hr. Sixty percent go to the chow-mein window, and 40% go to the rib window. Twenty percent of those who come from the chow-mein window with their order go next to the rib window; the other 80% leave the restaurant. Ten percent of those who purchase ribs then go to the chow-mein window, and the other 90% leave. It takes on average 4 min to fill a chow-mein order and 5 min to fill a sparerib order, the service times being exponential. How many on average are in the restaurant? What is the average wait at each window? If a person wants both chow mein and ribs, how long, on average, does he spend in the restaurant?

4.13. Verify, by substitution into Equation (4.14), that Equation (4.15) is a solution for single-server closed Jackson networks.

4.14. For a closed Jackson network, (a) generalize the steady-state balance equation set (4.14) to allow for c_i servers at each node, and (b) show that the solution given by (4.17) satisfies (4.14). [*Hint:* See the hint at Problem 4.9.]

4.15C. Consider the same problem as Example 4.4, but suppose management decides to add a second server at node 2, identical to the one already there. Find the steady-state availabilities that (a) both machines are operating, and (b) at least one is operating.

4.16C. Tonyexpress, a regional truck delivery service [one of its customers is Boobtube Boychicks (see Problem 4.11)], has a fleet of 50 trucks. Routine maintenance is done during off hours, so no trucks are unavailable for service because of routine maintenance. However, trucks do break down periodically and require repair. The breakdown process can be well approximated by a Poisson process, and repair records indicate a mean time to breakdown of 38 days. Sixty-eight percent of these breakdowns can be handled at Tonyexpress' own repair facility; the remainder must be handled by the manufacturer (the fleet consists of all the same make of trucks). Further, about 7% of those sent to local repair, must, after being worked on, be sent on to the manufacturer's repair facility. Repair times at the local repair facility are exponentially distributed with a mean of 2.75 days, and there are four repair bays, but only three operate at any given time, since there are only three repair crews available. Turnaround times for those sent to the factory have been found to be IID exponential random variables, with a mean of 10 days (since Tonyexpress is such a good customer, whenever a truck is received by the manufacturer, a mechanic is immediately put on the job). Since trucks are so well cared for, it is not unreasonable to assume that repaired trucks are as good as new. When trucks break down, Tonyexpress has a contract with TowsRus, a towing company with a very ample fleet of tow trucks. The mean time to pick up and tow a truck to the Tonyexpress facility is 0.15 days, and the mean time to tow a truck to the manufacturer (from the field or from Tonyexpress) is 0.75 days. Tony, the CEO of Tonyexpress, hires you to advise whether he should hire another repair crew. His major performance measure is truck availability—he is interested in the expected fraction of trucks available and the percentage of time that he has at least 45 trucks operational.

4.17C. Find the average number of customers at each node and the node delay time for a closed network with 35 customers circulating between seven nodes using the switch matrix given by

$$
\begin{pmatrix}
\frac{1}{3} & \frac{1}{4} & 0 & \frac{1}{4} & 0 & \frac{1}{6} & 0 \\
\frac{1}{3} & \frac{1}{4} & 0 & \frac{1}{4} & \frac{1}{6} & 0 & 0 \\
0 & 0 & \frac{1}{3} & \frac{1}{3} & \frac{1}{3} & 0 & 0 \\
\frac{1}{3} & 0 & \frac{1}{3} & 0 & \frac{1}{3} & 0 & 0 \\
0 & 0 & 0 & \frac{5}{6} & 0 & 0 & \frac{1}{6} \\
\frac{1}{6} & \frac{1}{6} & \frac{1}{6} & \frac{1}{6} & \frac{1}{6} & \frac{1}{6} & 0 \\
0 & \frac{1}{6} & \frac{1}{6} & \frac{1}{6} & \frac{1}{6} & \frac{1}{6} & \frac{1}{6}
\end{pmatrix}
$$

and assuming the same service rates as in Problem 4.10.

4.18. Show that for a single machine and a single repairman, the solution obtained from Equation (2.41) of Section 2.7 is the same as the solution obtained from Equation (4.31).

4.19. Generalize Equation (4.31) to handle multiple servers.

4.20C. Solve Problem 2.44 by looking at it as a cyclic queue, that is, a special case of a closed queueing network.

4.21C. Use the closed Jackson single-server software module to check out Example 4.5.

4.22C. Using the closed Jackson multiserver software module, (a) check the calculations for $p_3(n)$ of Example 4.4 given in the text section on Buzen's algorithm, and (b) calculate $p_2(n)$ and $p_1(n)$ also, by renumbering the nodes so that the node of interest is node k ($=3$ for this problem).

CHAPTER 5

Models with General Arrival or Service Patterns

For the models treated in this chapter, a Chapman–Kolmogorov analysis as in the previous chapters is not possible since we no longer have a Markov process because of the relaxation of the exponential assumption on interarrival times and/or service times. However, for many of the models considered here, while we no longer have a Markov process, there is, nevertheless, imbedded within this non-Markov stochastic process a Markov chain (referred to as an imbedded Markov chain; see Section 1.9). For these types of models, we can employ some of the theory of Markov chains from Section 1.9 for our analysis. The main sections of the chapter treat the cases of Poisson input with a general single server ($M/G/1$) and general input with a single exponential server ($G/M/1$).

5.1 SINGLE-SERVER QUEUES WITH POISSON INPUT AND GENERAL SERVICE ($M/G/1$)

Consider a queueing system immediately after a customer has departed and service is about to commence on the next customer in queue, with service times assumed to be independently and identically distributed random variables with an arbitrary probability distribution. We denote the cumulative distribution function (CDF) by $B(t)$, and the density function, if it exists, by $b(t)$. The arrival process is, as before, Poisson with parameter λ. The imbedded stochastic process $X(t_i)$, where X denotes the number in the system and $t_1, t_2, t_3, \ldots$ are the successive times of completion of service, can be shown to be Markovian as follows. Since t_i is the completion time of the ith customer, then $X(t_i)$ is the number of customers that the ith customer leaves behind upon departure. Since the state space is discrete, let us use a subscript notation so that X_i represents the number of customers remaining in the system as the ith customer departs. We can then write for all $n > 0$

that

$$X_{n+1} = \begin{cases} X_n - 1 + A_{n+1} & (X_n \geq 1), \\ A_{n+1} & (X_n = 0), \end{cases} \tag{5.1}$$

where X_n is the number in the system at the nth departure point and A_{n+1} is the number of customers who arrived during the service time $S^{(n+1)}$ of the $(n + 1)$st customer.

By assumption, the random variable $S^{(n+1)}$ is independent of previous service times and the length of the queue, so let it henceforth be denoted by S. Since arrivals are Poissonian, the random variable A_{n+1} depends only on S and not on the queue or on the time of service initiation, so let it henceforth be denoted by A. Then it follows that

$$\Pr\{A = a\} = \int_0^\infty \Pr\{A = a \mid S = t\} \, dB(t) \tag{5.2}$$

and

$$\Pr\{A = a \mid S = t\} = \frac{e^{-\lambda t}(\lambda t)^a}{a!}, \tag{5.3}$$

so that

$$\Pr\{X_{n+1} = j \mid X_n = i\} = \Pr\{A = j - i + 1\}$$

$$= \begin{cases} \int_0^\infty \dfrac{e^{-\lambda t}(\lambda t)^{j-i+1}}{(j-i+1)!} \, dB(t) & (j \geq i - 1, i \geq 1), \\ 0 & (j < i - 1, i \geq 1). \end{cases} \tag{5.4}$$

[For generality, we use the Stieltjes integral here. In cases where the density function exists, as in most cases encountered, $dB(t)$ can be replaced by $b(t)dt$, thus yielding the more familiar Riemann integral of elementary calculus.]

If a departing customer leaves an empty system, the system state remains zero until an arrival comes. Thus the transition probabilities for the case $i = 0$ are identical to those for $i = 1$. Hence we can readily see that the imbedded process is Markovian, since only the indices (i, j) are involved in (5.4), and furthermore, since the state variable is discrete, it is a Markov chain. This allows the utilization of Markov-chain theory in the analysis of the $M/G/1$ model. Prior to doing this, however, we derive the main measures of effectiveness through a direct expected-value argument.

5.1.1   Expected-Value Measures of Effectiveness:
The Pollaczek-Khintchine Formula

We begin here by deriving the expected steady-state system size L, using mean values. From this we can then obtain the other measures in the usual way. Again considering the queue just as a customer departs and assuming for the time being that a steady-state solution exists, we can rewrite (5.1) as

$$X_{n+1} = X_n - U(X_n) + A, \qquad (5.5)$$

where U is the unit step function

$$U(X_n) = \begin{cases} 1 & (X_n > 0), \\ 0 & (X_n = 0). \end{cases}$$

Since we are assuming a steady-state solution, we can take expected values of these random variables, noting that $E[X_{n+1}] = E[X_n] = L^{(D)}$ [the superscript (D) is used on L to denote that it represents the expected steady-state system size at departure points and not at any general time as does L]. Taking the expectation of both sides of (5.5) yields

$$L^{(D)} = L^{(D)} - E[U(X_n)] + E[A] \;\Rightarrow\; E[U(X_n)] = E[A].$$

But, from (5.2) and (5.3), we have

$$E[U(X_n)] = E[A] = \int_0^\infty E[A \mid S = t]\, dB(t)$$

$$= \int_0^\infty \lambda t\, dB(t) = \lambda E[S] = \frac{\lambda}{\mu} \equiv \rho. \qquad (5.6)$$

Next, squaring (5.5) gives

$$X_{n+1}^2 = X_n^2 + U^2(X_n) + A^2 - 2X_n U(X_n) - 2AU(X_n) + 2AX_n. \quad (5.7)$$

By taking expected values of (5.7), noting that $E[X_{n+1}^2] = E[X_n^2]$, we then have that

$$0 = E[U^2(X_n)] + E[A^2] - 2E[X_n U(X_n)] - 2E[AU(X_n)] + 2E[AX_n].$$

But $U^2(X_n) = U(X_n)$ and $X_n U(X_n) = X_n$. Then using (5.6) yields

$$0 = \rho + E[A^2] - 2L^{(D)} - 2\rho^2 + 2L^{(D)}\rho,$$

or

$$L^{(D)} = \frac{\rho - 2\rho^2 + E[A^2]}{2(1-\rho)}. \tag{5.8}$$

Now

$$E[A^2] = \text{Var}[A] + (E[A])^2 = \text{Var}[A] + \rho^2,$$

where (e.g., see Ross, 1996)

$$\text{Var}[A] = E\big[\text{Var}[A \mid S]\big] + \text{Var}\big[E[A \mid S]\big].$$

Therefore, from Equation (5.3), the equation above becomes

$$\text{Var}[A] = E[\lambda S] + \text{Var}[\lambda S] = \rho + \lambda^2 \sigma_S^2, \tag{5.9}$$

where σ_S^2 is the variance of the service-time distribution. Thus by use of (5.9) in (5.8), we get

$$L^{(D)} = \rho + \frac{\rho^2 + \lambda^2 \sigma_S^2}{2(1-\rho)}. \tag{5.10}$$

It can be shown (this is discussed more fully in a later section) that although $L^{(D)}$ is the expected steady-state system size at departure points, it is also equal to the expected steady-state system size at an arbitrary point in time, L. So we may write that

$$\boxed{L = \rho + \frac{\rho^2 + \lambda^2 \sigma_S^2}{2(1-\rho)}.} \tag{5.11}$$

Equation (5.11) is often referred to as the Pollaczek–Khintchine (PK) formula. From it, the expected wait W in the system can be obtained via Little's formula. Next, W_q can be obtained by the usual relation $W = W_q + 1/\mu$. Finally, L_q can be found by once again utilizing Little's formula, or the relation $L_q = L - \rho$. (It is common to find the resultant formulas for W, W_q, and L_q also referred to as the PK formula. All of these results are essentially equivalent in view of Little's formula and the fundamental properties of the single-server queue.)

It is only necessary, therefore, to know λ plus the mean and variance of the service-time distribution to obtain the expected-value measures of effectiveness of an $M/G/1$ queue. These are, indeed, powerful results, since this information concerning the service mechanism is usually readily available or can be estimated without great difficulty. For example, letting $\sigma_S^2 = 0$

gives results for the *M/D/*1 model. The reader can easily check this by comparing (5.11) with the variance set equal to 0 in L for the $M/E_k/1$ model of Section 3.3.3 with $k = \infty$.

Example 5.1
Consider a single-server, Poisson-input queue with mean arrival rate of 10/hr. Currently, the server works according to an exponential distribution with mean service time of 5 min. Management has a training course which will result in an improvement (decrease) in the variance of the service time but a slight increase in the mean. After completion of the course, it is estimated that the mean service time will increase to 5.5 min but the standard deviation will decrease from 5 min (the exponential case) to 4 min. Management would like to know whether they should have the server undergo further training.

To answer the question, we will compare L and W for each case, the first model being *M/M/*1 and the second *M/G/*1. For *M/M/*1, we can either use the results of Section 2.2 or the PK formula above with $\sigma_S = 1/\mu$; for the *M/G/*1, we use the PK results. The comparisons are presented in Table 5.1.

Thus it is not profitable to have the server "better" trained. Note that reducing the standard deviation twice as much as the increase in mean resulted in considerably poorer performance. Performance is much more sensitive to the mean service time than to its variation. In this example, the mean time increases only 10%, while the standard deviation decreases by 20% (a decrease in the variance of almost 36%). It is of interest to calculate the reduction in variance required to make up for the increase of 0.5 in the mean. We can do this by solving for σ_S^2 in (5.11) as follows:

$$L = 5 = \rho + \frac{\rho^2 + \lambda^2 \sigma_S^2}{2(1 - \rho)},$$

where $\rho = \lambda/\mu = 10(\frac{11}{2})(\frac{1}{60}) = \frac{11}{12}$. This yields a $\sigma_S^2 < 0$, which, of course, is not possible. What this means is that even with a service-time variance of 0 (deterministic service times), L is greater than 5, and, in fact, turns out to be

$$L = \rho + \frac{\rho^2}{2(1 - \rho)} \doteq 6.0.$$

Table 5.1 Comparison of Models

	Present (*M/M/*1)	After Training (*M/G/*1)
L	5	8.625
W	30 min	51.750 min

Problem 5.4 asks the reader to find the value of σ_S^2 required to yield the same L if the mean service time were increased to only 5.2 min after training.

5.1.2 Departure-Point Steady-State System-Size Probabilities

This subsection treats the development of the steady-state system-size probabilities at departure points. We let π_n represent the probability of n in the system at a departure point (a point of time just slightly after a customer has completed service) after steady state is reached. The probabilities $\{\pi_n\}$ are not in general the same as the previous steady-state system-size probabilities $\{p_n\}$, which were valid for any arbitrary point of time after steady state is reached. For the $M/G/1$ model, however, it turns out that the set $\{\pi_n\}$ and the set $\{p_n\}$ are identical; this will be verified in a later section in this chapter.

We have shown in Section 5.1 that the imbedded stochastic process at departure points is a Markov chain. We denote the transition probability matrix by

$$P = \{p_{ij}\},$$

where

$$
\begin{aligned}
p_{ij} &= \Pr\{\text{system size immediately after a departure point is } j \,| \\
&\quad\quad \text{system size after previous departure was } i\} \\
&= \Pr\{X_{n+1} = j \mid X_n = i\}.
\end{aligned}
$$

From (5.4) we have

$$p_{ij} = \int_0^\infty \frac{e^{-\lambda t}(\lambda t)^{j-i+1}}{(j-i+1)!}\, dB(t) \qquad (j \geq i-1, \quad i \geq 1). \qquad (5.12)$$

Simplification results on defining

$$k_n = \Pr\{n \text{ arrivals during a service time}\} = \int_0^\infty \frac{e^{-\lambda t}(\lambda t)^n}{n!}\, dB(t), \qquad (5.13)$$

so that p_{ij} can be seen to equal k_{j-i+1} and

$$
P = \{p_{ij}\} =
\begin{pmatrix}
k_0 & k_1 & k_2 & k_3 & \cdots \\
k_0 & k_1 & k_2 & k_3 & \cdots \\
0 & k_0 & k_1 & k_2 & \cdots \\
0 & 0 & k_0 & k_1 & \cdots \\
0 & 0 & 0 & k_0 & \cdots \\
\vdots & \vdots & \vdots & \vdots &
\end{pmatrix}. \qquad (5.14)
$$

Assuming steady state is achievable, the steady-state probability vector π

$= \{\pi_n\}$, can be found as the solution to the stationary equation (see Section 1.9)

$$\pi P = \pi. \tag{5.15}$$

This yields (see Problem 5.6)

$$\pi_i = \pi_0 k_i + \sum_{j=1}^{i+1} \pi_j k_{i-j+1} \qquad (i = 0, 1, 2, \ldots). \tag{5.16}$$

Now define the generating functions

$$\Pi(z) = \sum_{i=0}^{\infty} \pi_i z^i \quad \text{and} \quad K(z) = \sum_{i=0}^{\infty} k_i z^i \qquad (|z| \le 1). \tag{5.17}$$

Then multiplying (5.16) by z^i, summing over i, and solving for $\Pi(z)$ yields (see Problem 5.7)

$$\Pi(z) = \frac{\pi_0(1 - z)K(z)}{K(z) - z}. \tag{5.18}$$

Using the fact that $\Pi(1) = 1$, along with L'Hôpital's rule, and realizing that $K(1) = 1$ and $K'(1) = \lambda(1/\mu)$, we find that

$$\pi_0 = 1 - \rho \qquad (\rho \equiv \lambda E[\text{service time}]). \tag{5.19}$$

Hence

$$\Pi(z) = \frac{(1 - \rho)(1 - z)K(z)}{K(z) - z}. \tag{5.20}$$

Since by definition $\Pi'(1)$ is the expected system size, it is possible to derive the PK formula (see Problem 5.8) directly from Equation (5.20).

Equation (5.20) is as far as we can go in obtaining the $\{\pi_n\}$ (which we show, in the next section, are equivalent to the $\{p_n\}$) without making assumptions as to the specific service-time distribution. Problem 5.9 asks the reader to verify that Equation (5.20) reduces to the generating function for *M/M/*1 given by (2.8) when one assumes exponential service. We shall now present an example of an *M/G/*1 system with empirical service times.

Example 5.2
To illustrate how the results above can be utilized on an empirically determined service distribution, consider the following situation. The Bearing

Table 5.2 Service-Time Distribution

t(min)	$b(t)^a$	$B(t)$
9	2/3	2/3
12	1/3	1

[a] $E[S] = 1/\mu = 10$ min; $\sigma_S^2 = 2$ min^2.

Straight Corporation, a single-product company, makes a very specialized plastic bearing. The company is a high-volume producer, and the bearing undergoes a single machining operation, performed on a specialized machine. Because of the volume of sales, the company keeps a large number of machines in operation at all times (we may assume for practical purposes that the machine population is infinite).

Machines break down according to a Poisson process with a mean of 5/hr. The company has a single expert repairman, and the machine characteristics are such that the breakdowns are due to one of two possible malfunctions. Depending on which of the malfunctions caused the breakdown, it takes the expert repairman either 9 or 12 min to repair a machine. Since he is an expert and the machines are identical, any variation in these service times is minuscule and can be ignored. The type of malfunction which causes a breakdown occurs at random, but it has been observed that one-third of the malfunctions require the 12-min repair time. The company wishes to know the probability that more than three machines will be down at any time.

The service-time mechanism can be looked at as the two-point probability distribution given in Table 5.2. If management were interested in only the expected number of machines down, it would be an easy matter to obtain it from the PK formula, and it turns out that $L = 2.96$ (see Problem 5.10). However, to answer the question asked, it is necessary to find p_0, p_1, p_2, and p_3, since

$$\text{Pr\{more than 3 machines down\}} = \sum_{n=4}^{\infty} p_n = 1 - \sum_{n=0}^{3} p_n.$$

The probability that no machines are down is easily obtained as

$$p_0 = \pi_0 = 1 - \rho = 1 - \tfrac{5}{6} = \tfrac{1}{6}.$$

To find the others, we can make use of (5.16) iteratively. However, an alternative procedure which does not require successively calculating lower probabilities in order to obtain a particular p_n is first to calculate the generating function given in Equation (5.20) and then to expand it in a series to obtain $p_i = \pi_i$ ($i = 1, 2, 3$). To evaluate $\Pi(z)$, we must first find $K(z)$ from Equation (5.17), which necessitates finding k_n as given in (5.13). Since we have a two-point discrete distribution for service, the Stieltjes integral reduces to a summation, and we have

$$k_n = \frac{1}{n!} \{ e^{-5(3/20)} [5(\tfrac{3}{20})]^n (\tfrac{2}{3}) + e^{-5(1/5)} [5(\tfrac{1}{5})]^n (\tfrac{1}{3}) \}$$

$$= \frac{2}{3n!} e^{-3/4} (\tfrac{3}{4})^n + \frac{1}{3n!} e^{-1}.$$

Thus

$$K(z) = \tfrac{2}{3} e^{-3/4} \sum_{i=0}^{\infty} \frac{(3z/4)^i}{i!} + \tfrac{1}{3} e^{-1} \sum_{i=0}^{\infty} \frac{z^i}{i!}.$$

Although we can get $K(z)$ in closed form (since the sums are the infinite series expressions for $e^{3z/4}$ and e^z, respectively), it behooves us not to do so, as we ultimately desire a power series expansion for $\Pi(z)$ in order to pick off the probabilities. To make $K(z)$ easier to work with, define

$$c_i = \tfrac{2}{3} e^{-3/4} (\tfrac{3}{4})^i + \tfrac{1}{3} e^{-1}, \tag{5.21}$$

so that $K(z)$ can be written as

$$K(z) = \sum_{i=0}^{\infty} \frac{c_i z^i}{i!}.$$

Now from (5.20), we have

$$\Pi(z) = \frac{(1-\rho)(1-z) \sum_{i=0}^{\infty} c_i z^i / i!}{\sum_{i=0}^{\infty} c_i z^i / i! - z},$$

which gives

$$\Pi(z) = \frac{(1-\rho) \left(\sum_{i=0}^{\infty} \dfrac{c_i z^i}{i!} - \sum_{i=0}^{\infty} \dfrac{c_i z^{i+1}}{i!} \right)}{c_0 + (c_1 - 1)z + \sum_{i=2}^{\infty} \dfrac{c_i z^i}{i!}}$$

$$= \frac{(1-\rho) \left[1 + \sum_{i=1}^{\infty} \left(\dfrac{c_i}{c_0 i!} - \dfrac{c_{i-1}}{c_0 (i-1)!} \right) z^i \right]}{1 + \dfrac{c_1 - 1}{c_0} z + \sum_{i=2}^{\infty} \dfrac{c_i z^i}{c_0 i!}}. \tag{5.22}$$

It is necessary to have (5.22) in terms of a power series in z and not the ratio of two power series. For our example, we require coefficients of terms up to and including z^3. These can be obtained by long division, carefully keeping track of the needed coefficients. However, it can be seen by long division that the ratio of two power series is itself a power series, namely,

$$\frac{1 + \sum_{i=1}^{\infty} a_i z^i}{1 + \sum_{i=1}^{\infty} b_i z^i} = \sum_{i=0}^{\infty} d_i z^i,$$

where d_i can be obtained recursively from

$$d_i = \begin{cases} a_i - \sum_{j=1}^{i} b_j d_{i-j} & (i = 1, 2, \ldots), \\ 1 & (i = 0). \end{cases}$$

Thus it is only necessary to obtain d_1, d_2, and d_3, which, when multiplied by $1 - \rho$, give p_1, p_2, and p_3, respectively. In terms of the $\{c_i\}$, we get

$$p_1 = (1 - \rho) \left(\frac{1}{c_0} - 1 \right),$$

$$p_2 = (1 - \rho) \left(\frac{1 - c_1}{c_0} - 1 \right) \frac{1}{c_0}, \qquad (5.23)$$

$$p_3 = (1 - \rho) \frac{1}{c_0} \left[\frac{1 - c_1}{c_0} \left(\frac{1 - c_1}{c_0} - 1 \right) - \frac{c_2}{2c_0} \right].$$

Finally, using (5.21) to evaluate c_0, c_1, and c_2, and then substituting into (5.23), yields

$$p_1 \doteq 0.2143, \qquad p_2 \doteq 0.1773, \qquad p_3 \doteq 0.1293$$

$$\Rightarrow \quad \Pr\{>3 \text{ machines down}\} = 1 - \sum_{n=0}^{3} p_n \doteq 0.3124.$$

The results for the two-point service distribution can be rather easily generalized to a k-point service distribution, using a similar analysis (see Greenberg, 1973, and Problem 5.12). Let the probability that the service time is t_i be given by b_i. Then

$$K(z) = \sum_{i=0}^{\infty} \frac{c_i z^i}{i!} \quad \text{with} \quad c_i = \sum_{j=1}^{k} b_j e^{-\lambda t_j} (\lambda t_j)^i \qquad (5.24)$$

and

$$\Pi(z) = \frac{(1 - \rho)\left[1 + \sum_{i=1}^{\infty}\left(\frac{c_i}{c_0 i!} - \frac{c_{i-1}}{c_0(i-1)!}\right)z^i\right]}{1 + \frac{c_1 - 1}{c_0}z + \sum_{i=2}^{\infty}\frac{c_i z^i}{c_0 i!}}.$$

The above equation is identical to (5.22); however, the c_i are different. Using the same relationship for the quotient of two power series as given previously, the $\{d_i\}$ and hence the $\{p_i\}$ can be obtained and are identical to those given by (5.23), with c_i now given by (5.24). Of course, one could generate as many p_i as desired by continuing the procedure further, that is, by calculating more d_i.

Example 5.3
The Bearing Straight Company of Example 5.2 has been able to tighten up its repair efforts (largely by increased automation) to the point where all repairs can now safely be assumed to take precisely 6 min (λ is still equal to $\frac{1}{12}$/min). Thus we now have an *M/D/1* problem, which we can completely solve by the direct use of Equation (5.20).

Assuming that *all* service times are exactly $1/\mu$, we have from (5.13), (5.17), and (5.18) that

$$k_n = \frac{e^{-\rho}\rho^n}{n!} \qquad (\rho = \lambda/\mu),$$

$$K(z) = \sum_{i=0}^{\infty} \frac{e^{-\rho}\rho^i}{i!} z^i = e^{-\rho(1-z)},$$

and

$$\Pi(z) = \frac{(1 - \rho)(1 - z)}{1 - ze^{\rho(1-z)}}. \qquad (5.25)$$

Now, to obtain the individual probabilities, we next expand (5.25) as a geometric series, giving

$$\Pi(z) = (1 - \rho)(1 - z) \sum_{k=0}^{\infty} e^{k\rho(1-z)} z^k$$

$$= (1 - \rho)(1 - z) \sum_{k=0}^{\infty} e^{k\rho} \sum_{m=0}^{\infty} \frac{(-k\rho z)^m}{m!} z^k$$

$$= (1 - \rho)(1 - z) \sum_{k=0}^{\infty} \sum_{m=0}^{\infty} e^{k\rho} (-1)^m \frac{(k\rho)^m}{m!} z^{m+k}$$

$$= (1 - \rho)(1 - z) \sum_{k=0}^{\infty} \sum_{n=k}^{\infty} e^{k\rho} (-1)^{n-k} \frac{(k\rho)^{n-k}}{(n-k)!} z^n.$$

We now change the order of summation and get (graphing the region of summation on the n–k plane will easily verify this)

$$\Pi(z) = (1 - \rho)(1 - z) \sum_{n=0}^{\infty} \sum_{k=0}^{n} e^{k\rho} (-1)^{n-k} \frac{(k\rho)^{n-k}}{(n-k)!} z^n.$$

Finally, we complete the multiplication by the factor $1 - z$ and find that

$$\Pi(z) = (1 - \rho) \left(\sum_{n=0}^{\infty} \sum_{k=0}^{n} e^{k\rho} (-1)^{n-k} \frac{(k\rho)^{n-k}}{(n-k)!} z^n \right.$$
$$\left. - \sum_{n=1}^{\infty} \sum_{k=0}^{n-1} e^{k\rho} (-1)^{n-k-1} \frac{(k\rho)^{n-k-1}}{(n-k-1)!} z^n \right). \qquad (5.26)$$

Thus we have a power series in z and can pick off the $\{p_n\}$. For $n = 0$, we already know that $p_0 = 1 - \rho$. For $n = 1$,

$$p_1 = (1 - \rho)(e^\rho - 1).$$

For $n \geq 2$, the $k = 0$ term of both double summations in (5.26) is always 0 because of $k\rho$, so that both summations on k can start at $k = 1$. Hence we can write the final result as

$$p_n = (1 - \rho) \left(\sum_{k=1}^{n} e^{k\rho} (-1)^{n-k} \frac{(k\rho)^{n-k}}{(n-k)!} - \sum_{k=1}^{n-1} e^{k\rho} (-1)^{n-k-1} \frac{(k\rho)^{n-k-1}}{(n-k-1)!} \right).$$
$$(5.27)$$

The mean system size follows easily right from the PK formula with the variance set equal to 0 and $\rho = \frac{1}{2}$ as

$$L = \rho + \frac{\rho^2}{2(1 - \rho)} = \frac{1}{2} + \frac{1}{4} = \frac{3}{4}.$$

5.1.3   Proof that $\pi_n = p_n$

We now show that π_n, the steady-state probability of n in the system at a departure point, is equal to p_n, the steady-state probability of n in the system at an arbitrary point in time. We begin by considering a specific realization of the actual process over a long interval of time T and let $X(T)$ be the system size at time T. If $A_n(t)$ denotes the number of unit upward jumps or crossings (arrivals) from state n occurring in $(0, t)$, and $D_n(t)$ the number of unit downward jumps (departures) to state n in $(0, t)$, then since arrivals occur singly and customers are served singly, we must have

$$|A_n(T) - D_n(T)| \leq 1. \tag{5.28}$$

Furthermore, it must be true that the total number of departures, $D(T)$, relates to the total number of arrivals, $A(T)$, by

$$D(T) = A(T) + X(0) - X(T). \tag{5.29}$$

Now, the departure-point probabilities are found as

$$\pi_n = \lim_{T \to \infty} \frac{D_n(T)}{D(T)}. \tag{5.30}$$

When we add and subtract $A_n(T)$ from the numerator of (5.30) and use (5.29) in its denominator, we see that

$$\frac{D_n(T)}{D(T)} = \frac{A_n(T) + D_n(T) - A_n(T)}{A(T) + X(0) - X(T)}. \tag{5.31a}$$

Since $X(0)$ is finite and $X(T)$ must be too because of the assumption of stationarity, it follows from (5.28), (5.31a), and the fact that $A(T) \to \infty$ that

$$\lim_{T \to \infty} \frac{D_n(T)}{D(T)} = \lim_{T \to \infty} \frac{A_n(T)}{A(T)} \tag{5.31b}$$

with probability one. Since the arrivals occur at the points of a Poisson process operating independently of the state of the process, we invoke the PASTA property that Poisson arrivals find time averages. Therefore the general-time probability p_n is identical to the arrival-point probability $q_n = \lim_{T \to \infty}[A_n(T)/A(T)]$, which is, in turn, equal to the departure-point probability from (5.31b). Thus all three sets of probabilities are equal for the $M/G/1$ problem, and the desired result is shown.

5.1.4 Ergodic Theory

To find conditions for the existence of the steady state for the $M/G/1$ imbedded chain, we first obtain a sufficient condition from the theory of Markov chains in view of the Markovian nature of the departure-point process and the above-shown equality of departure-point and general-time probabilities. Then we show that this sufficient condition is also necessary by the direct use of the generating function $\Pi(z)$ given by Equation (5.18).

The Markov-chain proof relies heavily on two well-known theorems presented in Section 1.9 as Theorems 1.1 and 1.2. The former says that a discrete-time Markov chain has identical limiting and stationary distributions if it is irreducible, aperiodic, and positive recurrent, while the latter gives a sufficient condition for the positive recurrence of an irreducible and aperiodic chain.

The behavior of any single-channel queueing system is, at least, a function of the input parameters and the service-time distribution, say $B(t)$. Hence we would expect, as in $M/M/1$, that the existence of ergodicity depends upon the value of the utilization factor $\rho =$ (mean service time)/(mean interarrival time). The transition matrix P which characterizes the imbedded chain of the $M/G/1$ is given by Equation (5.14), remembering that

$$k_n = \Pr\{n \text{ arrivals during a service time } S = t\}$$

$$= \int_0^\infty \frac{e^{-\lambda t}(\lambda t)^n}{n!}\, dB(t),$$

with $\int t\, dB(t) = E[S]$. The problem then is to determine a fairly simple sufficient condition under which the $M/G/1$ has a steady state, and our past experience suggests that we try $\rho = \lambda E[S] < 1$.

We show in the following that the $M/G/1$ imbedded chain is indeed irreducible, aperiodic, and positive recurrent when $\rho < 1$, and hence, by Theorem 1.1, possesses a long-run distribution under this condition. The chain is clearly irreducible, since any state can be reached from any other state. It can next be observed directly from the transition matrix that $\Pr\{$passing from state k to state k in one transition$\} = p_{kk}^{(1)} > 0$ for all k, and therefore the period of each k, defined as the greatest common divisor of the set of all n such that $p_{kk}^{(n)} > 0$, is one. Hence the system is aperiodic. Notice that neither irreducibility nor aperiodicity depends upon ρ, but, rather, each is an inherent property of this chain. In fact, it is the positive recurrence which depends upon the value of ρ. It thus remains for us to show that the chain is positive recurrent when $\rho < 1$, and we could then apply the theorem to infer the existence of the steady-state distribution.

To obtain the required result, we employ Theorem 1.2 of Section 1.9.6 and show, using Foster's method (Foster, 1953), that the necessary inequality has the required solution when $\rho < 1$. An educated guess at this required solution is

$$x_j = \frac{j}{1-\rho} \qquad (j \geq 0).$$

Note that $\Sigma j p_{ij}$ is the mean "to" state starting from i.

Then, from the matrix P,

$$\sum_{j=0}^{\infty} p_{ij} x_j = \sum_{j=i-1}^{\infty} k_{j-i+1}\left(\frac{j}{1-\rho}\right)$$

$$= \frac{k_0(i-1)}{1-\rho} + \frac{k_1 i}{1-\rho} + \frac{k_2(i+1)}{1-\rho} + \cdots$$

$$= \frac{k_0(i-1)}{1-\rho} + \frac{k_1(i-1)}{1-\rho} + \frac{k_2(i-1)}{1-\rho} + \cdots + \frac{k_1}{1-\rho} + \frac{2k_2}{1-\rho} + \cdots$$

$$= (i-1)\sum_{j=0}^{\infty} \frac{k_j}{1-\rho} + \sum_{j=1}^{\infty} \frac{jk_j}{1-\rho}$$

$$= \frac{i-1}{1-\rho} + \sum_{j=1}^{\infty} \frac{jk_j}{1-\rho}.$$

But it should be noted that

$$\sum_{j=1}^{\infty} jk_j = \sum_{j=1}^{\infty} j\frac{1}{j!}\int_0^{\infty} (\lambda t)^j e^{-\lambda t}\, dB(t)$$

$$= \int_0^{\infty} e^{-\lambda t}\left(\sum_{j=1}^{\infty} \frac{1}{(j-1)!}(\lambda t)^j\right) dB(t)$$

$$= \int_0^{\infty} e^{-\lambda t}\lambda t e^{\lambda t}\, dB(t)$$

$$= \int_0^{\infty} \lambda t\, dB(t) = \lambda E[S] = \rho. \qquad (5.32)$$

So

$$\sum_{j=0}^{\infty} p_{ij}x_j = \frac{i-1+\rho}{1-\rho} = x_i - 1 \qquad (x_i \geq 0, \quad \text{since } 1-\rho > 0).$$

(Note that this implies that the expected single-step displacement of the chain from state $i > 0$ is $\rho - 1 < 0$.) Also

$$\sum_{j=0}^{\infty} p_{0j}x_j = \sum_{j=0}^{\infty} k_j x_j = \sum_{j=0}^{\infty} \frac{jk_j}{1-\rho} = \frac{\rho}{1-\rho} < \infty.$$

Hence it follows that the chain possesses identical stationary and long-run distributions when $\rho < 1$.

The proof of the necessity of $\rho < 1$ for ergodicity arises directly from the existence of the generating function $\Pi(z)$ over the interval $|z| < 1$,

$$\Pi(z) = \frac{\pi_0(1-z)K(z)}{K(z)-z}.$$

We know that $\Pi(1)$ must be equal to one; hence

$$1 = \lim_{z \to 1} \Pi(z)$$

$$= \pi_0 \frac{-K(1)}{K'(1)-1} \qquad \text{(using L'Hôpital's rule)}$$

$$= \pi_0 \frac{-1}{\rho-1} \qquad (\rho = \lambda E[S]).$$

But $\pi_0 > 0$ and therefore $\rho - 1 < 0$; thus $\rho < 1$ is necessary and sufficient for steady state.

5.1.5   Waiting Times

In this section we wish to present an assortment of important results concerning the delay times. It has already been shown that the average system wait is related to the average system size by Little's formula $W = L/\lambda$. A natural thing to require then is a possible relationship either between higher moments or between distribution functions [or equivalently between Laplace–Stieltjes transforms (LSTs)]. It turns out that this can be done with some extra effort.

To begin, we note that the stationary probability for the $M/G/1$ can always be written in terms of the waiting-time CDF as

$$p_n = \pi_n = \frac{1}{n!} \int_0^\infty (\lambda t)^n e^{-\lambda t}\, dW(t) \qquad (n \geq 0),$$

since the system size under FCFS will equal n at an arbitrary departure point if there have been n (Poisson) arrivals during the departure's system wait. If we multiply this equation by z^n, sum on n, and define the usual generating function, then it is found that

$$P(z) = \sum_{n=0}^\infty p_n z^n = \int_0^\infty e^{-\lambda t} \sum_{n=0}^\infty \frac{(\lambda t z)^n}{n!}\, dW(t)$$

$$= \int_0^\infty e^{-\lambda t(1-z)}\, dW(t) = W^*[\lambda(1-z)], \qquad (5.33)$$

where $W^*(s)$ is the LST of $W(t)$. The succession of moments of system size and delay can now be easily related to each other by repeated differentiation of the equality $P(z) = W^*[\lambda(1-z)]$. We therefore have by the chain rule that

$$\frac{d^k P(z)}{dz^k} = (-1)^k \lambda^k \frac{d^k W^*(u)}{du^k}\bigg|_{u=\lambda(1-z)}$$

$$= (-1)^k \lambda^k (-1)^k E[T^k e^{-Tu}]\big|_{u=\lambda(1-z)}.$$

Hence if $L_{(k)}$ is used to denote the kth factorial moment of the system size and W_k the regular kth moment of the system waiting time, then

$$L_{(k)} = \frac{d^k P(z)}{dz^k}\bigg|_{z=1} = \lambda^k W_k. \qquad (5.34)$$

This result provides a nice generalization of Little's formula, since the higher ordinary moments can be obtained from the factorial moments.

In the *M/M/*1 queue we were able to easily obtain a simple formula for the waiting-time distribution in terms of the service-time distribution (see Section 2.2.4 and Problem 2.8), namely,

$$W(t) = (1-\rho) \sum_{n=0}^\infty \rho^n B^{(n+1)}(t), \qquad (5.35)$$

where $B(t)$ is the exponential CDF and $B^{(n+1)}(t)$ its $(n+1)$st convolution. The derivation of this result absolutely required the memoryless property of the exponential service, since the arrivals catch the server in the midst of a serving period with probability equal to ρ. However, we have now lost the memoryless property and are therefore going to require an alternative approach to derive a comparable result for $M/G/1$.

To do so, we begin by deriving a simple relationship between the LSTs of the service and the waiting times, $B^*(s)$ and $W^*(s)$, respectively. From (5.33) we know that $P(z) = W^*[\lambda(1 - z)]$, and from (5.20) that $P(z) = (1 - \rho)(1 - z)K(z)/[K(z) - z]$. But, from (5.17) and (5.13),

$$K(z) = \int_0^\infty e^{-\lambda t} \sum_{n=0}^\infty \frac{(\lambda t z)^n}{n!} \, dB(t)$$

$$= \int_0^\infty e^{-\lambda t(1 - z)} \, dB(t) = B^*[\lambda(1 - z)]. \qquad (5.36)$$

Putting these three equations together, we find that

$$W^*[\lambda(1 - z)] = \frac{(1 - \rho)(1 - z)B^*[\lambda(1 - z)]}{B^*[\lambda(1 - z)] - z},$$

or

$$W^*(s) = \frac{(1 - \rho)sB^*(s)}{s - \lambda[1 - B^*(s)]}. \qquad (5.37a)$$

But, from the convolution property of transforms, $W^*(s) = W_q^*(s)B^*(s)$, since $T = T_q + S$. Thus

$$W_q^*(s) = \frac{(1 - \rho)s}{s - \lambda[1 - B^*(s)]}. \qquad (5.37b)$$

Expanding the right-hand side as a geometric series (since $(\lambda/s)[1 - B^*(s)] < 1$), we have

$$W_q^*(s) = (1 - \rho) \sum_{n=0}^\infty \left(\frac{\lambda}{s} [1 - B^*(s)] \right)^n$$

$$= (1 - \rho) \sum_{n=0}^\infty \left(\frac{\rho\mu}{s} [1 - B^*(s)] \right)^n.$$

But it can be seen that $\mu[1 - B^*(s)]/s$ is the LST of the residual-service-time CDF

$$R(t) = \mu \int_0^t [1 - B(x)] \, dx.$$

Therefore, it follows that

$$W_q^*(s) = (1 - \rho) \sum_{n=0}^{\infty} [\rho R^*(s)]^n,$$

which yields, after term-by-term inversion utilizing the convolution property, a result suprisingly similar to (5.35), namely,

$$W_q(t) = (1 - \rho) \sum_{n=0}^{\infty} \rho^n R^{(n)}(t). \tag{5.38}$$

Let us consider this result a little more carefully to see exactly what it is saying. First of all, what is the actual meaning of the function $R(t)$? It can, in fact, be shown using renewal theory (see Ross, 1996, for example) that $R(t)$ is nothing more than the CDF of the remaining service time of the customer being served at the instant the new customer arrives. Then Equation (5.38) says that if time is reordered with this remaining service time as the fundamental unit, any arrival in the steady state finds n such time units of potential service in front of it with probability $(1 - \rho)\rho^n$, giving a result remarkably like that for the *M/M/*1 queue. We can also use these results to derive an extension of the PK formula to relate iteratively the higher moments of the wait (call them $W_{q,k}$). First, rewrite the basic transform equation as

$$W_q^*(s)\{s - \lambda[1 - B^*(s)]\} = (1 - \rho)s.$$

Now apply Leibniz' rule for the derivative of a product to get

$$\sum_{i=0}^{k} \binom{k}{i} \frac{d^{k-i}W_q^*(s)}{ds^{k-i}} \frac{d^i\{s - \lambda[1 - B^*(s)]\}}{ds^i} = \frac{d^k[(1 - \rho)s]}{ds^k}.$$

We may assume $k > 1$, since it is the higher moments we seek. Then the right-hand side of the above equation is zero, and thus

$$0 = \frac{d^k W_q^*(s)}{ds^k}\{s - \lambda[1 - B^*(s)]\} + k\frac{d^{k-1}W_q^*(s)}{ds^{k-1}}[1 + \lambda B^{*\prime}(s)]$$

$$+ \sum_{i=2}^{k} \binom{k}{i} \frac{d^{k-i}W_q^*(s)}{ds^{k-i}} \lambda \frac{d^i B^*(s)}{ds^i}.$$

Now set $s = 0$, thus giving

$$0 = k(-1)^{k-1}W_{q,k-1}(1-\rho) + \sum_{i=2}^{k} \lambda \binom{k}{i}(-1)^{k-i}W_{q,k-i}E[S^i](-1)^i,$$

or

$$W_{q,k-1} = \frac{\lambda}{k(1-\rho)} \sum_{i=2}^{k} \binom{k}{i} W_{q,k-i}E[S^i].$$

We may rewrite this slightly by letting $K = k - 1$ and $j = i - 1$, from which we get the more familiar form,

$$W_{q,K} = \frac{\lambda}{1-\rho} \sum_{j=1}^{K} \binom{K}{j} W_{q,K-j} \frac{E[S^{j+1}]}{j+1}.$$

Example 5.4
Let us illustrate some of these results by going back to the Bearing Straight Corporation of Example 5.2. Bearing Straight has determined that it loses $5000 per hour that a machine is down, and that an additional penalty must be incurred because of the possibility of an excessive number of machines being down. It is decided to cost this variability at $10,000 \times$ (standard deviation of customer delay). Under such a total-cost structure, what is the total cost of their policy, using the parameters indicated in Example 5.2 and assuming that repair labor is a sunk cost?

Problem 5.10 asks us to show that $L \doteq 2.96$. But we are also going to need the variance of the system waits T, where we know using (5.34) that

$$E[N(N-1)] \equiv L_{(2)} = \lambda^2 W_2.$$

Thus

$$\text{Var}[T] = W_2 - W^2 = \frac{L_{(2)}}{\lambda^2} - \frac{L^2}{\lambda^2}.$$

To get $L_{(2)}$, the second derivative of $P(z)$ is found from (5.22) and then evaluated at $z = 1$ to be 14.50 (see Problem 5.16). Therefore $\text{Var}[T] = (14.50 - 8.75)/25 = 0.23 \text{ hr}^2$. Thus the total cost of Bearing's policy computes as $C = 5000L + 10,000\sqrt{0.23} \doteq \$19,596/\text{hr}$.

5.1.6 Busy-Period Analysis

The determination of the distribution of the busy period for an $M/G/1$ queue is a somewhat more difficult matter than finding that of the $M/M/1$, particularly in view of the fact that the service-time CDF must be carried as an unknown. But it is not too much of a task to find the LST of the busy-period CDF, from which we can easily obtain any number of moments.

To begin, let $G(x)$ denote the CDF of the busy period X of an $M/G/1$ with service CDF $B(t)$. Then we condition X on the length of the first service time inaugurating the busy period. Since each arrival during that service time will contribute to the busy period by having arrivals come during its service time, we can look at each arrival during the first service time of the busy period as essentially generating its own busy period. Thus we can write

$$G(x) = \int_0^x \Pr\{(\text{busy period generated by all arrivals during } t) \le x - t \mid \text{first service time} = t\} \, dB(t)$$

$$= \int_0^x \sum_{n=0}^{\infty} \frac{e^{-\lambda t}(\lambda t)^n}{n!} G^{(n)}(x - t) \, dB(t),$$

(5.39)

where $G^{(n)}(x)$ is the n-fold convolution of $G(x)$. Next let

$$G^*(s) = \int_0^{\infty} e^{-sx} \, dG(x)$$

be the LST of $G(x)$, and $B^*(s)$ be the LST of $B(t)$.

Then, by taking the transform of both sides of (5.39), it is found that

$$G^*(s) = \int_0^{\infty} \int_0^x \sum_{n=0}^{\infty} e^{-sx} \frac{e^{-\lambda t}(\lambda t)^n}{n!} G^{(n)}(x - t) \, dB(t) \, dx.$$

Changing the order of integration gives

$$G^*(s) = \int_0^{\infty} \sum_{n=0}^{\infty} \frac{e^{-\lambda t}(\lambda t)^n}{n!} \, dB(t) \int_t^{\infty} e^{-sx} G^{(n)}(x - t) \, dx,$$

and by the convolution property,

$$G^*(s) = \int_0^{\infty} \sum_{n=0}^{\infty} \frac{e^{-\lambda t}(\lambda t)^n}{n!} e^{-st} [G^*(s)]^n \, dB(t)$$

$$= \int_0^{\infty} e^{-\lambda t} e^{\lambda t G^*(s)} e^{-st} \, dB(t)$$

$$= B^*[s + \lambda - \lambda G^*(s)].$$

Hence the mean length of the busy period is found as

$$E[X] = -\left. \frac{dG^*(s)}{ds} \right|_{s=0} \equiv -G^{*\prime}(0),$$

where

$$G^{*\prime}(s) = B^{*\prime}[s + \lambda - \lambda G^*(s)][1 - \lambda G^{*\prime}(s)].$$

Therefore

$$E[X] = -B^{*\prime}[\lambda - \lambda G^*(0)][1 - \lambda G^{*\prime}(0)] = -B^{*\prime}(0) \cdot \{1 + \lambda E[X]\},$$

or

$$E[X] = -\frac{B^{*\prime}(0)}{1 + \lambda B^{*\prime}(0)}.$$

But $B^{*\prime}(0) = -1/\mu$; hence

$$E[X] = \frac{1/\mu}{1 - \lambda/\mu} = \frac{1}{\mu - \lambda},$$

which, surprisingly, is exactly the same result we obtained earlier for $M/M/1$. However, we note that the proof given for $M/M/1$ in Section 2.11 is perfectly valid for the $M/G/1$ model, since no use is made of the exponentiality of service.

5.1.7 Finite $M/G/1$ Queues

The analysis of the finite-capacity $M/G/1/K$ queue proceeds in a way very similar to that of the unlimited-waiting-room case. Let us thus examine each of the main results of $M/G/1/\infty$ for applicability to $M/G/1/K$.

The PK formula will not hold now, since the expected number of (joined) arrivals during a service period must be conditioned on the system size. The best way to get the new result is directly from the steady-state probabilities, since there are now only a finite number of them.

The single-step transition matrix must here be truncated at $K - 1$, so that

$$
P = \{p_{ij}\} = \begin{pmatrix}
k_0 & k_1 & k_2 & \cdots & 1 - \displaystyle\sum_{n=0}^{K-2} k_n \\[1.5em]
k_0 & k_1 & k_2 & \cdots & 1 - \displaystyle\sum_{n=0}^{K-2} k_n \\[1.5em]
0 & k_0 & k_1 & \cdots & 1 - \displaystyle\sum_{n=0}^{K-3} k_n \\[1.5em]
0 & 0 & k_0 & \cdots & 1 - \displaystyle\sum_{n=0}^{K-4} k_n \\[1.5em]
\vdots & \vdots & \vdots & & \vdots \\[1em]
0 & 0 & 0 & \cdots & 1 - k_0
\end{pmatrix},
$$

which implies that the stationary equation is

$$
\pi_i = \begin{cases}
\pi_0 k_i + \displaystyle\sum_{j=1}^{i+1} \pi_j k_{i-j+1} & (i = 0, 1, 2, \ldots, K - 2), \\[1.5em]
1 - \displaystyle\sum_{j=0}^{K-2} \pi_j & (i = K - 1).
\end{cases}
$$

These K (consistent) equations in K unknowns can then be solved for all the probabilities, and the average system size at points of departure is thus given by $L = \sum_{i=0}^{K-1} i\pi_i$. (Note that the maximum state of the Markov chain is not K, since we are observing just after a departure. We assume $K > 1$ because otherwise the resultant model $M/G/1/1$ is just a special case of the $M/G/c/c$ to be discussed shortly in Section 5.2.2.)

We also notice that the first portion of the stationary equation is identical to that of the unlimited $M/G/1$, and therefore deduce that the respective stationary probabilities, say $\{\pi_i\}$ for $M/G/1/K$ and $\{\pi_i^*\}$ for $M/G/1/\infty$, must be at worst proportional for $i \le K - 1$; that is, $\pi_i = C\pi_i^*$, $i = 0, 1, \ldots,$ $K - 1$. We find C by the usual use of the summability-to-one condition and get $C = 1/\sum_{i=0}^{K-1} \pi_i^*$.

Furthermore, we note that the probability distribution for the system size encountered by an arrival will be different from $\{\pi_i\}$, since now the state space must be enlarged to include K. Let q_n' then denote the probability that an arriving customer finds a system with n customers. (Here we are speaking about the distribution of arriving customers whether or not they join the queue, as opposed to only those arrivals who join, denoted by q_n. The q_n' distribution is often of interest in its own right.) Now, if we go back to Section 5.1.3 and the proof that $\pi_n = p_n$ for $M/G/1$, it is noted in that proof, essentially Equation (5.31), that the distribution of system sizes just prior to points of arrival (our $\{q_n\}$) is identical to the departure-point probabilities (here $\{\pi_n\}$) as long as arrivals occur singly and service is not in bulk. Such is also the case with q_n', except that the state spaces are different. This difference is easily taken care of by first noting that Equation (5.31) is really saying that

$$\pi_n = \text{Pr}\{\text{arrival finds } n \mid \text{customer does in fact join}\}$$

$$= q_n = \frac{q_n'}{1 - q_K'} \qquad (0 \le n \le K - 1).$$

Therefore

$$q_n' = (1 - q_K')\pi_n \qquad (0 \le n \le K - 1).$$

To get q_K' now, we use an approach similar to one mentioned earlier for simple Markovian models in Section 2.4, where we equate the effective arrival rate with the effective departure rate; that is,

$$\lambda(1 - q_K') = \mu(1 - p_0).$$

Therefore

$$q_n' = \frac{(1 - p_0)\pi_n}{\rho} \qquad (0 \le n \le K - 1),$$

$$q_K' = \frac{\rho - 1 + p_0}{\rho}.$$

But since the original arrival process is Poisson, $q_n' = p_n$ for all n. Thus

$$q_0' = p_0 = \frac{(1 - p_0)\pi_0}{\rho} \implies p_0 = \frac{\pi_0}{\pi_0 + \rho},$$

and finally

$$q_n' = \frac{\pi_n}{\pi_0 + \rho}.$$

5.1.8   Some Additional Results

In this subsection we present some assorted additional results for $M/G/1$ queues. First, a brief discussion of customer impatience is presented, followed in order by similarly brief discussions of the problems of priorities, output, and transience, with mention made of finite source and batching, all in the context of the *M/G/*1.

With respect to impatience, one can easily introduce balking into *M/G/*1 by prescribing a probability b that an arrival decides to actually join the system. Then the true input process becomes a (filtered) Poisson with mean $b\lambda t$, and Equation (5.13) thus has to be written as

$$k_n = \int_0^\infty \frac{e^{-b\lambda t}(b\lambda t)^n}{n!} \, dB(t).$$

But the rest of the analysis goes through parallel to that for the regular *M/G/*1, with the probability of idleness, p_0, now equal to $1 - b\lambda/\mu$. For a more comprehensive treatment of impatience in *M/G/*1, the reader is referred to Rao (1968), who treats both balking and reneging.

Next, we look at the problem of priorities in the *M/G/*1. It turns out that the nonpreemptive, many-priority model of Section 3.4.2 can be extended very nicely to situations with nonexponential service. The definitions and derivation are exactly the same, with the single exception of the value of $E[S_0]$, the expected time required to finish the item in service at the time of an arrival, the value of which must change in view of the loss of memoryless-ness by the service. But we have, in fact, actually done this calculation for *M/G/*1 as Problem 5.5 and found for S equal to the service time that

$$E[S_0] = \lambda E[S^2]/2.$$

Therefore we may write that the expected delay in line of the ith of r priorities is given by

$$W_q^{(i)} = \frac{\lambda E[S^2]/2}{(1 - \sigma_{i-1})(1 - \sigma_i)},$$

where we recall that

$$\sigma_i = \sum_{k=1}^i \rho_k \qquad (\rho_k = \lambda_k/\mu_k).$$

As far as output is concerned, we have already shown in Chapter 4 that the steady-state *M/M/*1 has Poisson output, but would now like to know whether there are any other *M/G/*1/∞ queues which also possess this property. The answer is in fact no (except for the pathological case where service is 0 for all customers), and this follows from the simple observation that such queues are never reversible, so that their output processes cannot probabilistically match their inputs.

But what is the distribution of an arbitrary *M/G/*1 interdeparture time in the steady state? To derive this, define $C(t)$ as the CDF of the interdeparture times; then, for $B(t)$ equal to the service CDF, it follows that

$C(t) \equiv \Pr\{\text{interdeparture time} \leq t\}$

$\quad = \Pr\{\text{system experienced no idleness during interdeparture period)}$
$\quad\quad \times \Pr\{\text{interdeparture time} \leq t \mid \text{no idleness}\}$
$\quad\quad + \Pr\{\text{system experienced some idleness during interdeparture period}\}$
$\quad\quad \times \Pr\{\text{interdeparture time} \leq t \mid \text{some idleness}\}$

$$= \rho B(t) + (1 - \rho) \int_0^t B(t - u)\lambda e^{-\lambda u}\, du,$$

since the length of an interdeparture period with idleness is the sum of the idle time and service time. Problem 5.22(b) asks to show from the above equation that the exponentiality of $C(t)$ implies the exponentiality of $B(t)$.

Remember that the fact that the $M/M/1$ is the only $M/G/1$ with exponential output has serious negative implications for the solution of series models since we have shown that the output of a first stage will be exponential, which we would like it to be, only if it is $M/M/1$. However, small $M/G/1$ series problems can be handled numerically by appropriate utilization of the above formula for the CDF of the interdeparture times, $C(t)$.

By putting a capacity restriction on the $M/G/1$ at $K = 1$, it can be seen that such queues also have IID interdeparture times. This is so because the successive departure epochs are identical to the busy cycles, which are found as the sums of each idle time paired with an adjacent service time.

To get any transient results for $M/G/1$, we would have to appeal directly to the theory of Markov chains and the CK equation

$$p_j^{(m)} = \sum_k p_k^{(0)} p_{kj}^{(m)},$$

where $p_j^{(m)}$ is then the probability that the system state is in state j just after the mth customer has departed. The necessary matrix multiplications must be done with some caution, since we are dealing with an $\infty \times \infty$ matrix in the unlimited-waiting-room case. But this can indeed be done by carefully truncating the transition matrix at an appropriate point where the entries have become very small (see, e.g., Neuts, 1973).

To close this section, we make brief mention of two additional problem types, namely, finite-source and bulk queues. The finite-source $M/G/1$ is essentially the machine-repair problem with arbitrarily distributed repair times and has been solved in the literature, again using an imbedded-Markov-chain approach. Any interested reader is referred to Takács (1962) for a fairly detailed discussion of these kinds of problems. The bulk-input $M/G/1$, denoted by $M^{[X]}/G/1$, and the bulk-service $M/G/1$, denoted by $M/G^{[Y]}/1$, can also be solved with the use of Markov chains. The bulk-input model is presented in the next section, but the bulk-service problem is quite a bit more messy and therefore is not treated in this book. However, the reader is referred to Prabhu (1965b) for the details of this latter model.

5.1.9   The Bulk-Input Queue ($M^{[X]}/G/1$)

The $M^{[X]}/G/1$ queueing system can be described in the following manner.

1. Customers arrive as a Poisson process with parameter λ in groups of random size C, where C has the distribution

$$\Pr\{C = n\} = c_n \qquad (n > 1)$$

and the generating function (which will be assumed to exist) is

$$C(z) = \mathrm{E}[z^C] = \sum_{n=1}^{\infty} c_n z^n \qquad (|z| \le 1).$$

The probability that a total of n customers arrive in an interval of length t is thus given by

$$p_n(t) = \sum_{k=0}^{n} e^{-\lambda t} \frac{(\lambda t)^k}{k!} c_n^{(k)} \qquad (n \ge 0),$$

where $\{c_n^{(k)}\}$ is the k-fold convolution of $\{c_n\}$ with itself (that is, the n arrivals form a compound Poisson process),

$$c_n^{(0)} \equiv \begin{cases} 1 & (n = 0), \\ 0 & (n > 0). \end{cases}$$

2. The customers are served singly by one server on a FCFS basis.
3. The service times of the succession of customers are IID random variables with CDF $B(t)$ and LST $B^*(s)$.

Let us make a slight change here from $M/G/1$ which will help us later: the imbedded chain we shall use will be generated by the points (therefore called regeneration points) at which either a departure occurs *or* an idle period is ended. This process will be called $\{X_n, n = 1, 2, \ldots | X_n = \text{number of customers in the system immediately after the } n\text{th regeneration point}\}$, with transition matrix given by

$$
\begin{pmatrix}
0 & c_1 & c_2 & \cdots \\
k_0 & k_1 & k_2 & \cdots \\
0 & k_0 & k_1 & \cdots \\
0 & 0 & k_0 & \cdots \\
\vdots & \vdots & \vdots &
\end{pmatrix},
$$

where

$k_n = \Pr\{n \text{ arrivals during a full service period}\}$

$$
= \int_0^\infty p_n(t)\, dB(t) = \int_0^\infty \sum_{k=0}^n e^{-\lambda t} \frac{(\lambda t)^k}{k!} c_n^{(k)}\, dB(t) = p_{i,n+i-1} \qquad (i > 0).
$$

The application of Theorem 1.2, Section 1.9.6 in a fashion similar to that of Section 5.1.4 for $M/G/1$ shows that this chain is ergodic and hence possesses identical long-run and stationary distributions when

$$
\rho \equiv \sum_{n=1}^\infty n k_n = \mathrm{E}[\text{arrivals during a service time}] = \frac{\lambda \mathrm{E}[C]}{\mu} < 1.
$$

If the steady-state distribution $\{\pi_i\}$ is to exist for the chain, then it is the solution of the system (for all $j \geq 0$)

$$
\sum_{i=0}^\infty p_{ij} \pi_i = \pi_j \quad \text{and} \quad \sum_{i=0}^\infty \pi_i = 1.
$$

(Keep in mind that these $\{\pi_i\}$ are slightly different from those that would have resulted had we restricted ourselves to departure points only, as we did for $M/G/1$.) From the transition matrix (as in Section 5.1.2),

$$
\pi_j = c_j \pi_0 + \sum_{i=1}^{j+1} k_{j-i+1} \pi_i \qquad (c_0 \equiv 0),
$$

If the above stationary equation is multiplied by z^j and then summed on j, it is found that

$$
\sum_{j=0}^\infty \pi_j z^j = \sum_{j=0}^\infty c_j \pi_0 z^j + \sum_{j=0}^\infty \sum_{i=1}^{j+1} k_{j-i+1} \pi_i z^j .
$$

If the usual generating functions are defined as

$$\Pi(z) = \sum_{j=0}^{\infty} \pi_j z^j \quad \text{and} \quad K(z) = \sum_{j=0}^{\infty} k_j z^j,$$

then we find after reversing the order of summation in the final term that

$$\Pi(z) = \pi_0 C(z) + \frac{K(z)}{z} \sum_{i=1}^{\infty} \pi_i z^i$$

$$= \pi_0 C(z) + \frac{K(z)}{z} [\Pi(z) - \pi_0],$$

or

$$\Pi(z) = \frac{\pi_0 [K(z) - z C(z)]}{K(z) - z}.$$

Furthermore, it can be shown for $|z| \le 1$ that

$$K(z) = \sum_{j=0}^{\infty} \int_0^{\infty} \sum_{k=0}^{j} e^{-\lambda t} \frac{(\lambda t)^k}{k!} c_j^{(k)} \, dB(t) z^j$$

$$= \int_0^{\infty} e^{-\lambda t} \sum_{k=0}^{j} \frac{(\lambda t)^k}{k!} [C(z)]^k \, dB(t)$$

$$= \int_0^{\infty} e^{-\lambda t + \lambda t C(z)} \, dB(t) = B^*[\lambda - \lambda C(z)].$$

These results have all been derived without much difficulty in a manner similar to the approach for *M/G/*1. However, there now exists a problem that we have not faced before, namely, that the results derived for the imbedded Markov chain do not directly apply to the total general-time stochastic process, $\{X(t), t \ge 0 | X(t) = \text{number in the system at time } t\}$. In order to relate the general-time steady-state probabilities, say $\{p_n\}$, to $\{\pi_n\}$, we must appeal to some results from semi-Markov processes, which are presented in Chapter 6, Section 6.3; a reference on this subject for anyone with further interest in this material is Heyman and Sobel (1982).

5.1.10 Departure-Point State Dependence, Decomposition, and Server Vacations

In Section 2.8 we treated queues with state dependences such that the mean service rate was a function of the number of customers in the system. Whenever the number in the system changed (arrival or departure), the mean service rate would itself change accordingly. But in many situations, it might not be possible to change the service rate at any time a new arrival may come, but rather only upon the initiation of a service (or, almost equivalently, at the conclusion of a service time). For example, in many cases where the service is a man–machine operation and the machine is capable of running at various speeds, the operator would set the speed only prior to the actual commencement of service. Once service had begun, the speed could not be changed until that service was completed, for to do otherwise would necessitate stopping work to alter the speed setting, and then restarting and/or repositioning the work. Further, stopping the operation prior to completion might damage the unit. This type of situation, where the mean service rate can be adjusted only prior to commencing service or at a customer departure point, is what we refer to as departure-point state dependence and is considered in this section.

We assume the state-dependent service mechanism is as follows. Let $B_i(t)$ be the service-time CDF of a customer who enters service after the most recent departure left i customers behind, and

$$k_{ni} = \Pr\{n \text{ arrivals during a service time} \mid$$
$$i \text{ in the system at latest departure}\}$$

$$= \int_0^\infty e^{-\lambda t} \frac{(\lambda t)^n}{n!} \, dB_i(t). \tag{5.40}$$

Then the transition matrix is given as

$$\boldsymbol{P} = \{p_{ij}\} = \begin{pmatrix} k_{00} & k_{10} & k_{20} & k_{30} & \cdots \\ k_{01} & k_{11} & k_{21} & k_{31} & \cdots \\ 0 & k_{02} & k_{12} & k_{22} & \cdots \\ 0 & 0 & k_{03} & k_{13} & \cdots \\ \vdots & \vdots & \vdots & \vdots & \end{pmatrix}. \tag{5.41}$$

A sufficient condition for the existence of a steady-state solution (see Crabill, 1968) is

$$\lim \sup\{\rho_j \equiv \lambda \mathrm{E}[S_j] < 1\}, \tag{5.42}$$

which says that all but a finite number of the ρ_j must be less than 1. Thus, assuming this condition is met, we can find the steady-state probability distribution by solving the usual stationary equation $\boldsymbol{\pi P} = \boldsymbol{\pi}$. Although this gives the departure-point state probabilities, we have shown in Section 5.1.3 for non-state-dependent service that these are equivalent to the general-time probabilities. The addition of state dependence of service does not alter the proof in any way, so that here, also, the departure-point and general-time state probabilities are equivalent.

Thus the use of the stationary equation (5.15) results in

$$p_j = \pi_j = \pi_0 k_{j,0} + \pi_1 k_{j,1} + \pi_2 k_{j-1,2} + \pi_3 k_{j-2,3} + \cdots + \pi_{j+1} k_{0,j+1} \quad (j \ge 0).$$
$$(5.43)$$

Again, we define the generating functions

$$\Pi(z) = \sum_{j=0}^{\infty} \pi_j z^j \quad \text{and} \quad K_i(z) = \sum_{j=0}^{\infty} k_{ji} z^j.$$

Then multiplying both sides of (5.43) by z^j and summing over all j, we get

$$\Pi(z) = \pi_0 K_0(z) + \pi_1 K_1(z) + \pi_2 z K_2(z) + \cdots + \pi_{j+1} z^j K_{j+1}(z) + \cdots.$$
$$(5.44)$$

This is as far as we are able to proceed in general. No closed-form expression for $\Pi(z)$ in terms of the $K_i(z)$ is obtainable, so we now present a specific case for $B_i(t)$ to illustrate the typical solution procedure and give another specific case as an exercise (see Problem 5.21).

Consider the case where customers beginning a busy period get exceptional service at a rate μ_0 unequal to the rate μ offered to everybody else. Thus we have

$$B_n(t) = \begin{cases} 1 - e^{-\mu_0 t} & (n = 0), \\ 1 - e^{-\mu t} & (n > 0). \end{cases}$$

From (5.40), it follows that

$$k_{n0} = \int_0^\infty \frac{e^{-\lambda t}(\lambda t)^n}{n!} \mu_0 e^{-\mu_0 t}\, dt = \frac{\mu_0 \lambda^n}{(\lambda + \mu_0)^{n+1}},$$
$$k_n = \int_0^\infty \frac{e^{-\lambda t}(\lambda t)^n}{n!} \mu e^{-\mu t}\, dt = \frac{\mu \lambda^n}{(\lambda + \mu)^{n+1}}. \qquad (5.45)$$

Now using (5.44), with $K(z) \equiv K_i(z)$ for $i > 0$, we thus have that

$$\Pi(z) = \pi_0 K_0(z) + \frac{K(z)}{z}[\Pi(z) - \pi_0],$$

so that

$$\Pi(z) = \frac{\pi_0 [z K_0(z) - K(z)]}{z - K(z)}. \qquad (5.46)$$

To get π_0, we take the limit of $\Pi(z)$ as $z \to 1$, recognizing that $K(1) = K_0(1) = 1$, $K_0'(1) = \rho_0 = \lambda/\mu_0$ and $K'(1) = \rho = \lambda/\mu$, and then use L'Hôpital's rule to get

$$1 = \frac{\pi_0(1 + \rho_0 - \rho)}{1 - \rho} \quad \Rightarrow \quad \pi_0 = \frac{1 - \rho}{1 + \rho_0 - \rho}.$$

The remaining probabilities may be found by iteration on the stationary equation (5.43), and we find after repeated calculations and verification by induction that [see Problem 5.21(b)]

$$\pi_n = \left(\frac{\rho_0^n}{(\rho_0 + 1)^n} + \sum_{k=0}^{n-1} \frac{\rho_0^{n-k} \rho^{k+1}}{(\rho_0 + 1)^{n-k}} \right) \pi_0.$$

Now, particularly interesting things happen to the above state-dependent-service model when the probability generating function $K_0(z)$ can be expressed as the product $K(z)D(z)$, where $D(z)$ is also a probability generating function defined over the nonnegative integers (see Harris and Marchal, 1988). Then it follows from (5.46) that

$$\Pi(z) = \frac{\pi_0 K(z)[1 - zD(z)]}{K(z) - z}.$$

What happens now is that an appropriate rewriting leads to a decomposition of $\Pi(z)$ itself into the product of two generating functions as

$$\Pi(z) = \frac{\pi_0(1 - z)K(z)}{K(z) - z} \times \frac{1 - zD(z)}{1 - z}. \tag{5.47}$$

We immediately notice that the first factor on the right-hand side is the precise probability generating function for the steady-state system size of the $M/G/1$, as given in (5.18) (though the exact value of the $M/G/1$'s π_0 will not be the same as that of the state-dependent model). Now, it turns out that the second factor can also be algebraically shown to be a legitimate probability generating function. Hence it follows from the product decomposition of $\Pi(z)$ in (5.47) that the queue's system sizes are the sum of two random variables defined on the nonnegative integers, the first coming from an ordinary $M/G/1$, while the second is associated with a random variable having generating function proportional to $[1 - zD(z)]/[1 - z]$.

We can connect the above development to a particular type of *server-vac-ation* queue. Suppose that, after each busy period is concluded, the server takes a vacation away from its work. If, upon return from vacation, the server finds that the system is still empty, it goes away again, etc. Therefore, it follows that k_{i0} is the probability that $i + 1$ arrivals occur during the service time combined with the final vacation, with probability generating function

$$K_0(z) = K(z) \times \frac{K_V(z)}{z}, \qquad (5.48)$$

where $K_V(z)$ is the generating function for the number of arrivals during an arbitrary vacation time. Thus the role of the $D(z)$ of (5.47) is played by the quotient $K_V(z)/z$, and the second factor on the right-hand side is now $[1 - K_V(z)]/[1 - z]$. By analogy with the LST of the residual service time of Section 5.1.5, as we show below, this latter quotient is proportional to the generating function of the number of arrivals coming in during a residual vacation time. So, taken all together, we are claiming that the system state of the vacation queue is the sum of the size of an ordinary, nonvacation $M/G/1$ and the number of arrivals coming in during a random residual vacation time.

Now we verify the claim that $[1 - K_V(z)]/[1 - z]$ is proportional to the generating function of the number of arrivals coming in during a residual vacation time, where $K_V(z)$ is the generating function of (5.17) based on a k_n value derived using the vacation distribution function $V(t)$ in place of the service-time CDF $B(t)$ in Equation (5.13). From Section 5.1.5, the residual distribution function associated with the vacation CDF $V(t)$ is given as

$$R_V(t) = \frac{1}{\nu} \int_0^t [1 - V(x)] \, dx,$$

where $\bar{\nu}$ is the mean vacation length. Then, from Section 5.1.5's results for the ordinary $M/G/1$, including Equation (5.36) and the material leading to (5.38), we see that

$$K_V(z) = V^*[\lambda(1 - z)] \quad \text{and} \quad R_V^*(s) = \frac{1 - V^*(s)}{\bar{\nu} s}.$$

Therefore, the probability generating function for the number of arrivals during a residual vacation time may be written as

$$K_{R-V}(z) = \frac{1 - V^*[\lambda(1 - z)]}{\bar{\nu}\lambda(1 - z)} = \frac{1 - K_V(z)}{\bar{\nu}\lambda(1 - z)}.$$

The result is thus shown.

Example 5.5

The Hugh Borum Company has a production process which involves drilling holes into castings. The interarrival times of the castings at the single drill press were found to be governed closely by an exponential distribution with mean 4 min, while there was such a variety of hole dimensions that service times also appeared to follow exponentiality, with mean 3 min. However, the very high speed at which the press operated necessitated a cooling-off period, which was done whenever the queue emptied, for a fixed amount of time equal to 2 min. Management wanted to know what penalty they paid in lengthening the average system size because of the breather that the equipment had to take.

Since we have shown in (5.47) that the generating function of the system size is the product of two distinct generating functions, it follows that the system size is the sum of two independent random variables, with the first coming from an ordinary $M/G/1$ ($G = M$ in this problem) and the second equal to the number of arrivals during a residual vacation time. Thus it follows that the mean system size is the sum of the $M/M/1$ mean $\rho/(1 - \rho) = 3$ and the mean number expected to arrive during a residual vacation time. Since the vacation time is the constant 2 min, it follows that the mean remaining vacation time at a random point is merely $2/2 = 1$ min. Thus the required answer for the Hugh Borum management is $\lambda \times 1 = 0.25$ customers, a value with which they are quite comfortable in view of the base average of 3 customers.

5.2   MULTISERVER QUEUES WITH POISSON INPUT AND GENERAL SERVICE

We begin here with an immediate disadvantage from the point of view of being able to derive necessary results, since $M/G/c/\infty$ and the $M/G/c/K$ loss-delay system do not possess imbedded Markov chains in the usual sense. This is so at departure points because the number of arrivals during any interdeparture period is dependent upon more than just the system size at the immediate departure predecessor, due to the presence of multiple servers. There are, however, some special $M/G/c$'s which do possess enough structure to get fairly complete results, including, of course, $M/M/c$. Another such example is the $M/D/c$, which will be taken up in detail in Chapter 6. What we wish then to do in this section is to try to get some general results for both $M/G/c/\infty$ and $M/G/c/c$ which can easily be applied by merely specifying G.

5.2.1 Some Results for $M/G/c/\infty$

For $M/G/c/\infty$, the main general result that may be found is a line version of the relationship between the kth factorial moment of system size and the regular kth moment of system delay given by (5.34), namely,

$$L_{(k)} = \lambda^k W_k.$$

The proof of this result follows.

To begin, we know from our earlier work on waiting times for $M/G/1$ (Section 5.1.5) that

$$\pi_n = \Pr\{n \text{ in system just after a departure}\} = \frac{1}{n!} \int_0^\infty (\lambda t)^n e^{-\lambda t}\, dW(t).$$

But this equation is also valid in modified form for $M/G/c$ if we consider everything in terms of the queue and not the system. Then it is true that

$$\pi_n^q \equiv \Pr\{n \text{ in queue just after a departure}\} = \frac{1}{n!} \int_0^\infty (\lambda t)^n e^{-\lambda t}\, dW_q(t),$$

with the mean queue length at departure points, say $L_q^{(D)}$, given by

$$L_q^{(D)} = \sum_{n=1}^\infty n\pi_n^q = \int_0^\infty \lambda t\, dW_q(t) = \lambda W_q,$$

which is Little's formula. Now suppose we use $L_{q(k)}^{(D)}$ to denote the kth factorial moment of the departure-point queue size. Then

$$L_{q(k)}^{(D)} = \sum_{n=1}^\infty n(n-1)\cdots(n-k+1)\pi_n^q$$

$$= \int_0^\infty dW_q(t) \sum_{n=1}^\infty \frac{n(n-1)\cdots(n-k+1)(\lambda t)^n e^{-\lambda t}}{n!}.$$

But the summand is nothing more than the kth factorial moment of the Poisson, which can be shown to be $(\lambda t)^k$. Hence

$$L_{q(k)}^{(D)} = \lambda^k W_{q,k}, \tag{5.49}$$

where $W_{q,k}$ is the ordinary kth moment of the line waiting time. This is now a generalization of Equation (5.34) to line waits for an $M/G/c$.

5.2.2   The *M/G/∞* Queue and *M/G/c/c* Loss System

We begin this section with a derivation of two key results for the *M/G/∞* model, namely, the transient distribution for the number of customers in the system at time t (as we did for *M/M/∞*), and the transient distribution for the number of customers who have completed service by time t, that is, the departure counting process. To start, let the overall system-size process be called $N(t)$, the departure process $Y(t)$, and the input process $X(t) = Y(t) + N(t)$. By the laws of conditional probability, we find that

$$\Pr\{N(t) = n\} = \sum_{i=n}^{\infty} \Pr\{N(t) = n \mid X(t) = i\} \frac{e^{-\lambda t}(\lambda t)^i}{i!},$$

since the input is Poisson. The probability that a customer who arrives at time x will still be present at time t is given by $1 - B(t - x)$, $B(u)$ being the service-time CDF. Hence it follows that the probability that an arbitrary one of these customers is still in service is given by

$$q_t = \int_0^t \Pr\{\text{service time} > t - x \mid \text{arrival at time } x\} \Pr\{\text{arrival at } x\} \, dx.$$

Since the input is Poisson, $\Pr\{\text{arrival at } x\}$ is uniform on $(0, t)$; hence it is $1/t$, from Equation (1.16). Thus

$$q_t = \frac{1}{t} \int_0^t [1 - B(t - x)] \, dx = \frac{1}{t} \int_0^t [1 - B(x)] \, dx,$$

and is independent of any other arrival. Therefore by the binomial law,

$$\Pr\{N(t) = n \mid X(t) = i\} = \binom{i}{n} q_t^n (1 - q_t)^{i-n} \qquad (n \geq 0),$$

and the transient distribution is

$$\Pr\{N(t) = n\} = \sum_{i=n}^{\infty} \binom{i}{n} \frac{q_t^n (1 - q_t)^{i-n} e^{-\lambda t}(\lambda t)^i}{i!}$$

$$= \frac{(\lambda q_t t)^n e^{-\lambda t}}{n!} \sum_{i=n}^{\infty} \frac{[\lambda t(1 - q_t)]^{i-n}}{(i - n)!}$$

$$= \frac{(\lambda q_t t)^n e^{-\lambda t} e^{\lambda t - \lambda q_t t}}{n!} = \frac{(\lambda q_t t)^n e^{-\lambda q_t t}}{n!},$$

namely, nonhomogeneous Poisson with mean $\lambda q_t t$.

To derive the equilibrium solution, take the limit as $t \to \infty$ of this transient answer. It is thereby found that

$$\lim_{t \to \infty}(\lambda q_t t) = \lambda \int_0^\infty [1 - B(x)] \, dx = \frac{\lambda}{\mu},$$

and hence the equilibrium solution is Poisson with mean $\lambda E[S] = \lambda/\mu$. We make special note of this result and its similarity to the steady-state probabilities we derived in Chapter 2 for the $M/M/\infty$—the importance of this observation will become more apparent below in our discussion of the $M/G/c/c$ problem.

The distribution of the departure-counting process $Y(t)$ can be found by exactly the same argument as above, using $1 - q_t = \int_0^t B(x) \, dx/t$ instead of q_t. The result, as expected, is

$$\Pr\{Y(t) = n\} = \frac{[\lambda(1 - q_t)t]^n e^{-\lambda(1 - q_t)t}}{n!}.$$

In the limit as $t \to \infty$, we see that q_t goes to zero, and thus the interdeparture process is Poisson in the steady state, which is precisely the same as the arrival process.

Note how this output result compares with our discussion in Chapter 4, where we showed that the output of an $M/M/c$ is Poisson for any value of c. Furthermore, in Section 5.1.8 we pointed out the converse result that $M/G/1$ is the only $M/G/1$ with Poisson output [Problem 5.22(b)]. More generally, we can say that $M/M/c$ is the only $M/G/c$ with Poisson output, though we must now, as a result of the calculations of this section, add the caveat that c must be finite.

Now, getting back to $M/G/c/c$, we wish to look into a result that we quoted earlier in Section 2.5. It is the surprising fact that the steady-state system-size distribution given by (2.39), namely, the truncated Poisson

$$p_n = \frac{(\lambda/\mu)^n/n!}{\sum_{i=0}^c (\lambda/\mu)^i/i!} \qquad (0 \leq n \leq c),$$

is valid for *any* $M/G/c/c$, independent of the form of G. The specific value from this for p_c, as noted in Section 2.5, is called Erlang's loss or B formula, and, as noted just above in this section, the result extends to $M/G/\infty$, where

$p_n = e^{-\lambda/\mu}(\lambda/\mu)^n/n!$ for any form of G. We now sketch a proof of the general assertion for $M/G/c/c$.

The $c = 1$ case (i.e., $M/G/1/1$) is very simple, since we have essentially already noted this fact with the observation (from Section 2.4) that $p_0 = 1 - \rho_{\text{eff}} = 1 - p_1$ for any $G/G/1/1$ queue. Since $\rho_{\text{eff}} = \lambda(1 - p_1)/\mu$, it follows that

$$p_1 = \frac{\lambda/\mu}{1 + \lambda/\mu}, \qquad p_0 = \frac{1}{1 + \lambda/\mu},$$

which is precisely the result needed.

For the general problem $c > 1$, the formula follows from a set of more complex observations involving reversibility, product-form solutions, and multidimensional Markov processes, much as was done in Chapter 4 for the modeling of Jackson, product-form networks. In order to invoke these theories, it is critical to define a Markov process in the context of the $M/G/c/c$, since, as we have already noted, such systems are not Markovian by definition. This is done by expanding the model state space from n to the multi-dimensional vector $(n, u_1, u_2, \ldots, u_c)$, where $0 \le u_1 \le u_2 \le \cdots \le u_c$ are the c ordered service ages (i.e., completed service times so far, ranked smallest to largest, recognizing that $u_1, \ldots, u_{c-n}$ will be zero when the system state is n). That such a vector is Markovian can be seen from the fact that its future state is clearly a function of only its current position, with change over infinitesimal intervals depending on the fact that the instantaneous probability of a service completion when the service age is u depends solely on u. Then, altogether, it turns out that this augmented process is *reversible*—we have already had a major hint that this is true from our observation above that the output process of an $M/G/\infty$ is Poisson. More detailed discussions of the precise nature of the reversibility can be found in Ross (1996) and Wolff (1989).

From the reversibility, then, comes the critical observation that the limiting joint distribution of $(n, u_1, u_2, \ldots, u_c)$ has the proportional product form

$$p_n(u_1, u_2, \ldots, u_c) = Ca_n\bar{B}(u_1)\bar{B}(u_2) \cdots \bar{B}(u_c)/n!, \qquad (5.50)$$

where $u_1, \ldots, u_{c-n}$ are zero, C is proportional to the zero-state probability, a_n is proportional to the probability that n servers are busy, $B(u_i)$ is the service-time distribution function [thus $\bar{B}(u)$ is the probability that a service time is at least equal to u], and the $n!$ accounts for all of the rearrangements possible for the n order statistics within the $\{u_1, \ldots, u_c\}$ corresponding to the n busy servers. From the boundary conditions of the problem and the Poisson input assumption, we see that

$$C \equiv p_n(0, 0, \ldots, 0) = \lambda p_{n-1}(0, 0, \ldots, 0) = \lambda[\lambda p_{n-2}(0, 0, \ldots, 0)]$$

$$= \cdots = \lambda^n p_0(0, 0, \ldots, 0) = \lambda^n p_0.$$

Thus it follows from (5.50) that this is a completely separable product form, since we know from Section 5.1.5 that $\mu \bar{B}(u)$ is a legitimate density function (viz., of the residual service time) and each such term will independently integrate out to 1. The resultant marginal system-size probability function is

$$p_n = p_0 \frac{(\lambda/\mu)^n}{n!}.$$

It then follows that

$$p_0 = \left(\sum_{n=0}^{c} \frac{(\lambda/\mu)^n}{n!} \right)^{-1}$$

and

$$p_n = \frac{(\lambda/\mu)^n/n!}{\sum_{i=0}^{c} (\lambda/\mu)^i/i!} \qquad (0 \le n \le c),$$

which is precisely what we got for $M/M/c/c$ in Chapter 2.

Note that the fact that the steady-state probabilities for the $M/G/c/c$ are insensitive to the choice of G means that these probabilities will always satisfy the $M/M/c/c$ birth–death recursion

$$\lambda p_n = (n + 1)\mu p_{n+1} \qquad (n = 0, \ldots, c - 1).$$

It also turns out that we can retain the insensitivity with respect to the service distribution G even when the arrival process is generalized from a Poisson to a state-dependent birth process with rate λ_n depending on the system state (Wolff, 1989). In this case, the left-hand side of the birth–death recursion is merely changed to $\lambda_n p_n$.

We summarize in Table 5.3 the results we have just discussed on the insensitivity of $M/G/c/K$ models to the form of the service-time distribution G.

Table 5.3 Insensitivity Results for $M/G/c/K$ Models

$M/G/c/c$ vs. $M/M/c/c$	Steady-state probabilities and output process independent of form of G
$M/G/\infty$ vs. $M/M/\infty$	Steady-state probabilities and output process independent of form of G
$M/G/c/\infty$ vs. $M/M/c/\infty$	Output processes equal if and only if $G \equiv M$

5.3 GENERAL INPUT AND EXPONENTIAL SERVICE

In this section we concentrate on the queueing situation where service times are assumed to be exponential and no specific assumption is made concerning the arrival pattern other than that successive interarrival times are IID. For this case, results can be obtained for c parallel servers using an analysis similar to that for the $c = 1$ case with a slight increase in complexity in certain calculations. So we first consider $c = 1$ and then generalize to c servers.

5.3.1 Arrival-Point Steady-State System-Size Probabilities for G/M/1

We examine first a single server with mean service rate μ and exponential service times. We assume that the mean arrival rate is λ, that arrivals come singly, and, as mentioned above, that successive interarrival times are IID. An imbedded-Markov-chain approach is also utilized here to obtain our results.

Following the type of analysis used in Section 5.1 for $M/G/1$, we now consider the system just prior to an arrival, and X_i will represent the number in the system that the ith arrival sees upon joining the system ($X_i = 0, 1, 2, \ldots$). We then have

$$X_{n+1} = X_n + 1 - B_n \qquad (B_n \le X_n + 1, \quad X_n \ge 0),$$

where B_n is the number of customers served during the interarrival time $T^{(n)}$ between the nth and $(n + 1)$st arrivals. Since the interarrival times are assumed independent, the random variable $T^{(n)}$ can be denoted by T, and we denote its CDF by $A(t)$. Since service is exponential, the random variable B_n depends on only the length of the interval and not on the extent of the service the present customer has already received. Thus we can drop the time-dependent subscript and henceforth denote B_n as B. We have, then, that

$$Pr\{B = b\} = \int_0^\infty Pr\{B = b \mid T = t\} \, dA(t)$$

$$= \int_0^\infty \frac{e^{-\mu t}(\mu t)^b}{b!} \, dA(t), \qquad (5.51)$$

so that

$$p_{ij} \equiv \Pr\{X_{n+1} = j \mid x_n = i\} = \begin{cases} \Pr\{B = i + 1 - j\} & (i + 1 \geq j \geq 1), \\ 0 & (i + 1 < j) \end{cases}$$

$$= \begin{cases} \displaystyle\int_0^\infty \frac{e^{-\mu t}(\mu t)^{i+1-j}}{(i+1-j)!}\, dA(t) & (i + 1 \geq j \geq 1), \\ 0 & (i + 1 < j). \end{cases} \tag{5.52}$$

(Note that the case $j = 0$ must be treated separately, since idle time results over a portion of the interarrival time T in this case, and it is not sufficient to say that $i + 1 - j$ are to be served during T, as they could have been served in a time less than T.) Since p_{i0} must be $1 - \Sigma_{j=1}^{i+1} p_{ij}$, we see from (5.52) that all p_{ij} depend only on i and j; thus we have a Markov chain.

Continuing now in a manner similar to Section 5.1.2, we introduce the following simplifying notation:

$$b_n = \Pr\{n \text{ services during an interarrival time}\} = \Pr\{B = n\}$$

$$= \int_0^\infty \frac{e^{-\mu t}(\mu t)^n}{n!}\, dA(t), \tag{5.53}$$

so that we can now obtain the imbedded, single-step transition probability matrix

$$\mathbf{P} = \{p_{ij}\} = \begin{pmatrix} 1 - b_0 & b_0 & 0 & 0 & 0 & \cdots \\ 1 - \displaystyle\sum_{k=0}^{1} b_k & b_1 & b_0 & 0 & 0 & \cdots \\ 1 - \displaystyle\sum_{k=0}^{2} b_k & b_2 & b_1 & b_0 & 0 & \cdots \\ \cdot & \cdot & \cdot & \cdot & \cdot & \\ \cdot & \cdot & \cdot & \cdot & \cdot & \\ \cdot & \cdot & \cdot & \cdot & \cdot & \end{pmatrix}. \tag{5.54}$$

Assuming that a steady-state solution exists (this will be taken up later) and denoting the probability vector that an arrival finds n in the system by $q = \{q_n\}$, $n = 0, 1, 2, \ldots$, we have the usual stationary equations

$$qP = q \quad \text{and} \quad qe = 1, \tag{5.55}$$

which yield

$$q_i = \sum_{k=0}^{\infty} q_{i+k-1} b_k \qquad (i \geq 1),$$

$$q_0 = \sum_{j=0}^{\infty} q_j \left(1 - \sum_{k=0}^{j} b_k \right).$$

(5.56)

We note here that a major difference between (5.56) and its counterpart for the $M/G/1$, Equation (5.16), is that the equations of (5.56) have an infinite summation, whereas each equation had a finite summation in (5.16). It turns out that this works to our advantage, and we now employ the method of operators on (5.56). (Problem 5.27 asks the reader to derive the same results using generating functions.)

Letting $Dq_i = q_{i+1}$, we find for $i \geq 1$ that (5.56) can be written as

$$q_i - (q_{i-1} b_0 + q_i b_1 + q_{i+1} b_2 + \cdots) = 0,$$

so that

$$q_{i-1}(D - b_0 - Db_1 - D^2 b_2 - D^3 b_3 - \cdots) = 0.$$

Thus, for a nontrivial solution,

$$D - \sum_{n=0}^{\infty} b_n D^n = 0.$$

(5.57)

But note that since b_n is a probability, the second term on the left is merely the probability generating function of the $\{b_n\}$ [call it $\beta(z)$], so that (5.57) becomes

$$\boxed{\beta(z) = z.}$$

(5.58)

Completely analogous to Equation (5.36), it can be shown that $\beta(z) = A^*[\mu(1 - z)]$, where $A^*(z)$ is the LST of the interarrival-time CDF (see Problem 5.27), so that (5.58) may also be written as

$$z = A^*[\mu(1 - z)].$$

(5.58a)

We shortly prove that Equation (5.58) has only a single root with absolute value less than 1 (it must thus be real and positive). Denoting this root by r_0, we therefore have that

$$q_i = Cr_0^i \qquad (i \geq 0).$$

(5.59)

The constant C, as usual, is to be determined from the summability-to-one boundary condition, and its value is $1 - r_0$.

To show that there is one and only one positive root of (5.58) between 0 and 1, we consider the two sides of the equation separately, as

$$y = \beta(z) \quad \text{and} \quad y = z. \tag{5.60}$$

First, we observe the facts that

$$0 < \beta(0) = b_0 < 1 \quad \text{and} \quad \beta(1) = \sum_{n=0}^{\infty} b_n = 1.$$

We can also easily show that $\beta(z)$ is both monotonically nondecreasing and convex, because

$$\beta'(z) = \sum_{n=1}^{\infty} nb_n z^{n-1} \geq 0,$$

$$\beta''(z) = \sum_{n=2}^{\infty} n(n-1)b_n z^{n-2} \geq 0.$$

Furthermore, since the service times are exponential, all the b_n, $n > 2$, are greater than 0 and $\beta(z)$ is thus strictly convex. We graph Equations (5.60) in Figure 5.1 to observe their intersections. There are two possible cases: either no roots between 0 and 1, or exactly one root between 0 and 1. The left-hand case (no roots between 0 and 1) occurs if

$$\beta'(1) = E[\text{number served during interarrival time}] = \mu \frac{1}{\lambda} \leq 1.$$

This implies that $\mu/\lambda > 1$, or equivalently $\rho = \lambda/\mu < 1$, is necessary and sufficient for ergodicity, since otherwise $\Sigma q_i = \infty$. Hence when a steady-state

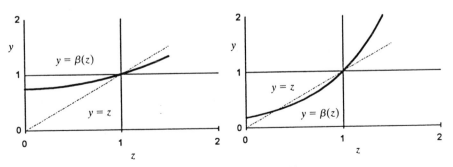

Fig. 5.1 Plot of Equation (5.60).

solution exists, there is exactly one root (r_0) of Equation (5.58) between 0 and 1. Finding this root generally involves numerical procedures, but it is readily obtainable. For example, the method of successive substitution,

$$z^{(k+1)} = \beta(z^{(k)}) \qquad (k = 0, 1, 2, \dots, \quad 0 < z^{(0)} < 1), \qquad (5.61)$$

is guaranteed to converge because of the shape of $\beta(z)$. More on the numerical aspects of the problem will follow later.

When these results are all put together, we can write that the steady-state arrival-point distribution is given by

$$\boxed{q_n = (1 - r_0)r_0^n \qquad (n \geq 0, \quad \rho < 1).} \qquad (5.62)$$

It is informative to note the analogy between (5.62) and the $M/M/1$ steady-state probability given by $p_n = (1 - \rho)\rho^n$. We can therefore use all the expected-value measures of effectiveness results obtained for $M/M/1$ by merely replacing ρ with r_0. However, it must be pointed out that q_n is the steady-state probability of n in the system just prior to an arrival and not the general-time steady-state probability p_n, so that the expected-value measures apply only at arrival points. Unlike the $M/G/1$ model, it is not true here that $q_n = p_n$. In fact, it turns out that the equality holds for the current model if, and only if, the arrivals are Poisson; that is, $q_n = p_n$ for $G/M/1$ if, and only if, $G = M$. However, q_n and p_n can be related; more will be said about this in Chapter 6, Section 6.3.

In light of these circumstances, we use a superscript (A) to denote the fact that a particular measure of effectiveness is taken relative to arrival points only, and thus write from (5.62) that

$$L^{(A)} = \frac{r_0}{1 - r_0} \quad \text{and} \quad L_q^{(A)} = \frac{r_0^2}{1 - r_0}. \qquad (5.63)$$

The line-delay and system-waiting-time distribution functions, $W_q(t)$ and $W(t)$, can also be obtained from the $M/M/1$ with r_0 replacing ρ, that is,

$$\boxed{\begin{aligned} W_q(t) &= 1 - r_0 e^{-\mu(1 - r_0)t} \qquad (t \geq 0), \\ W(t) &= 1 - e^{-\mu(1 - r_0)t} \qquad (t \geq 0), \end{aligned}} \qquad (5.64)$$

with mean values given by

$$W_q = \frac{r_0}{\mu(1 - r_0)} \quad \text{and} \quad W = \frac{1}{\mu(1 - r_0)}. \tag{5.65}$$

These waiting-time results are exact, since if one refers back to the development of Section 2.2.4, it can be seen that the steady-state probabilities used in the development leading to Equation (2.21) were in reality the q_n. In that case it was necessary to justify using p_n by noting that the two were equivalent because of Poisson arrivals (PASTA).

Example 5.6
Suppose we know in a single-server queueing situation that the service time is exponential with mean μ but have no theoretical basis for assuming the input to be Poisson or Erlang. From past history, we have determined a k-point probability distribution for interarrival times, so that

$$\Pr\{\text{interarrival time} = t_i\} = a(t_i) = a_i \quad (1 \le i \le k).$$

We must first determine the root r_0 from (5.58), written here as

$$z = A^*[\mu(1 - z)] = \sum_{i=1}^{k} a_i e^{-\mu t_i(1 - z)}.$$

To illustrate numerically, consider the case where the interarrival-time distribution is given in Table 5.4 and $1/\mu = 2$ min. We must solve for the root of

$$A^*(z) = \beta(z) = 0.2e^{-(1-z)} + 0.7e^{-1.5(1-z)} + 0.1e^{-2(1-z)} = z, \tag{5.66}$$

which is a relatively easy equation to solve. We shall use successive substitution on (5.66), beginning from $z^{(0)} = 0.5$, and the resulting sequence of

Table 5.4   Interarrival-Time Probability Distribution*

t (min)	$a(t)$	$A(t)$
2	0.2	0.2
3	0.7	0.9
4	0.1	1.0

* $1/\lambda = 2.9$ min.

Table 5.5   Successive Substitution Steps

k	$z^{(k)}$	$\beta(z^{(k)})$
1	0.500	0.489
2	0.489	0.481
3	0.481	0.476
4	0.476	0.472
5	0.472	0.470
6	0.470	0.468
7	0.468	0.467
8	0.467	0.467

values is as shown in Table 5.5. We see that it took only eight iterations for the process to converge to three decimal places. With r_0 thus estimated as 0.467, we now compute the measures of effectiveness as

$$q_n \doteq 0.533(0.467)^n \qquad (n \geq 0),$$

$$L^{(A)} \doteq \frac{r_0}{1 - r_0} = 0.876, \qquad L_q^{(A)} \doteq \frac{r_0^2}{1 - r_0} \doteq 0.409,$$

$$W_q \doteq \frac{r_0}{\mu(1 - r_0)} \doteq 1.752 \text{ min}, \qquad W \doteq W_q + 1/\mu \doteq 3.752 \text{ min}.$$

(Remember that Little's formula cannot be used to relate mean waiting times with mean queue sizes measured at arrival points.)

Thus we see from this exercise that it is not difficult to obtain results for empirical distributions. This was true also for the $M/G/1$ as illustrated by Example 5.2. It turns out that this is quite a useful result, since any probability distribution can be approximated by a finite discrete distribution of k points, through the use of an approximating histogram.

5.3.2   Arrival-Point Steady-State System-Size Probabilities for $G/M/c$

When we move to generalize to c servers, much of the derivation remains the same, with the major exception of the value of b_n and its effect on the imbedded matrix and the rootfinding problem. The mean system service rate is now going to be either $n\mu$ or $c\mu$, depending on the state, so that b_n will now depend on both i and j. This leads to a very different-looking transition matrix for the Markov chain, and its derivation follows.

To begin, we note that it is still true that $p_{ij} = 0$ for all $j > i + 1$. Now, for $j \leq i + 1$ but $\geq c$, the system serves at the mean rate $c\mu$, since all servers are busy; so, as with $G/M/1$,

$$p_{ij} = b_{i+1-j} \qquad (c \le j \le i + 1),$$

though here

$$b_n = \int_0^\infty \frac{e^{-c\mu t}(c\mu t)^n}{n!} \, dA(t). \tag{5.67}$$

As a result, the transition probability matrix for the multiserver problem has the same kind of layout from its cth column (starting the count from the 0th one) on out to the right as the $G/M/1$ does from its 1st column on out to the right, as seen in Equation (5.54). Furthermore, it turns out that the rootfinding problem remains the same except that now the number of servers must enter the calculation as

$$\beta(z) = A^*[c\mu(1 - z)] = z.$$

The two cases which cause some difficulties are $(i \ge c, j < c)$, for which the system service rate varies from $c\mu$ down to $j\mu$, and $(i < c, j < c)$, for which the system service rate varies from $i\mu$ down to $j\mu$. Therefore, there are now c columns replacing the first one of (5.54), numbered 0 to $c - 1$.

Consider now the case when $j \le i + 1 \le c$. Here, everyone is being served, and the probability that anyone has completed service by a time t is the CDF of an individual server, namely, $1 - e^{-\mu t}$. To go from i to j, there must be $i + 1 - j$ service completions by time t; hence, using the binomial distribution, we have

$$p_{ij} = \int_0^\infty \binom{i+1}{i+1-j} (1 - e^{-\mu t})^{i+1-j} e^{-\mu t j} \, dA(t) \qquad (j \le i + 1 \le c). \tag{5.68}$$

Lastly, it remains to consider the case $i + 1 > c > j$. Here, the system starts out with all servers busy, since $i \ge c$, and sometime during the interarrival time T, servers start to become idle, until finally only j servers are busy. Let us assume that at a time V after the arrival comes $(0 < V < T)$, it goes into service (all prior customers have left), with $H(v)$ the CDF of V. Thus to get from state i to j in time T, we must have $c - j$ service completions from V to T, or during a time interval of length $T - V$. Using the binomial distribution again and realizing that service time is memoryless, we have

$$p_{ij} = \int_0^\infty \int_0^t \binom{c}{c-j} (1 - e^{-\mu(t-v)})^{c-j} e^{-\mu(t-v)j} \, dH(v) \, dA(t). \tag{5.69}$$

The random variable V is merely the time until $i - c + 1$ people have been served with all c servers working, which is the $(i - c + 1)$-fold convolution of the exponential distribution with parameter $c\mu$. Hence $h(v) = dH(v)/dv$ is Erlang type $i - c + 1$, namely,

$$h(v) = \frac{c\mu(c\mu v)^{i-c}e^{-c\mu v}}{(i-c)!}.$$

Substituting for $dH(v)$ gives us for $j < c < i + 1$ that

$$p_{ij} = \binom{c}{c-j}\frac{(c\mu)^{i-c+1}}{(i-c)!}\int_0^\infty\int_0^t(1-e^{-\mu(t-v)})^{c-j}e^{-\mu(t-v)j}v^{i-c}e^{-c\mu v}\,dv\,dA(t).$$

(5.70)

Now, the stationary equation is

$$q_j = \sum_{i=0}^\infty p_{ij}q_i \qquad (j \geq 0),$$

where the $\{p_{ij}\}$ are as given throughout the preceding discusion. However, for $j \geq c$, we have

$$q_j = \sum_{i=0}^{j-2} 0 \cdot q_i + \sum_{i=j-1}^\infty b_{i+1-j}q_i$$

$$= \sum_{k=0}^\infty b_k q_{j+k-1} \qquad (j \geq c). \tag{5.71}$$

Equation (5.71) is identical to the first line of (5.56), and hence, using analyses like those for $c = 1$, we have

$$q_j = Cr_0^j \qquad (j \geq c), \tag{5.72}$$

where r_0 is the root of $\beta(z) = A^*[c\mu(1-z)] = z$.

The constant C and the q_j $(j = 0, 1, \ldots, c-1)$ must be determined from the boundary condition $\sum_{i=0}^\infty q_j = 1$ and the first $c - 1$ of the stationary equations, using the various formulas for the $\{p_{ij}\}$ given above. This is not particularly easy to do, since the $c + 1$ equations in $c + 1$ unknowns are all infinite summations. We can, however, get an expression for C in terms of $q_1, q_2, \ldots, q_c$ and r_0, and then develop a recursive relation among the q_j. The procedure is as follows.

The boundary condition yields

$$1 = \sum_{j=0}^{\infty} q_j = \sum_{j=0}^{c-1} q_j + \sum_{j=c}^{\infty} Cr_0^j.$$

Hence

$$C = \frac{1 - \sum_{j=0}^{c-1} q_j}{\sum_{j=c}^{\infty} r_0^j} = \frac{1 - \sum_{j=0}^{c-1} q_j}{r_0^c(1 - r_0)^{-1}}. \tag{5.73}$$

Now, a recursive relation for q_j when $j < c$ can be obtained:

$$q_j = \sum_{i=0}^{\infty} p_{ij}q_i = \sum_{i=0}^{c-1} p_{ij}q_i + \sum_{i=c}^{\infty} p_{ij}Cr_0^i$$

$$= \sum_{i=j-1}^{c-1} p_{ij}q_i + C\sum_{i=c}^{\infty} p_{ij}r_0^i \qquad (1 \le j \le c - 1),$$

since $p_{ij} = 0$ for $j > i + 1$. Then, rewriting, we have

$$q_{j-1} = \frac{q_j - \sum_{i=j}^{c-1} p_{ij}q_i - C\sum_{i=c}^{\infty} p_{ij}r_0^i}{p_{j-1,j}} \qquad (1 \le j \le c - 1).$$

Dividing through by C and letting $q_j' = q_j/C$ gives

$$q_{j-1}' = \frac{q_j' - \sum_{i=j}^{c-1} p_{ij}q_i' - \sum_{i=c}^{\infty} p_{ij}r_0^i}{p_{j-1,j}} \qquad (1 \le j \le c - 1). \tag{5.74}$$

From the stationary equation, we may also write that

$$q_c = \sum_{i=0}^{\infty} p_{ic}q_i = \sum_{i=c-1}^{c} p_{ic}q_i + \sum_{i=c+1}^{\infty} b_{i+1-c}Cr_0^i.$$

Hence

$$q_{c-1} = \frac{(1 - p_{cc})q_c - C\sum_{i=c+1}^{\infty} b_{i+1-c}r_0^i}{p_{c-1,c}}$$

and

$$q_{c-1}' = \frac{(1 - p_{cc})q_c' - \sum_{i=c+1}^{\infty} b_{i+1-c}r_0^i}{p_{c-1,c}}$$

$$= \frac{(1 - b_1)q_c' - \sum_{i=c+1}^{\infty} b_{i+1-c}r_0^i}{b_0}.$$

But we also have, using (5.72), that $q_c' = r_0^c$, so that q_{c-1}' can be determined using the earlier formulas for b_n and p_{ij}. Then $q_{c-1}', q_{c-2}', \ldots, q_0'$ can be obtained by repeated use of (5.74). Now writing (5.73) in terms of q'_i finally gives

$$C = \frac{1 - C\sum_{j=0}^{c-1} q'_j}{r_0^c(1 - r_0)^{-1}} = \left(\sum_{j=0}^{c-1} q'_j + \frac{r_0^c}{1 - r_0}\right)^{-1}. \tag{5.75}$$

Although we have just presented a complete derivation of a very specific approach to obtaining the initial c arrival-point probabilities for the $G/M/c$ model, namely, $\{q_j, j = 0, 1, 2, \ldots, c - 1\}$, there is a computational recursion due to Takács (1962) which is better suited to the kind of spreadsheet computations we have developed for our software. Its derivation, however, is extremely long, so we just present the steps of this process, precisely as we have implemented them for the $G/M/c$ queues presented in the software.

As before, we let r_0 be the unique real solution in $(0, 1)$ of $z = A^*[c\mu(1 - z)]$, where A^* is the LST of the interarrival distribution. Then the Takács algorithm is used to define and calculate, in sequence:

$$A_j^* \equiv A^*(j\mu) \qquad \text{for} \quad j = 0, 1, 2, \ldots, c,$$

$$C_j \equiv \frac{A_1^*}{1 - A_1^*} \cdot \frac{A_2^*}{1 - A_2^*} \cdots \frac{A_j^*}{1 - A_j^*} \qquad \text{for } j = 1, 2, \ldots, c, \qquad C_0 = 1$$

and

$$D_j \equiv \sum_{k=j+1}^{c} \binom{c}{k} \frac{c(1 - A_k^*) - k}{[C_k(1 - A_k^*)][c(1 - r_0) - k]} \qquad \text{for} \quad j = 0, 1, 2, \ldots, c - 1,$$

$$M \equiv \left(\frac{1}{1 - r_0} + D_0\right)^{-1}.$$

Then it can be proved that

$$q_j = \begin{cases} \sum_{i=j}^{c-1} (-1)^{i-j} \binom{i}{j} MC_i D_i & (j = 0, 1, \ldots, c - 1), \\ Mr_0^{j-c} & (j \geq c). \end{cases}$$

Note that it follows by comparing the above with Equation (5.72) that the constant $M = q_c$.

To determine the line-delay distribution function $W_q(t)$, we recognize that there should be a direct analogy back to the derivation of $W_q(t)$ for the $M/M/c$ in Section 2.3, leading to Equation (2.31). Since both $G/M/c$ and $M/M/c$ have exponential service, the only difference in the respective delay distribution derivations would be the values used for the arrival-point probabilities. For the $G/M/c$, we need to use the probabilities $\{q_j, \ j \geq c\}$. By rewriting (2.31) as

$$W_q(t) = 1 - \frac{p_c}{1 - \rho} e^{-c\mu(1-\rho)t},$$

we are able to conclude that the line delay distribution of the $G/M/c$ is given for $t \geq 0$ by

$$W_q(t) = 1 - \frac{q_c}{1 - r_0} e^{-c\mu(1-r_0)t} = 1 - \frac{Cr_0^c}{1 - r_0} e^{-c\mu(1-r_0)t}. \qquad (5.76)$$

The mean delay is easily seen to be

$$W_q = \frac{q_c}{c\mu(1 - r_0)^2} = \frac{Cr_0^c}{c\mu(1 - r_0)^2}.$$

Just as in $M/G/1$, there are numerous other $G/M/1$-type problems we might want to consider, such as busy periods, $G/M/1/K$, impatience, priorities, output, and transience. Due to space limitations, we are not able to pursue any of these topics at length, and shall instead make a few comments on each and indicate a number of references.

Cohen (1982) is probably the most comprehensive reference for nearly all of these problems. Of course, specific references in the open literature may be better for particular subjects. For example, it is not difficult to get the expected length of the busy period for any $G/M/1$ queue, and there is a very nice approach to this in Ross (1970) using renewal theory. We supply some of the details of this argument in the next chapter in the context of the $G/G/1$ queue.

The approach to the truncation of the queue would be very similar to that described early in Section 5.1.7 for the imbedded chain of $M/G/1$, while impatience could be nicely introduced by permitting some departures from the system to occur before customers reach service. This can essentially be accomplished by changing the parameter of the exponential service to $\mu + r$, r now being the probability of such a renege. If, in addition, we desire to make reneges functions of queue size, as they probably should be, then we have a problem with state-dependent departures, such that

$$b_{mn} = \Pr\{m \text{ services during an interarrival time} \mid$$
$$n \text{ in system at latest departure}\}$$
$$= \int_0^\infty \frac{e^{-[\mu+r(n)]t}\{[\mu + r(n)]t\}^m}{m!} \, dA(t),$$

where $r(n)$ is defined to be the reneging rate during interarrival periods which began with n in the system. The analysis would then proceed in a way similar to the departure-point state dependence of Section 5.1.10.

As far as priorities are concerned, when the assumption of Poisson inputs is relaxed, it becomes very difficult to obtain any results. One possible way of approaching the problem, suggested by Jaiswal (1968), is to use the technique of supplementary variables to gain some "Markovianness." How-ever, one supplementary variable would be required for each priority, hence at least two for any such problem. Even the near-Markovian assumption of Erlang input is messy, though some work has appeared on this problem for a small number of priorities.

For output, we have already indirectly obtained some results. It was noted in our discussion of series queues that the limiting output of an $M/G/1$ is Poisson if, and only if, G is exponential. Likewise, it can be shown that the limiting output $G/M/1$ is Poisson if, and only if, G is exponential (see Problem 5.35).

Bulk services $(G/M^{[Y]}/1)$ can be handled in a way comparable to the manner in which we solve the $M^{[X]}/G/1$ problem in Section 5.1.9. Some results are also possible for these extensions of the basic $G/M/1$ models when considering c-channels.

Finally, we close with a few comments about transient analysis. As for $M/G/1$, we have to again appeal to the CK equation

$$p_j^{(m)} = \sum_k p_k^{(0)} p_{kj}^{(m)},$$

where $p_j^{(m)}$ is the probability that the system state is j just before the mth customer has arrived. The necessary matrix multiplications must be done with some caution, since we are still dealing with an $\infty \times \infty$ matrix when there is unlimited waiting room. But this can be done (albeit carefully) by truncating the transition matrix at an appropriate point (see Neuts, 1973).

PROBLEMS

*Whenever a problem is best solved by use of this book's accompanying software, we have added a boldface **C** to the right of the problem number.*

5.1.   Calculate the imbedded transition probabilities for an $M/G/1$ where service is uniformly distributed on (a, b).

5.2C.   Do Problem 3.18 without making any assumptions concerning the service-time distribution, that is, utilizing only the mean and variance of the service-time data.

5.3. (a) Find L, L_q, W_q, and W for an $M/G/1$ with service times which are beta-distributed.

(b) Arrivals to an $M/G/1$ queue occur at the mean rate of one-third per unit time. Find the average steady-state waiting times in system and queue when the distribution function of the service time is an Erlang type 2.

5.4. For Example 5.1, find the σ_S^2 necessary to yield $L = 5$ if the mean service time after training increases to 5.2 min.

5.5. Show that the expected amount of time required to finish the item in service encountered by an arbitrary arrival is given by $(\lambda/2)E[S^2]$ for any $M/G/1$ when S is the total service time of an item.

5.6. Verify Equation (5.16) for $i = 0, 1, 2, 3$.

5.7. Derive the generating function $\Pi(z)$ for $M/G/1$ as given by Equation (5.18).

5.8. From Equation (5.20), derive the PK formula (5.11) by using the fact that $L = \Pi'(1)$. [*Hint:* Use l'Hôpital's rule twice.]

5.9. Use Equation (5.20) for the $M/G/1$ queue with G assumed exponential and show that it reduces to the generating function of $M/M/1$ given by Equation (2.8).

5.10C. Calculate L for Example 5.2.

5.11C. A certain assembly-line operation is assumed to be of the $M/G/1$ type, with input rate 5/hr and service times with mean 9 min and variance 90 min^2. Find L, L_q, W, and W_q. Is the operation improved or degraded if the service times are forced to be exponential with the same mean?

5.12. Derive the generating function $\Pi(z)$ following Equation (5.24) for the $M/G/1$ k-point service-time model following the analysis of Example 5.2.

5.13. Consider an $M/D/1$ queue with service time equal to b time units. Suppose further that one is only able to determine the system size when time is a multiple of b. Show that the stochastic process $\{X_n, n = 0, 1, 2, \ldots \mid X_n$ equal to the system size at time $t = nb\}$ is a Markov chain, and find its transition matrix.

5.14C. Determine the lowest value of the Erlang parameter k which will allow you to approximate the mean system waiting time of an $M/D/1$ queue to no worse than 0.5% by an $M/E_k/1$ when $\lambda = 4$ and $\mu = 4.5$.

5.15. Find the LST of the waiting-time-in-line distributions for $M/E_2/1$ using (5.37b), and then invert the result using partial fractions.

5.16. Verify the computation for Example 5.4 that $P''(1) = 14.50$.

5.17. Find the third and fourth ordinary moments of the system delay for the Bearing Straight problem, given that the third and fourth moments of the system size are 149.2 and 1670.6, respectively.

5.18. Find the variance of the $M/G/1$ busy period from the Laplace–Stieltjes transform of its CDF.

5.19. Consider a single-server queue to which customers arrive according to a Poisson process with parameter $\lambda = 0.04$/min and where the service times of all customers are fixed at 10 min. When there are three units in line, the system becomes saturated and all additional arrivals are turned away. The instants of departure give rise to an imbedded Markov chain with states 0, 1, 2, and 3. Find the one-step transition matrix of this chain and the resultant stationary distribution. Then compare this answer with the result you would have gotten without truncation.

5.20. Given the two-point service distribution of Table 5.2, find the output CDF for that $M/G/1$ queue.

5.21. (a) Find the steady-state probabilities for an $M/G/1$ state-dependent queue where

$$B_i(t) = \begin{cases} 1 - e^{-\mu_1 t} & (i = 1), \\ 1 - e^{-\mu t} & (i > 1), \end{cases}$$

that is, where there are two possible mean service rates, μ_1 if there is no queue and μ if there is a queue.
(b) Verify the steady-state probability result following Equation (5.46).

5.22. (a) Prove that the distribution of system sizes just prior to arrivals is equal to that after departures in any $G/G/c$ queue.
(b) Use the result of Section 5.1.8 for the output CDF $C(t)$ of the

interdeparture process of the $M/G/1$ to show that the $M/M/1$ is the only $M/G/1$ with Poisson output.

5.23. Derive the stationary system-size probabilities for $M/G/2/2$ using only Little's formula and fundamental steady-state identities.

5.24. The Mutual Exclusive Life Insurance Company (MELIC) is building a new headquarters in downtown Burbank. The telephone company wishes to determine the number of lines to feed into the building to assure MELIC of no more than 5% loss in calls due to busy circuits. Find the number of lines when it is estimated that the calling stream is Poisson with mean 100/hr throughout the day and that the mean call duration is 2 min.

5.25C. When an AIDS case is first diagnosed in the United States by a physician, it is required that a report be filed (i.e., serviced) on the case with the Centers for Disease Control in Atlanta, Georgia. It takes a random amount of time (with an expected value of approximately 3 months) for a doctor to finish the report and send it into the CDC (distribution unknown). An OR team has analyzed the report arrival stream and believes that new patients come to the doctors all over the nation as a Poisson process. If 50,000 new reports are completed each year, what is the mean number of reports in process by doctors at any one instant (assuming steady state)?

5.26. Derive the equivalent to Erlang's loss formula (2.39) for the case in which the input source is *finite* with rate proportional to the remaining source size, service times are general, and there are two channels. (The resultant answer is an example of a so-called Engset formula.)

5.27. Derive Equation (5.58) using the method of generating functions on Equation (5.56), $i \geq 1$. Then show that $\beta(z) = A^*[\mu(1-z)]$.

5.28. (a) Find the solution, r_0, in closed form, of $\beta(z) = z$, the generating-function equation for a $G/M/1$ queue when the interarrival times follow the hyperexponential (a special, balanced mixture of two exponentials) probability distribution

$$a(t) = 2q^2 \lambda e^{-2q\lambda t} + 2(1-q)^2 \lambda e^{-2(1-q)\lambda t}$$
$$(t>0, \quad 0<q\leq 0.5, \quad \lambda>0).$$

(b)C Find the steady-state arrival-time distribution for a $G/M/1$ queue with $\mu = 2$ and interarrival density function

$$a(t) = 1.0481e^{-1.8387t} + 0.5962e^{-1.3871t}.$$

5.29. Show that if $G = M$, the root r_0 of Equation (5.58) is ρ, and that Equation (5.62) yields the familiar $M/M/1$ result.

5.30. (a) Prove that r_0 is always greater than $e^{-1/\rho}$.

(b) You are observing two different $G/M/1$ queues, and know that one interarrival CDF (say A_1) is everywhere larger than the second, A_2. Show that $\beta_1(z) \le \beta_2(z)$, that $r_0^{(1)} \le r_0^{(2)}$, and thus that $\Sigma_{i=0}^n q_i^{(1)} \ge \Sigma_{i=0}^n q_i^{(2)}$.

5.31. Find $\beta(z)$ for $G/M/1$ in the event that G is (a) deterministic, (b) Erlang type 2. Then, under the assumption that the mean input rate λ is 3 and that the mean service time is $\frac{1}{5}$, find the steady-state arrival-point distribution for each of the two cases.

5.32C. Consider a $G/M/1$ system whose solution depends on the real root of the nonlinear equation

$$z = \frac{12}{31 - z} + \tfrac{3}{5}e^{-10(1 - z)/3}.$$

Write your own spreadsheet program to use successive substitution to find the root, starting from an initial guess of $z^{(0)} = 0.4$ and continuing until successive values of z are the same when rounded to three decimal places. Then use this answer to determine the line-delay CDF under the assumption that the service rate is 1.

5.33C. Suppose in Example 5.6 that there are two servers, each working at a mean rate of 0.25/min with their times being exponential. Find the arrival-point steady-state system-size probabilities, the expected system size and queue length at arrival points, and the expected time in queue and in system.

5.34. Suppose it is known in a $D/M/1$ queue that some customers are reneging before reaching service and that the reneging pattern is such that $\Pr\{\text{renege in } (t, t + \Delta t)\} = r \, \Delta t + o(\Delta t)$. Find the resultant imbedded Markov chain.

5.35. Show (a) for $M/G/1$ queues that stationary system waiting times are exponential if, and only if, $G = M$, and (b) that the stationary output of a $G/M/1$ queue is Poisson if, and only if, G is exponential. [*Hint:* Note for (b) that the idle time the queue undergoes in an arbitrary interdeparture period (called the virtual idle time) has a CDF given by $F(u) = A(u) + \int_u^\infty e^{-\mu(1-r_0)(t-u)} \, dA(t)$. Then use the fact that each departure time is a sum of a virtual idle time and a service time.]

5.36C. Determine the lowest value of the Erlang parameter k which will allow you to approximate the mean line delay of a $D/M/c$ queue to no worse than 2.0% by an $E_k/M/c$ when $\lambda = 4$, $\mu = 1.5$, and $c = 3$.

5.37C. Arrivals occur to a $G/M/4$ system from either of two sources. Any particular interarrival time is twice as likely to come from the first source as the second, and will be exponentially distributed with parameter $\lambda_1 = 0.5$ per unit time when coming from the first source and exponential with parameter $\lambda_2 = 0.25$ when coming from the second. It is further known that the mean service time is 9 time units. Find the mean system size and expected system waiting time.

5.38C. The arrival stream described in Problem 5.37 is now observed to come equally likely from the two sources, and the four servers have been merged into one having a mean service time of 2.25. The resultant $G/M/1$ queue does not match up with any of the prescribed models available in the text's software (why?). Use your own spreadsheet software (or adapt the text's) to find the root that is needed in the problem's solution, and then compute the waiting-time value beyond which only 5% of the arriving customer delays will fall.

CHAPTER 6

More General Models and Theoretical Topics

In this chapter we provide some assorted additional results. As a rule, these results were not included earlier because the models were either not Markovian or inappropriate for the discussions in Chapter 5 dealing with $M/G/1$ and $G/M/c$. Most of this new material follows in a logical way from previous material in the sense that it ties up some loose theoretical ends and should also help to provide a more complete picture of the kinds of models which may occur in real life.

6.1 $G/E_k/1$, $G^{[k]}/M/1$, AND $G/PH_k/1$

Remember from Chapter 5 that the waiting-time distribution function for the stationary, general single-server problem with exponential service (i.e., $G/M/1$) requires the single real root on $(0, 1)$ for the characteristic equation

$$z = A^*[\mu(1 - z)] = \beta(z),$$

where A^* is the Laplace–Stieltjes transform (LST) of the interarrival times, and $\beta(z)$ is the probability generating function (pgf) of the number of service completions during interarrival times, with $\rho = \lambda/\mu < 1$. The pgf (defined and analytic at least on the complex unit circle) is easily shown to be monotone nondecreasing and convex for real z, and thus the root is readily obtainable. For example, the well-known method of successive substitution is guaranteed to converge from any nonnegative starting point less than 1 because of the shape of $\beta(z)$.

The problem becomes more interesting and potentially quite useful when Erlang-k service times are used instead. Here the roots need to be generally located and then found from within the interior of the unit circle for

$$z^k = A^*[\mu(1 - z)] = \beta(z), \tag{6.1}$$

266

where $A^*(z)$ is once more the LST of the interarrival distribution function evaluated at $\mu(1 - z)$. A simple example of the form of (6.1) may be found in the characteristic equation of the $M/E_k/1$ queue from Chapter 3, namely,

$$\mu z^{k+1} - (\lambda + \mu)z^k + \lambda = 0,$$

which may be rewritten after some simple algebra as $z^k = \lambda/[\lambda + \mu(1 - z)]$. If we now let $z = r e^{i\theta}$, then we can rewrite (6.1) in the useful form

$$r^k e^{i\theta k} = A^*[\mu(1 - r e^{i\theta})]e^{2\pi n i} \qquad (6.1a)$$

for $n = 1, 2, \ldots, k$, where the final exponential factor is equal to 1 for all n and is added to the equation in preparation for taking the complex kth root. There is clearly one solution to (6.1) at $z = 1$, and by Rouché's theorem, we can show that there are k others strictly inside the unit circle $|z| = 1$ when the traffic intensity $k\lambda/\mu < 1$ (i.e., the phase completion rate is μ, while the overall service rate is μ/k). For later reference, it is useful to reformulate this equation by taking the kth root of both sides:

$$r e^{i\theta} = \{A^*[\mu(1 - r e^{i\theta})]\}^{1/k} e^{2\pi n i/k}. \qquad (6.2)$$

Given that we know that this model has k roots inside the unit circle, we now attempt to find their precise locations. The following result from Chaudhry et al. (1990) provides an important sufficient condition under which the roots will be distinct and offers some key information on determining their location. Then we use this and a related result to get a more complete feel for the circumstances under which the characteristic equation associated with an arbitrary $G/E_k/1$ queue will have distinct roots. The Chaudhry–Harris–Marchal theorem is that:

One of the k roots of the characteristic equation of the $G/E_k/1$ model (or, equivalently, the $G^{[k]}/M/1$) is real and in $(0, 1)$ for all values of k, and there is a second real root in $(-1, 0)$ only when k is even. In addition, the other roots ($k - 1$ for k odd, $k - 2$ for k even) are distinct if the Laplace–Stieltjes transform $A^*(s)$ of the interarrival-time distribution can be written as $[A_1^*(s)]^k$ where $A_1^*(s)$ is itself a legitimate LST.

The proof of this assertion is straightforward. For simplicity, let us use (6.1) in the form $z^k = \beta(z)$. Then by a geometric argument essentially the same as that for the $G/M/1$ used in Figure 5.1, it follows that there exists a unique *real* root in $(0, 1)$ for *all* k when

$$\beta'(1) = \frac{\mu}{\lambda} > \left.\frac{dz^k}{dz}\right|_{z=1} = k,$$

which is equivalent to $k\lambda/\mu < 1$, which is true from ergodicity.

For k even, we see that there is an additional real root in $(-1, 0)$ with a smaller modulus than the positive root, since z^k is a symmetric function and $0 < \beta(-z) < \beta(z)$ for $z \in (0, 1)$. For distinctness, the assumption of the power form for $A^*(s)$ (which is the property of *infinite divisibility*) permits us to rewrite the characteristic equation as

$$z^k = \beta(z) = [\beta_1(z)]^k,$$

where the kth root $\beta_1(z)$ of $\beta(z)$ is a unique and analytic pgf. Thus when we take the kth root, it follows that

$$z = \beta_1(z)e^{2\pi ni/k} \quad (n = 1, 2, \ldots, k)$$
$$\Rightarrow r\,e^{i\theta} = \beta_1(r\,e^{i\theta})e^{2\pi ni/k}. \tag{6.3}$$

Then, by an argument virtually identical to that used to show that there is one characteristic root for the $G/M/1$ inside the unit circle, we can show that the above equation has a unique root in the unit circle for each $n = 1, \ldots, k$ as long as $\beta_1'(1) = \mu/k\lambda > 1$ or $k\lambda/\mu < 1$. Finally, no two of these k roots can be equal, since that would imply the equality of their respective roots of unity, thus requiring a contradiction of the uniqueness we just showed.

The sufficient condition of this theorem can be weakened by asking that $\beta(z)$ be nonzero but not necessarily infinitely divisible. The kth root function is then analytic, and a proof can be devised (see Chaudhry et al., 1990) using Rouché's theorem and not requiring that the function $\beta(z)$ be a legitimate probability generating function.

Unfortunately, there are some $\beta(z)$ associated with $G/E_k/1$ models which have zeros inside the unit circle. However, we can feel comfortable knowing that a good number of queues encountered in practice will have infinitely divisible interarrival-time distributions, since the exponential and Erlang are infinitely divisible and distributions built up from them by convolutions (such as generalized Erlang) will be also. For all other distributions, one should always first try to determine whether $\beta(z)$ is ever zero, for if not, the kth-root approach of the infinitely divisible case will work. If there is a zero of $\beta(z)$ in the unit disk and it is isolated (as it will be in most cases), it is fairly easy to program any numerical procedure to go around the difficulties. Furthermore, there are examples for which $\beta(z)$ has zeros, but where the characteristic equations still have distinct roots. Because of the distinctness of the roots inside the unit circle, we see that the waiting-time distribution function of Section 1.4 for the line delays will be a linear but possibly nonconvex combination of negative exponential functions with possibly complex parameters.

The algorithm we recommend for this problem is successive substitution, though this time we are to work in the complex domain and must thus solve k different problems as we move n from 1 to k. A priori, we know that each of these problems has a distinct, complex-valued solution. To illustrate this approach to the rootfinding, let us consider an $E_3/E_3/1$ problem, with $\lambda = 1$ and $\mu = 4$ ($p = k\lambda/\mu = 3/4$). Then, given that $A^*(s) = [j\lambda/(j\lambda + s)]^3$ here, the resultant problem is to find the roots of

$$z^3 = A^*[\mu(1-z)] = \left(\frac{j\lambda}{j\lambda + \mu(1-z)}\right)^3 = \frac{27}{(7-4z)^3}.$$

We see that the problem is thus equivalent to finding the roots of a sixth-degree polynomial, with one root at 1, one real root in $(0, 1)$, two complex conjugate roots with absolute values less than 1, and two possibly complex roots with absolute values more than 1. But rather than turning to a polynomial rootfinder, we shall use successive substitution on (6.3) as

$$z = \frac{3}{7-4z} \exp\left(\frac{2\pi ni}{3}\right) \qquad (n = 1, 2, 3). \tag{6.4}$$

First, for $n = 3$, we get the real root on $(0, 1)$, and the solution is easily found by the quadratic formula to be 0.75, since (6.4) simplifies in this case to $(4z - 3)(1 - z) = 0$.

Now, let $n = 2$ in (6.4), so that we are to solve

$$z = \frac{3}{7-4z} \exp\left(\frac{4\pi i}{3}\right),$$

where $\exp(4\pi i/3)$ is the complex number $\cos(4\pi/3) + i\sin(4\pi/3) = -0.5 - (\sqrt{3}/2)i$, which we use as the starting point for iterations defined by

$$z^{(m+1)} = \frac{3}{7-4z^{(m)}} \exp\left(\frac{4\pi i}{3}\right).$$

Table 6.1   Successive Substitution Steps

m	$z^{(m)}$	Right-Hand Side
1	$(-.500, -.866)$	$(-.242, -.196)$
2	$(-.242, -.196)$	$(-.218, -.305)$
3	$(-.218, -.305)$	$(-.236, -.294)$
4	$(-.236, -.294)$	$(-.233, -.293)$
5	$(-.233, -.293)$	$(-.233, -.293)$

The results are displayed in Table 6.1, where we note a very rapid convergence. The roots in question are therefore $-0.233 \pm 0.293i$, each of which is clearly less than 1 in absolute value. Thus, in total analogy with the $G/M/1$ model, we may write the steady-state arrival-point probabilities as

$$q_n = C_1(0.75)^n + C_2(-0.233 - 0.293i)^n + C_3(-0.233 + 0.293i)^n,$$

where C_3 must equal the complex conjugate of C_2 in order for the complex arithmetic to lead to real-valued answers. Furthermore, it also follows that the line-delay distribution function has the form

$$W_q(t) = 1 - (K_1 e^{-\mu(1-0.75)t} - K_2 e^{-\mu(1+0.233+0.293i)t} - \overline{K}_2 e^{-\mu(1+0.233-0.293i)t})$$
$$= 1 - (K_1 e^{-0.25\mu t} - K_2 e^{-(1.233+0.293i)\mu t} - \overline{K}_2 e^{-(1.233-0.293i)\mu t}),$$

where the complex constants (with positive real parts) need to be found from boundary conditions.

At this point, we are temporarily going to put aside the question of how to find the remaining constants in these solutions. These steps turn out to be more clearly explained by two alternative methods for solving this problem, and discussions of these follow in the subsequent Sections 6.1.1 and 6.1.2.

6.1.1   Matrix Geometric Solutions for $G/E_k/1$ and $G/PH_k/1$

By use of a clever device largely due to Neuts (1981), the fundamental arrival-point argument used for the $G/M/1$ may be extended to the $G/PH_k/1$. The idea is to establish a transition matrix whose form is similar to (5.54) except that the entries become matrices. To show how this results, consider a $G/E_2/1$ system. Denoting the state of the system in the usual way for Erlang service $[(n, i) = n$ in system, customer in phase i of service], we have the transition matrix for the imbedded discrete parameter chain at arrival epochs as

$$P = \begin{array}{c c} & \begin{array}{ccccccccc} 0 & 1,2 & 1,1 & 2,2 & 2,1 & 3,2 & 3,1 & 4,2 & 4,1 & 5,2 \quad \cdots \end{array} \\ \begin{array}{c} 0 \\ 1,2 \\ 1,1 \\ 2,2 \\ 2,1 \\ 3,2 \\ 3,1 \\ 4,2 \\ 4,1 \\ \vdots \end{array} & \left(\begin{array}{ccccccccccc} 1-\Sigma b_j & b_0 & b_1 & 0 & \cdots \\ 1-\Sigma b_j & b_2 & b_3 & b_0 & b_1 & 0 & \cdots \\ 1-\Sigma b_j & b_1 & b_2 & 0 & b_0 & 0 & \cdots \\ 1-\Sigma b_j & b_4 & b_5 & b_2 & b_3 & b_0 & b_1 & 0 & \cdots \\ 1-\Sigma b_j & b_3 & b_4 & b_1 & b_2 & 0 & b_0 & 0 & \cdots \\ 1-\Sigma b_j & b_6 & b_7 & b_4 & b_5 & b_2 & b_3 & b_0 & b_1 & 0 & \cdots \\ 1-\Sigma b_j & b_5 & b_6 & b_3 & b_4 & b_1 & b_2 & 0 & b_0 & 0 & \cdots \\ 1-\Sigma b_j & b_8 & b_9 & b_6 & b_7 & b_4 & b_5 & b_2 & b_3 & b_0 & \cdots \\ 1-\Sigma b_j & b_7 & b_8 & b_5 & b_6 & b_3 & b_4 & b_1 & b_2 & 0 & \cdots \\ \vdots & \vdots & \vdots & \vdots & \vdots & \vdots & \vdots & \vdots & \vdots & \vdots \end{array} \right) \end{array},$$

where now $b_n = \Pr\{n \text{ phases of service completed during an interarrival time}\}$.

Denoting by B_i the matrix

$$\begin{bmatrix} b_{2i} & b_{2i+1} \\ b_{2i-1} & b_{2i} \end{bmatrix},$$

by B_{01} the vector (b_0, b_1), by B_{00} the scalar $1 - b_0 - b_1$, and by B_{i0} the column vector

$$\begin{pmatrix} 1 - \Sigma b_j \\ 1 - \Sigma b_j \end{pmatrix},$$

this can be rewritten as follows:

$$P = \begin{array}{c c} & \begin{array}{ccccccc} 0 & 1 & 2 & 3 & 4 & 5 & \cdots \end{array} \\ \begin{array}{c} 0 \\ 1 \\ 2 \\ 3 \\ \vdots \end{array} & \left(\begin{array}{ccccccc} B_{00} & B_{01} & 0 & \cdots \\ B_{10} & B_1 & B_0 & 0 & \cdots \\ B_{20} & B_2 & B_1 & B_0 & 0 & \cdots \\ B_{30} & B_3 & B_2 & B_1 & B_0 & 0 & \cdots \\ \vdots & \vdots & \vdots & \vdots & \vdots & \vdots \end{array} \right) \end{array}. \qquad (6.5)$$

The P matrix of Equation (6.5) looks very similar to the P matrix of the $G/M/1$ queue given in (5.54), but with 2×2 matrices replacing scalars. A slight variation exists in the first row with its 1×2 matrix and the first column with its 2×1 matrices. This same pattern would exist for E_k service with k replacing 2; that is, the B_i would be $k \times k$ matrices, and so on. Further, a similar kind of structural pattern can be obtained for PH_k service. Differences exist only in the first row and first few columns at most.

Neuts further showed that the similarity does not end here, but that the solution is totally analogous to that for the $G/M/1$ and leads to what Neuts has called a matrix-geometric solution. We of course wish to find the invariant, nonnegative probability vector q satisfying $qP = q$ and $qe = 1$. If we partition the vector q into a sequence of contiguously laid-out vectors $q_0, q_1, \ldots,$ where q_j is k-dimensional, $j \geq 1$, and q_0 is of dimension one, then, as in (5.56), the stationary equations become

$$q_j = \sum_{i=0}^{\infty} q_{j+i-1} B_i \qquad (j > 1),$$

$$q_0 = \sum_{i=0}^{\infty} q_i A_i \qquad [A_0 = (B_{00}, B_{01}), \quad A_i = (B_{i0}, B_i), \quad i \geq 1]$$

with $\sum q_i e = 1$.

The form of the solution to this turns out to be $q_n = CR^n$, where R is a special nonnegative and irreducible $k \times k$ matrix. It is in fact the minimal nonnegative solution (in the sense that all its entries are individually the smallest in the set of solving matrices Z) to the matrix equation

$$Z = \sum_{i=0}^{\infty} Z^i B_i = B(Z). \qquad (6.6)$$

This last relationship is the matrix equivalent of the fundamental generating-function equation for the $G/M/1$ given by (5.58). The answer is unique for $\rho < 1$, and the actual computation of R must be numerical. One method of finding the matrix R is a numerical iterative procedure motivated by the process of successive substitution we used for getting the root of (5.58). This time, we write

$$Z^{(k+1)} = B(Z^{(k)}) \qquad (k = 0, 1, 2, \ldots).$$

The reader is referred to Neuts (1981) and Kao (1991) for further details on this extension of $G/M/1$ to $G/PH_k/1$. This current discussion also carries through to $G/PH_k/c$, though the size of the state space increases rapidly, making computations formidable unless c and k are small. Since the Erlang type k is a special type of phase-type distribution, we note that the matrix-geometric approach presents an alternative solution method to the complex-plane rootfinding procedure introduced earlier in this chapter for the $G/E_k/1$ and $G/E_k/c$ problems.

The matrix geometric $\{q_n\}$ determined for PH service as described above can be used to find line waiting-time distribution functions, by setting up an appropriate continuous-parameter Markov chain with an absorbing state and finding the time to absorption, much as we did in Section 3.3.2. The line delay distribution of an arriving customer can be obtained beginning from conditional waiting-time probabilities, given that an arrival finds n in the system, and then summing over all states. Each conditional problem requires the simultaneous solution of a system of linear differential equations in the unknown conditional waiting-time distributions, which are then multiplied by their respective q_n. Thus, obtaining waiting-time distributions in this fashion requires solving large numbers of differential equations using carefully chosen numerical procedures (see Neuts, 1981).

6.1.2 Quasi-Birth–Death Processes

An interesting special case of $G/PH_k/1$ occurs when the input is Poisson. The infinitesimal generator of such a queue has a matrix structure composed of tridiagonal elements which are themselves matrices, thus generating a matrix analog of the standard birth–death process. This structure arises because the allowable transitions from state (n, i) (i.e., n in system, customer in phase i of service) are to the set $\{(n, i - 1), (n + 1, i), (n - 1, k)\}$, where k is the "first" stage of service, with successive service stages labeled with decreasing numbers.

An illustrative $M/PH_2/1$ generator matrix Q with mean service rates μ_1 and μ_2 is as follows, analogous to that provided for the $M/E_2/1$ of Section 6.1.1:

$Q =$

	0	1,2	1,1	2,2	2,1	3,2	3,1	4,2	4,1	5,2	...
0	$-\Sigma b_0$	λ	0	0	...						
1,2	0	$-\Sigma b_1$	μ_2	λ	0	0	...				
1,1	μ_1	0	$-\Sigma b_2$	0	λ	0	0	...			
2,2	0	0	0	$-\Sigma b_3$	μ_2	λ	0	0	...		
2,1	0	μ_1	0	0	$-\Sigma b_4$	0	λ	0	0	...	
3,2	0	0	0	0	0	$-\Sigma b_5$	μ_2	λ	0	0	...
3,1	0	0	0	μ_1	0	0	$-\Sigma b_6$	0	λ	0	...
4,2	0	0	0	0	0	0	0	$-\Sigma b_7$	μ_2	0	...
4,1	0	0	0	0	0	μ_1	0	0	$-\Sigma b_8$	0	...
⋮	⋮	⋮	⋮	⋮	⋮	⋮	⋮	⋮	⋮	⋮	

,

where Σb_i denotes the sum of the quantities in row i. When this matrix is rewritten in block form, we find that

$$
Q = \begin{array}{c} \\ 0 \\ 1 \\ 2 \\ 3 \\ \vdots \end{array}
\begin{array}{cccccccc}
0 & 1 & 2 & 3 & 4 & 5 & \cdots \\
\end{array}
\left(\begin{array}{ccccccc}
B_{00} & B_{01} & 0 & \cdots & & & \\
B_{10} & B_1 & B_0 & 0 & \cdots & & \\
0 & B_2 & B_1 & B_0 & 0 & \cdots & \\
0 & 0 & B_2 & B_1 & B_0 & 0 & \cdots \\
\vdots & \vdots & \vdots & \vdots & \vdots & \vdots & \vdots
\end{array}\right).
$$

where

$$
B_{00} = -\sum b_0 = -\lambda, \qquad B_{01} = [\lambda \quad 0], \qquad B_{10} = [0, \quad \mu_1],
$$

$$
B_0 = \begin{bmatrix} \lambda & 0 \\ 0 & \lambda \end{bmatrix}, \qquad
B_1 = \begin{bmatrix} -\sum b_i & \mu_2 \\ 0 & -\sum b_i \end{bmatrix}, \qquad
B_0 = \begin{bmatrix} 0 & 0 \\ \mu_1 & 0 \end{bmatrix}.
$$

Note that this block-matrix representation of the Q matrix has a tridiagonal structure reminiscent of the usual birth–death generator seen back in Section 1.9.3; thus the name *quasi-birth–death* (QBD) *process*.

After formulating the Q values for a specific problem, the solution can be determined analytically via the techniques mentioned in Section 6.1.1 or computed through successive substitution in the equation $0 = \pi Q$. The QBD approach is useful in analyzing priority queues, as in the previously mentioned work of Miller (1981)—as noted in 3.4.1—or in problems like those with one server and two queues, with priority for the longer line. Such problems are not $M/PH_k/1$ queues, but they have multiple classes of Poisson arrivals coupled with exponential service that leads to a QBD form.

6.2 GENERAL INPUT, GENERAL SERVICE (G/G/1)

Although nearly completely void of specific structure, we are nonetheless able to get some results for single-server queues with general (i.e., arbitrary) input and service. The major things we are able to do follow from an integral equation of the Wiener–Hopf type for the stationary distribution of the waiting time in queue of an arbitrary customer. This equation is largely due to Lindley (1952) and goes under his name. Further details on the $G/G/1$ may be found in the literature (e.g., Cohen, 1982), but a good portion of it would be beyond the level of this text.

We begin by observing that the relationship between the line waiting times, say $W_q^{(n)}$ and $W_q^{(n+1)}$, of the nth and $(n + 1)$st customers, which we noted earlier in the text as Equation (1.5), is totally valid for the arbitrary $G/G/1$ problem. This recurrence is given by

$$W_q^{(n+1)} = \begin{cases} W_q^{(n)} + S^{(n)} - T^{(n)} & (W_q^{(n)} + S^{(n)} - T^{(n)} > 0), \\ 0 & (W_q^{(n)} + S^{(n)} - T^{(n)} \leq 0), \end{cases}$$

or

$$W_q^{(n+1)} = \max(0, W_q^{(n)} + S^{(n)} - T^{(n)}),$$

where $S^{(n)}$ is the service time of the nth customer and $T^{(n)}$ is the time between the arrivals of the two customers. We can immediately note that the stochastic process $\{W_q^{(n)}, n = 0, 1, 2, \ldots\}$ is a discrete-time Markov process, since the behavior of $W_q^{(n+1)}$ is only a function of the stochastically determined value of $W_q^{(n)}$ and is independent of prior waiting-time history.

Now, from basic probability arguments, we have

$$W_q^{(n+1)}(t) \equiv \Pr\{[\text{line delay } W_q^{(n+1)} \text{ of } (n+1)\text{st customer}] \leq t\}$$

$$= \Pr\{W_q^{(n+1)} = 0\} + \Pr\{0 < W_q^{(n+1)} \leq t\}$$

$$= \Pr\{W_q^{(n)} + S^{(n)} - T^{(n)} \leq 0\} + \Pr\{0 < W_q^{(n)} + S^{(n)} - T^{(n)} \leq t\}$$

$$= \Pr\{W_q^{(n)} + S^{(n)} - T^{(n)} \leq t\}.$$

If we now define the random variable $U^{(n)} \equiv S^{(n)} - T^{(n)}$ with CDF $U^{(n)}(x)$, then, making use of the convolution formula,

$$W_q^{(n+1)}(t) = \int_{-\infty}^{t} W_q^{(n)}(t - x)\, dU^{(n)}(x) \qquad (0 \leq t < \infty).$$

In the steady state ($\rho < 1$), the two waiting-time CDFs must be identical; hence using $W_q(t)$ to denote the stationary delay distribution, we find *Lindley's equation* as

$$\begin{aligned} W_q(t) &= \begin{cases} \int_{-\infty}^{t} W_q(t - x)\, dU(x) & (0 \leq t < \infty), \\ 0 & (t < 0) \end{cases} \\ &= -\int_{0}^{\infty} W_q(y)\, dU(t - y) \quad (0 \leq t < \infty), \end{aligned} \qquad (6.7)$$

where $U(x)$ is the equilibrium $U^{(n)}(x)$ and is given by the convolution of S and $-T$:

$$U(x) = \int_{\max(0,x)}^{\infty} B(y)\, dA(y - x). \qquad (6.8)$$

The usual approach to the solution of a Wiener–Hopf integral equation such as (6.7) (see Feller, 1971) begins with the definition of a new function as

$$W_q^-(t) \equiv \begin{cases} \int_{-\infty}^t W_q(t-x)\, dU(x) & (t<0), \\ 0 & (t \geq 0). \end{cases} \tag{6.9}$$

Hence it follows from (6.7) that

$$W_q^-(t) + W_q(t) = \int_{-\infty}^t W_q(t-x)\, dU(x) \qquad (-\infty < t < \infty). \tag{6.10}$$

Note that $W_q^-(t)$ is the portion of the CDF associated with the negative values of $W_q^{(n)} + S - T$ when there is idle time between the nth and the $(n+1)$st customer.

It turns out to be easiest to try to obtain $W_q(t)$, $t>0$ (i.e., the positive part), since $W_q(t)$ is not continuous at 0, but has a jump equal to the arrival-point probability q_0, so that $W_q(0) = q_0$. Start this by denoting the two-sided Laplace transforms (LTs) of $W_q(t)$ and $W_q^-(t)$ as

$$\overline{W}_q(s) = \int_{-\infty}^{\infty} e^{-st}\, W_q(t)\, dt = \int_0^{\infty} e^{-st}\, W_q(t)\, dt$$

and

$$\overline{W}_q^-(s) = \int_{-\infty}^{\infty} e^{-st}\, W_q^-(t)\, dt = \int_{-\infty}^0 e^{-st}\, W_q^-(t)\, dt.$$

In addition, we shall use $U^*(s)$ as the (two-sided) LST of $U(t)$.

We then take the two-sided Laplace transform of both sides of (6.10). The transform of the right-hand side is found to be

$$\mathcal{L}_2\left\{ \int_{-\infty}^t W_q(t-x)\, dU(x) \right\} = \int_{-\infty}^{\infty} \int_{-\infty}^t e^{-(t-x)s}\, W_q(t-x)\, e^{-sx}\, dU(x)\, dt.$$

Since $W_q(t-x) = 0$ for $x \geq t$, we can write

$$\mathcal{L}_2 = \int_{-\infty}^{\infty} \int_{-\infty}^{\infty} e^{-(t-x)s}\, W_q(t-x)\, e^{-sx}\, dU(x)\, dt$$

$$= \left(\int_{-\infty}^{\infty} e^{-su}\, W_q(u)\, du \right) \left(\int_{-\infty}^{\infty} e^{-sx}\, dU(x) \right)$$

$$= \overline{W}_q(s)\, U^*(s).$$

But U is the CDF of the difference of the interarrival and service times and hence by the convolution property must have (two-sided) LST equal to the product of the interarrival transform $A^*(s)$ evaluated at $-s$ and the service transform $B^*(s)$, since $A(t)$ and $B(t)$ are both zero for $t < 0$. Hence $U^*(s) = A^*(-s)B^*(s)$, and thus from (6.10),

$$\overline{W_q^-}(s) + \overline{W_q}(s) = \overline{W_q}(s)A^*(-s)B^*(s)$$

$$\Rightarrow \quad \overline{W_q}(s) = \frac{\overline{W_q^-}(s)}{A^*(-s)B^*(s) - 1}. \tag{6.11}$$

Therefore, given any pair $\{A(t), B(t)\}$ for the *G/G*/1, we can theoretically find the Laplace transform of the line delay. The determination of $\overline{W_q^-}(s)$ is the primary difficulty in this computation, often requiring advanced concepts from the theory of complex variables.

To show how this all may work, let us consider the *M/M*/1 problem and then check the answer against our earlier result from Chapter 2. Here

$$B(t) = 1 - e^{-\mu t}, \qquad B^*(s) = \frac{\mu}{\mu + s},$$

$$A(t) = 1 - e^{-\lambda t},$$

and

$$A^*(-s) = \frac{\lambda}{\lambda - s}.$$

So, from (6.8),

$$U(x) = \begin{cases} \displaystyle\int_0^\infty (1 - e^{-\mu y})\lambda e^{-\lambda(y-x)}\,dy & (x < 0), \\[2ex] \displaystyle\int_x^\infty (1 - e^{-\mu y})\lambda e^{-\lambda(y-x)}\,dy & (x \geq 0) \end{cases}$$

$$= \begin{cases} \dfrac{\mu e^{\lambda x}}{\lambda + \mu} & (x < 0), \\[2ex] 1 - \dfrac{\lambda e^{-\mu x}}{\lambda + \mu} & (x \geq 0). \end{cases} \tag{6.12}$$

Thus

$$W_q^-(t) = \int_{-\infty}^{t} W_q(t - x) \, dU(x) \qquad (t < 0)$$

$$= \frac{\lambda \mu}{\lambda + \mu} \int_{-\infty}^{t} W_q(t - x) e^{\lambda x} \, dx \qquad (t < 0).$$

Letting $u = t - x$ yields

$$W_q^-(t) = \frac{\lambda \mu}{\lambda + \mu} \int_0^{\infty} W_q(u) e^{-\lambda(u-t)} \, du$$

$$= \frac{\lambda \mu e^{\lambda t}}{\lambda + \mu} \int_0^{\infty} W_q(u) e^{-\lambda u} \, du$$

$$= \frac{\lambda \mu e^{\lambda t} \overline{W}_q(\lambda)}{\lambda + \mu}. \qquad (6.13)$$

Now $\overline{W}_q(\lambda)$ may be easily found from some of our earlier work. In Section 5.2.1, we noted that

$$\pi_n^q = \Pr\{n \text{ in queue just after a departure}\}$$

$$= \frac{1}{n!} \int_0^{\infty} (\lambda t)^n e^{-\lambda t} \, dW_q(t)$$

for any $M/G/c$. Hence if $G = M$ and $c = 1$, we find that

$$\pi_0^q = \int_0^{\infty} e^{-\lambda t} \, dW_q(t).$$

Integration by parts then gives

$$\pi_0^q = e^{-\lambda t} W_q(t) \big|_0^{\infty} + \lambda \int_0^{\infty} e^{-\lambda t} W_q(t) \, dt.$$

But $\lim_{t \to \infty} e^{-\lambda t} = 0$, and since we are only concerned computationally with $W_q(t)$ for $t > 0$, let us make $W_q(0) = 0$ for the time being to simplify the analysis henceforth. In the end, we shall simply set $W_q(0) = p_0$, since it is true for all $M/G/1$ (which always satisfy Little's formula) that $W_q(0) = p_0 = 1 - \lambda/\mu$. Therefore

$$\pi_0^q = \lambda \overline{W}_q(\lambda).$$

We also know that π_0^q must be equal to $p_0 + p_1 = (1 - \rho)(1 + \rho)$, since this is an *M/M/*1. Hence

$$\overline{W}_q(\lambda) = \frac{(1 - \rho)(1 + \rho)}{\lambda},$$

and, from (6.13),

$$W_q^-(t) = \frac{e^{\lambda t}(1 - \rho)(1 + \rho)}{1 + \rho} = e^{\lambda t}(1 - \rho),$$

with transform

$$\overline{W}_q^-(s) = \frac{1 - \rho}{\lambda - s}.$$

Putting everything together, we find, using (6.11), that

$$\overline{W}_q(s) = \frac{(1 - \rho)/(\lambda - s)}{\lambda \mu/[(\lambda - s)(\mu + s)] - 1}$$

$$= \frac{(1 - \rho)(\mu + s)}{s(\mu - \lambda + s)}$$

$$= \frac{1 - \rho}{s} + \frac{\lambda(1 - \rho)}{s(\mu - \lambda + s)},$$

which inverts to

$$W_q(t) = 1 - \rho + \frac{\lambda(1 - \rho)(1 - e^{-(\mu - \lambda)t})}{\mu - \lambda}$$

$$= 1 - \rho e^{-\mu(1 - \rho)t} \qquad (t > 0).$$

Now, realizing that $W_q(0)$ equals $p_0 = 1 - \rho$, the result is thus the same as obtained in (2.21).

We illustrate the use of Lindley's equation by the following example.

Example 6.1
Our friend the hair-salon operator, H. R. Cutt, has decided that she would like to find the distribution of the line waits her customers undergo. She realizes that under the new priority rules she recently designated (see Example 3.7, Section 3.4.1), service times may be either of two possibilities: exponential with mean 5 min (for the trims, used one-third of the time), and exponential with mean 12.5 min (for the others). The arrival stream remains Poisson with parameter $\lambda = 5$/hr. If we assume that there are no priorities,

then Cutt's system is an $M/G/1$ queue, with service times given to be the mixed exponential (i.e., the weighted average of two the exponentials) having

$$b(t) = \tfrac{1}{3}\tfrac{1}{5}e^{-t/5} + \tfrac{2}{3}\tfrac{2}{25}e^{-2t/25},$$

$$B(t) = 1 - (\tfrac{1}{3}e^{-t/5} + \tfrac{2}{3}e^{-2t/25}),$$

$$B^*(s) = \frac{1}{15s + 3} + \frac{4}{75s + 6}.$$

So, from (6.9), noting that $a(t) = \tfrac{1}{12}e^{-t/12}$, we find that

$$W_q^-(t) = \int_{-\infty}^t W_q(t - x)\, dU(x) \qquad (t < 0),$$

where, from (6.8),

$$
U(x) =
\begin{cases}
\displaystyle\int_0^\infty \left(1 - \frac{e^{-y/5} + 2e^{-2y/25}}{3}\right)\frac{e^{-(y-x)/12}}{12}\, dy & (x < 0), \\[2em]
\displaystyle\int_x^\infty \left(1 - \frac{e^{-y/5} + 2e^{-2y/25}}{3}\right)\frac{e^{-(y-x)/12}}{12}\, dy & (x \geq 0)
\end{cases}
$$

$$
=
\begin{cases}
\displaystyle e^{x/12}\left(1 - \frac{\tfrac{1}{36}}{\tfrac{1}{12} + \tfrac{1}{5}} - \frac{\tfrac{1}{18}}{\tfrac{1}{12} + \tfrac{2}{25}}\right) & (x < 0), \\[2em]
\displaystyle 1 - \frac{\tfrac{1}{36}e^{x/5}}{\tfrac{1}{12} + \tfrac{1}{5}} - \frac{\tfrac{1}{18}e^{-2x/25}}{\tfrac{1}{12} + \tfrac{2}{25}} & (x \geq 0).
\end{cases}
$$

Thus

$$W_q^-(t) = \frac{1}{12} \times \frac{468}{833}\int_{-\infty}^t W_q(t - x)e^{x/12}\, dx \qquad (t < 0)$$

$$= \tfrac{39}{833}e^{t/12}\overline{W}_q(\tfrac{1}{12}).$$

We must therefore next find $\overline{W}_q(\lambda)$, $\lambda = \tfrac{1}{12}$. Since the result we quoted earlier in this section from Section 5.2.1 regarding π_n^q was valid for any $M/G/1$, it is certainly true in this current case. Hence, again,

$$\overline{W}_q(\lambda) = \frac{\pi_0^q}{\lambda}.$$

But here π_0^q will clearly be different from its value for $M/M/1$. We know that $\pi_0^q = \pi_0 + \pi_1$, where π_0 and π_1 refer to the departure-point probabilities of 0 and 1 in the system, respectively. Recalling the analysis of Chapter 5, we have $\pi_0 = 1 - \rho$, and from Equation (5.16), $\pi_1 = \pi_0(1 - k_0)/k_0$. But

$$k_0 = \int_0^\infty e^{-\lambda t}\, dB(t)$$

$$= \int_0^\infty e^{-t/12}\left(\frac{e^{-t/5}}{15} + \frac{4e^{-2t/25}}{75}\right) dt = \frac{468}{833}.$$

Since

$$\rho = \tfrac{1}{12}[\tfrac{1}{3}(5) + \tfrac{2}{3}(\tfrac{25}{2})] = \tfrac{5}{6},$$

we get $\pi_0 = \frac{1}{6}$ and

$$\pi_1 = \tfrac{1}{6}\left(\tfrac{365}{468}\right) = \tfrac{365}{2808}.$$

So

$$\pi_0^q \doteq 0.297 \quad\text{and}\quad \overline{W}_q(\lambda) \doteq \frac{0.297}{\frac{1}{12}} = 3.564.$$

Therefore

$$W_q^-(t) \doteq \tfrac{39}{833}(3.564)e^{t/12} \quad\text{and}\quad \overline{W}_q^-(s) \doteq \frac{2.00}{1 - 12s}.$$

Since $A^*(-s) = 1/(1 - 12s)$, we now have, from (6.11), that

$$\overline{W}_q(s) \doteq \cfrac{\cfrac{2.00}{1 - 12s}}{\cfrac{1}{1 - 12s}\left(\cfrac{1}{3 + 15s} + \cfrac{4}{6 + 75s}\right) - 1}$$

$$= \frac{2.00(18 + 315s + 1125s^2)}{36s + 2655s^2 + 13{,}500s^3}$$

$$\doteq \frac{1}{s} - \frac{0.010}{s + 0.18} - \frac{0.82}{s + 0.015}.$$

Finally, inversion of the LT yields

$$W_q(t) \doteq 1 - 0.01e^{-0.18t} - 0.82e^{-0.015t}.$$

6.2.1 $GE_j/GE_k/1$

The $G/G/1$ problem can be greatly simplified if it can be assumed that the interarrival and service distributions can be expressed individually as convolutions of independent and not necessarily identical exponential random variables. Such a form is called the *generalized Erlang*, and of course the regular Erlang is just a special case. This is not a particularly strong restriction, since it can be shown that any CDF can be approximated to almost any degree of accuracy by such a convolution, using an argument similar to that used to show the completeness of the polynomials in function space.

So we may write that

$$A^*(s) = \prod_{i=1}^{j} \frac{\lambda_i}{\lambda_i + s}, \qquad B^*(s) = \prod_{i=1}^{k} \frac{\mu_i}{\mu_i + s}.$$

Thus from (6.11),

$$\overline{W}_q(s) = \frac{\overline{W_q^-}(s)}{\prod_{i=1}^{j} \dfrac{\lambda_i}{\lambda_i - s} \prod_{i=1}^{k} \dfrac{\mu_i}{\mu_i + s} - 1}$$

$$= \frac{\overline{W_q^-}(s) \prod_{i=1}^{j} (\lambda_i - s) \prod_{i=1}^{k} (\mu_i + s)}{\prod_{i=1}^{j} \lambda_i \prod_{i=1}^{k} \mu_i - \prod_{i=1}^{j} (\lambda_i - s) \prod_{i=1}^{k} (\mu_i + s)}. \tag{6.14}$$

The denominator of $\overline{W}_q(s)$ is clearly a polynomial of degree $j + k = n$; its roots will be denoted by $s_1, \ldots, s_n$, where s_1 is easily seen to be 0. Furthermore, it can be shown from the form of the polynomial and Rouché's theorem that there are exactly $j - 1$ roots, $s_2, \ldots, s_j$, whose real parts are positive, and thus k roots, $s_{j+1}, \ldots, s_n$, with negative real parts. Thus we must be able to write that

$$\prod_{i=1}^{j} \lambda_i \prod_{i=1}^{k} \mu_i - \prod_{i=1}^{j} (\lambda_i - s) \prod_{i=1}^{k} (\mu_i + s)$$

$$= s(s - s_2) \cdots (s - s_j)(s - s_{j+1}) \cdots (s - s_n).$$

Hence, letting $z_i = s_{j+i}$,

$$\overline{W}_q(s) = \frac{\overline{W_q^-}(s) \, \Pi_{i=1}^{j} (\lambda_i - s) \, \Pi_{i=1}^{k}(\mu_i + s)}{s \, \Pi_{i=2}^{j} (s - s_i) \, \Pi_{i=1}^{k} (s - z_i)}.$$

But the numerator must also have the roots $s_2, \ldots, s_j$ to preserve analyticity for $\text{Re}(s) > 0$ and hence may be rewritten as $Cf(s) \, \Pi^j_{i=2}(s - s_i)$, where C is a constant to be found later. So now, after cancellation,

$$\overline{W}_q(s) = \frac{Cf(s)}{s \, \Pi^k_{i=1}(s - z_i)}.$$

The last key step is to show that $f(s)$ is itself also a polynomial. This can, in fact, be done again using concepts from the theory of complex variables. The final form of $\overline{W}_q(s)$ is therefore determined by finding $Cf(s)$. It turns out that $f(s)$ cannot have any roots with positive real parts and, in fact, has the roots $-\mu_1, \ldots, -\mu_k$. Hence $\overline{W}_q(s)$ may be written as

$$\overline{W}_q(s) = \frac{C \, \Pi^k_{i=1}(s + \mu_i)}{s \, \Pi^k_{i=1}(s - z_i)}.$$

To get C now we note that

$$\mathscr{L}\{W'_q(t)\} = s\overline{W}_q(s) - W_q(0)$$

$$= C \prod^k_{i=1} \frac{s + \mu_i}{s - z_i} - q_0.$$

But the value of the transform of $W'_q(t)$ as $s \to 0$ is equal to $1 - q_0$, since there is a jump in the CDF $W_q(t)$ equal to q_0 at the origin. Thus

$$1 - q_0 = C \prod^k_{i=1} \frac{\mu_i}{-z_i} - q_0 \quad \Rightarrow \quad C = \prod^k_{i=1} \frac{-z_i}{\mu_i}.$$

So

$$\overline{W}_q(s) = \frac{\Pi^k_{i=1}(-z_i/\mu_i)(s + \mu_i)}{s \, \Pi^k_{i=1}(s - z_i)}. \tag{6.15}$$

The above result is an extremely useful simplification, since it now puts $\overline{W}_q(s)$ into an easily invertible form. A partial-fraction expansion is then performed (assuming distinct z_i) to give the result

$$\overline{W}_q(s) = \frac{1}{s} - \sum^k_{i=1} \frac{C_i}{s - z_i},$$

which inverts to the generalized exponential mixture

$$W_q(t) = 1 - \sum_{i=1}^{k} C_i e^{z_i t},$$

where the $\{z_i\}$ have *negative* real parts and the values of the C_i would be determined in the usual way from the partial-fraction expansion.

While conceptually this is a simplification, in practice it is sometimes difficult to estimate the parameters λ_i $(i = 1, 2, \ldots, j)$ and μ_i $(i = 1, 2, \ldots, k)$, since j and k generally must be large in order for this method to be accurate. In the event that the $\{\mu_i\}$ are all equal to begin with, the model becomes a $G/E_k/1$, which therefore has "mixed" exponential waiting times for any interarrival-time distribution. Recognize also that each generalized Erlang (GE_k) is of phase type and further that any GE_k with distinct parameters may be written as a linear combination of exponentials with possibly negative mixing parameters.

As a check, let us verify these results for $M/M/1$. From (6.14), we need the roots of the polynomial denominator of (6.14),

$$0 = \lambda\mu - (\lambda - s)(\mu + s) = s^2 - (\lambda - \mu)s$$

with negative real parts. There is clearly one, and it is $z_1 = \lambda - \mu$. Hence, from (6.15),

$$\overline{W}_q(s) = \frac{(1 - \rho)(s + \mu)}{s(s - \lambda + \mu)},$$

which completely agrees with the result obtained earlier.

Example 6.2

Ms. W. A. R. Mup of the Phill R. Upp Company of Example 3.4 would like to know the percentage of her customers who wait more than 4 hr for oil delivery. She has already computed the average delay as 2 hr, but is now particularly concerned that too many customers are excessively delayed. We know that $\lambda = \frac{6}{5}$/hr and $\mu = \frac{3}{2}$/hr for this $M/E_2/1$ queue, and the results of this section now permit us to obtain the necessary service-time distribution $W_q(t)$.

Since there are two identical exponential stages in the service process, we see that $\mu_1 = \mu_2 = 2\mu = 3$/hr. Hence the LST of the waiting times is computed from (6.14) as

$$\overline{W}_q(s) = \frac{(s + 3)^2 \prod_{i=1}^{2} z_i}{3^2 s \prod_{i=1}^{2} (s - z_i)};$$

the quantities z_1 and z_2 are the roots with negative real parts of the polynomial denominator of (6.14),

$$0 = \lambda \mu_1 \mu_2 - (\lambda - s)(\mu_1 + s)(\mu_2 + s)$$
$$= s(5s^2 + 24s + 9).$$

Both roots of the quadratic factor are found to be real, with values -4.39 and -0.410. Hence,

$$\overline{W}_q(s) = \frac{(s + 3)^2}{5s(s + 4.39)(s + 0.410)}$$

$$= \frac{1}{s} + \frac{0.0221}{s + 4.39} - \frac{0.8221}{s + 0.410}.$$

Therefore

$$W_q(t) = 1 + 0.0221e^{-4.39t} - 0.8221e^{-0.410t}$$

$$\Rightarrow \quad \Pr\{T_q > 4\} \doteq 0.1595.$$

6.2.2   Discrete G/G/1

From the foregoing, it should be readily apparent that actual results for waiting times using Lindley's equation when interarrival and service times are continuous are often difficult to obtain. This is so primarily because of the complexity in obtaining $W_q^-(t)$. However, if the interarrival and service times are discrete (recall that any continuous distribution can be approximated by a k-point distribution), then Equation (6.7) can be used iteratively (since its right-hand side becomes a sum and we know that $\Sigma \, w_q(t_i) = 1$) to obtain the values of $W_q(t)$ at all realizable values of the discrete random variable t.

Since the state space is discrete here, the Markovian waiting-time process $\{W_q^{(n)}\}$ now clearly becomes a Markov chain. To illustrate briefly, let us assume that interarrival and service times can take on only two values each, namely, (a_1, a_2) and (b_1, b_2), respectively. Then $U = S - T$ can have at most four values—let them be $U_1 = -2$, $U_2 = -1$, $U_3 = 0$, and $U_4 = 1$. Such a system is ergodic whenever $E[U] < 0$, which will be true if we assign equal probabilities of $\frac{1}{4}$ to these values. Let us denote the steady-state probability of a wait of j as w_j; it can be found by solving the stationary equation $w_j = \Sigma_i \, w_i p_{ij}$.

Given the above values, the transition probabilities are all derived simply as

$$p_{00} = \Pr\{U = -2, -1 \text{ or } 0\} = \tfrac{3}{4}, \qquad p_{01} = \Pr\{U = 1\} = \tfrac{1}{4},$$
$$p_{10} = \Pr\{U = -2, \text{ or } -1\} = \tfrac{1}{2}, \qquad p_{11} = \Pr\{U = 0\} = \tfrac{1}{4},$$
$$p_{12} = \Pr\{U = 1\} = \tfrac{1}{4},$$

and, for $i > 1$,

$$p_{i,i-2} = p_{i,i-1} = p_{i,i} = p_{i,i+1} = \tfrac{1}{4}.$$

Thus the $\{w_j\}$ are found as the solution to

$$w_0 = \tfrac{3}{4}w_0 + \tfrac{1}{2}w_1 + \tfrac{1}{4}w_2,$$

$$w_j = \sum_{i=j-1}^{j+2} \frac{w_i}{4} \quad (j \geq 1),$$

$$1 = \sum_{j=0}^{\infty} w_j.$$

A more complete expansion of these ideas will be presented in Chapter 7 as a means of finding approximate $G/G/1$ solutions.

6.3 MULTICHANNEL QUEUES WITH POISSON INPUT AND CONSTANT SERVICE ($M/D/c$)

In the event that an $M/G/c$ is found to have deterministic service, there is sufficient special structure available to permit the obtaining of the stationary probability generating function for the distribution of the queue size, something which is not possible generally for $M/G/c$. The approach we use is fairly typical, and a similar one may be found in Saaty (1961), which is essentially originally due to Crommelin (1932).

We begin by rescaling time so that the constant service time (say $b = 1/\mu$) is now the basic unit of time; then λ becomes λ/b and μ becomes 1. For ease of notation, let us henceforth use P_n to denote the CDF of the system size in the steady state. We are then able to observe (in the spirit of Problem 5.13) that the queueing process is indeed Markovian and therefore

$p_0 = \Pr\{c$ or less in system at arbitrary instant in steady state$\}$
  $\times \Pr\{0$ arrivals in subsequent unit of time$\}$,

$p_1 = \Pr\{c$ or less in system$\} \Pr\{1$ arrival$\}$
  $+ \Pr\{c + 1$ in system$\} \Pr\{0$ arrivals$\}$,

$p_2 = \Pr\{c$ or less in system$\} \Pr\{2$ arrivals$\}$
  $+ \Pr\{c + 1$ in system$\} \Pr\{1$ arrival$\}$
  $+ \Pr\{c + 2$ in system$\} \Pr\{0$ arrivals$\}$,

$$\vdots$$

$p_n = \Pr\{c$ or less in system$\} \Pr\{n$ arrivals$\}$
  $+ \Pr\{c + 1$ in system$\} \Pr\{n - 1$ arrivals$\}$
  $+ \cdots + \Pr\{c + n$ in system$\} \Pr\{0$ arrivals$\}$,

or

$$p_0 = P_c e^{-\lambda},$$

$$p_1 = P_c \lambda e^{-\lambda} + p_{c+1} e^{-\lambda},$$

$$p_2 = \frac{P_c \lambda^2 e^{-\lambda}}{2} + p_{c+1} \lambda e^{-\lambda} + p_{c+2} e^{-\lambda}, \qquad (6.16)$$

$$\vdots$$

$$p_n = \frac{P_c \lambda^n e^{-\lambda}}{n!} + \frac{p_{c+1} \lambda^{n-1} e^{-\lambda}}{(n-1)!} + \cdots + p_{c+n} e^{-\lambda}.$$

When we define the usual generating function $P(z) \equiv \Sigma_{n=0}^{\infty} p_n z^n$, Equation (6.16) leads, after multiplying the *i*th row by z^i and then summing over all rows, to (see Problem 6.9)

$$P(z) = \frac{\Sigma_{n=0}^{c} p_n z^n - P_c z^c}{1 - z^c e^{\lambda(1-z)}} = \frac{\Sigma_{n=0}^{c-1} p_n (z^n - z^c)}{1 - z^c e^{\lambda(1-z)}}. \qquad (6.17)$$

Our first step is to prove that the poles of (6.17) are distinct. This is done by showing that no value which makes the denominator of (6.17) vanish can do the same for its derivative (see Problem 6.10(a)]. Then to get rid of the *c* unknown probabilities in the numerator of (6.17), we invoke the usual arguments, employing Rouché's theorem and the fact that $P(1) = 1$. Since $P(z)$ is analytic and bounded within the unit circle, all the zeros of the denominator within and on the unit circle must also make the numerator vanish. Rouché's theorem will tell us that there are $c - 1$ zeros of the numerator inside $|z| = 1$, while the *c*th is clearly $z = 1$. Since the numerator is a polynomial of degree *c*, it may be written as

$$N(z) = K(z - 1)(z - z_1) \cdots (z - z_{c-1}),$$

where $1, z_1, z_2, \ldots, z_{c-1}$ are the coincident roots of the denominator and numerator. To get *K*, we use the fact that $P(1) = 1$. By L'Hôpital's rule,

$$1 = \lim_{z \to 1} P(z) = \frac{K(1 - z_1) \cdots (1 - z_{c-1})}{\lambda - c} \quad \Rightarrow \quad K = \frac{\lambda - c}{(1 - z_1) \cdots (z - z_{c-1})}.$$

Hence

$$P(z) = \frac{\lambda - c}{(1 - z_1) \cdots (1 - z_{c-1})} \frac{(z - 1)(z - z_1) \cdots (z - z_{c-1})}{1 - z^c e^{\lambda(1-z)}}. \qquad (6.18)$$

The roots $\{z_i, i = 1, \ldots, c - 1\}$ can be obtained exactly like those of the

$G/E_k/1$ model from

$$z^c = e^{-\lambda(1-z)}.$$

The first step in obtaining the $\{p_n\}$ is to find p_0 by evaluating the generating function of (6.18) at $z = 0$, from which we find that

$$p_0 = \frac{(c - \lambda)(-1)^{c-1} \Pi_{i=1}^{c-1} z_i}{\Pi_{i=1}^{c-1}(1 - z_i)} \qquad (c \geq 2). \qquad (6.19)$$

Next, we get $\{p_1, \ldots, p_{c-1}\}$ from a $(c-1) \times (c-1)$ complex-valued linear system of equations created from the numerator of (6.17) set equal to zero at each of the $c-1$ roots $\{z_i\}$. The resultant matrix of coefficients has (i, j) entry given by

$$a_{ij} = z_i^j - z_i^c \qquad (i, j = 1, \ldots, c-1)$$

and right-hand-side elements equal to

$$b_i = p_0(z_i^c - 1) \qquad (i = 1, \ldots, c-1).$$

The remaining probabilities are found by recursion on (6.16), first obtaining p_c from $p_0 = P_c e^{-\lambda}$ as

$$p_c = (e^\lambda - 1)p_0 - \sum_{i=1}^{c-1} p_i.$$

Various other measures of effectiveness may also be obtained for this model. The reader is referred to Saaty (1961) and/or Crommelin (1932) for the computation of the probability of no delay and the waiting-time distribution, as well as some approximate methods for computing measures of effectiveness which are useful when extreme accuracy is not required.

6.4   SEMI-MARKOV AND MARKOV RENEWAL PROCESSES IN QUEUEING

In this section we treat a discrete-valued stochastic process which transits from state to state according to a Markov chain, but in which the time required to make each transition may be a random variable which is a function of both the "to" and "from" states. In the special cases where the transition times are independent exponential random variables with parameters dependent only on the "from" state, the semi-Markov process reduces to a continuous-parameter Markov chain. Let us henceforth refer notationally to the general semi-Markov process (SMP) as $\{X(t) \mid X(t) =$ state of the process at time $t \geq 0\}$, with the state space to be of course the nonnegative integers. There are numerous references for semi-Markov processes that the authors could suggest as further reading, two of which are Ross (1996, 1970) and Heyman and Sobel (1982).

We define $Q_{ij}(t)$ to be the joint conditional probability that, given we begin in state i, the next transition will be to state j in an amount of time less than or equal to t. Then it should be clear that the (imbedded) Markov chain which underlies the SMP has transition probabilities which are given by $p_{ij} = Q_{ij}(\infty)$. Thus the conditional transition-time distribution function for the time to transit from i to j, given i and j specified, can be found from the definition of conditional probability to be

$$F_{ij}(t) = \frac{Q_{ij}(t)}{p_{ij}} \tag{6.20}$$

whenever $p_{ij} > 0$, and is arbitrary otherwise. In addition, we shall denote the joint marginal distribution of i and t, namely, $\Sigma_{j=0}^{\infty} Q_{ij}(t)$, by $G_i(t)$, it is the CDF of the time to the next transition given that we started at state i. Let us also use $\{X_n \mid n = 0, 1, 2, \ldots\}$ to denote the imbedded Markov chain, $\{R_i(t) \mid t > 0\}$ to denote the number of transitions into state i occurring in $(0, t)$, and $R(t)$ to denote the vector whose ith component is $R_i(t)$.

The stochastic process $R(t)$ is called a Markov renewal process (MRP), and though arising from the same probabilistic situation as the semi-Markov process, the MRP should be carefully distinguished from the SMP. The MRP is a counting process which keeps track of the total number of visits to each state, while the SMP only records at each point in time the single state at which the process finds itself. But for all intents and purposes, the one always determines the other.

To illustrate these concepts, let us display three well-known examples. The first of these is the $M/M/1/\infty$ queue, which we have seen to be a Markov chain. Here the queueing process $\{N(t) = \text{number in system at time } t\}$ is clearly also semi-Markovian, with

$$Q_{01}(t) = 1 - e^{-\lambda t},$$

$$Q_{i,i-1}(t) = \text{Pr\{one occurrence (service or arrival) by } t\}$$
$$\times \text{Pr\{occurrence is a service completion\}}$$

$$= (1 - e^{-(\lambda+\mu)t}) \frac{\mu}{\lambda + \mu} \qquad (i \geq 1),$$

$$Q_{i,i+1}(t) = \text{Pr\{one occurrence (service or arrival) by } t\}$$
$$\times \text{Pr\{occurrence is an arrival\}}$$

$$= (1 - e^{-(\lambda+\mu)t}) \frac{\lambda}{\lambda + \mu} \qquad (i \geq 1).$$

So,

$$p_{01} = 1, \qquad p_{i,i-1} = \frac{\mu}{\lambda + \mu}, \quad p_{i,i+1} = \frac{\lambda}{\lambda + \mu} \quad (i \geq 1),$$

$$F_{01} = 1 - e^{-\lambda t}, \qquad F_{i,i-1} = F_{i,i+1} = 1 - e^{-(\lambda + \mu)t} \quad (i \geq 1).$$

In fact, it should also be clear that any Markov chain, whether in continuous time or not, is an SMP as well. The Markov chain in discrete time can be interpreted as an SMP whose transition times are the same constant, while the chain in continuous time has transition times which must all be exponential to maintain the memoryless property. So a simple second example of an SMP in queueing would be any birth–death model, such as the $M/M/c/\infty$ queue. There,

$$F_{01} = 1 - e^{-\lambda t},$$

$$F_{i,i-1} = F_{i,i+1} = \begin{cases} 1 - e^{-(\lambda + i\mu)t} & (1 \leq i \leq c), \\ 1 - e^{-(\lambda + c\mu)t} & (c < i < \infty). \end{cases}$$

For a third example of the appearance of SMPs in queueing, we turn to $M/G/1/\infty$. Here the system size is, in fact, neither Markovian nor semi-Markovian. But if we define a new stochastic process $\{X(t)\}$ as the number of customers who were left behind in the system by the most recent departure (this is not to be confused with the number in the system at the instant a customer departs, which is an imbedded Markov chain), then this new process is indeed semi-Markovian, and it is often called an imbedded SMP because it lies within the total general-time process.

The main use of SMPs in queueing is as a means to quantify systems which are non-Markovian, but which either are semi-Markovian or possess an imbedded SMP. The existence of semi-Markovian structure is quite advantageous and often easily permits the obtaining of some of the fundamental relationships required in queueing. To derive some of the key SMP results, we are going to need the following notation, remembering that $G_i(t)$ is the CDF of the time to the next transition given we started at state i:

$H_{ij}(t) = $ CDF of time until first transition into j beginning from i,

$$m_{ij} = \text{mean first passage time from } i \text{ to } j = \int_0^\infty t \, dH_{ij}(t),$$

$$m_i = \text{mean time spent in state } i \text{ during each visit} = \int_0^\infty t \, dG_i(t),$$

$$\eta_{ij} = \text{mean time spent in state } i \text{ before going to } j = \int_0^\infty t \, dF_{ij}(t).$$

From the definition of $G_i(t)$ given previously and Equation (6.20), we are able to immediately deduce the intuitive result that

$$m_i = \sum_{j=0}^{\infty} p_{ij} \eta_{ij}.$$

The key results that we desire are those which determine the limiting probabilities of the SMP and then relate these to the general-time queueing process in the event that the SMP is imbedded and does not possess the same distribution as the general process. These relationships are all fairly intuitive but of course require proof. A reference for the direct SMP results is Ross (1996), and Fabens (1961) gives a presentation of the relationship between the imbedded SMP and the general-time process. The key points of interest to us are presented as follows, much of it built up from the material of Section 1.9.

1. If state i communicates with state j in the imbedded Markov chain of an SMP, $F_{ij}(t)$ is not lattice (i.e., is not discrete with all of its outcomes multiples of one of the values it takes), and $m_{ij} < \infty$, then

   $v_j \equiv$ the steady-state probability that the SMP is in state j
   given that it started in state i

   $$= \lim_{t \to \infty} \Pr\{X(t) = j \mid X(0) = 0\} = \frac{m_j}{m_{jj}}.$$

2. Under the assumption that the underlying Markov chain of an SMP is irreducible and positive recurrent, and that $m_j < \infty$ for all j, if π_j is the stationary probability of j in the imbedded Markov chain, then

   $$v_j = \frac{\pi_j m_j}{\sum_{i=0}^{\infty} \pi_i m_i}.$$

3. If $\{X(t)\}$ is an aperiodic SMP in continuous time with $m_i < \infty$, then for $\delta(t)$ defined to be the time back to the most recent transition looking from t,

   $$\lim_{t \to \infty} \Pr\{\delta(t) \le \delta \mid X(t) = i\} = \int_0^{\infty} \frac{1 - G_i(x)}{m_i} dx \equiv R_i(\delta).$$

4. The general-time equilibrium probability of n in the system is found

to be

$$p_n = \sum_i \nu_i \int_0^\infty \Pr\{\text{required changes in } t \text{ to bring state from } i \text{ to } n\}\, dR_i(t)$$

$$(n \geq 0).$$

We now proceed to illustrate the use of the above results on the $G/M/1$ queue, whose analysis we were not quite able to finish in Chapter 5 because this model contains a semi-Markov process. Now, with this current theory, we are able to obtain the general-time state probabilities starting from the geometric arrival-point distribution $q_i = (1 - r_0)r_0^i$.

For the $G/M/1$, we recall that the imbedded chain measures the system size just before arrival occurrences, so that the unconditional wait T_i of the SMP in state i has the same CDF as the interarrival distribution, namely, $A(t)$, with mean $1/\lambda$. Hence it follows that the imbedded SMP has probabilities

$$\nu_i = \frac{q_i/\lambda}{\sum_{i=0}^\infty q_i/\lambda} \qquad (i \geq 0)$$

$$= q_i = (1 - r_0)r_0^i,$$

which is as expected, since the interoccurrence times between the regeneration points of the SMP are IID.

Now, the times back to the latest transition have CDFs $\{R_i(t)\}$, which must be for all i

$$R_i(t) = \lambda \int_0^t [1 - A(x)]\, dx \equiv R(t).$$

So

$$p_n = \sum_{i=n-1}^\infty \nu_j \int_0^\infty \Pr\{i - n + 1 \text{ departures in } t\}\, \lambda[1 - A(t)]\, dt \qquad (n > 0),$$

$$p_0 = \sum_{i=0}^\infty \nu_i \int_0^\infty \Pr\{\text{at least } i + 1 \text{ departures in } t\}\, \lambda[1 - A(t)]\, dt.$$

We already know that p_0 must equal $1 - \lambda/\mu = 1 - \rho$, since we have long ago shown that this is true for all $G/G/1$ queues. Therefore, we focus on $n > 0$, and it follows from the above that

$$p_n = \lambda \sum_{i=n-1}^{\infty} (1 - r_0) r_0^i \int_0^\infty \frac{e^{-\mu t}(\mu t)^{i-n+1}}{(i-n+1)!} [1 - A(t)] \, dt \qquad (n > 0).$$

Letting $j = i - n + 1$, we have

$$p_n = \lambda(1 - r_0) r_0^{n-1} \int_0^\infty e^{-\mu t}[1 - A(t)] \sum_{j=0}^{\infty} \frac{(r_0 \mu t)^j}{j!} \, dt$$

$$= \lambda(1 - r_0) r_0^{n-1} \int_0^\infty e^{-\mu t(1-r_0)}[1 - A(t)] \, dt.$$

Now integration by parts yields

$$p_n = \frac{\lambda}{\mu} r_0^{n-1} \left[1 - \int_0^\infty e^{-\mu t(1 - r_0)} \, dA(t) \right].$$

But, from Equation (5.58) and Problem 5.27,

$$\int_0^\infty e^{-\mu t(1-r_0)} \, dA(t) = \beta(r_0) = r_0;$$

hence for $n > 0$,

$$p_n = \frac{\lambda}{\mu} r_0^{n-1}(1 - r_0) = \frac{\lambda q_{n-1}}{\mu} \qquad (n > 0). \tag{6.21}$$

Note that we have now shown that the *general-time* probabilities for all $G/M/1$ queues are of the familiar geometric pattern which we have shown earlier for the $M/M/1$ and $E_k/M/1$ systems.

For completeness, we provide without proof the comparable results for the $G/M/c$ queue:

$$p_0 = 1 - \frac{\lambda}{c\mu} - \frac{\lambda}{\mu} \sum_{j=1}^{c-1} q_{j-1} \left(\frac{1}{j} - \frac{1}{c} \right),$$

$$p_n = \frac{\lambda q_{n-1}}{n\mu} \qquad (1 \le n < c),$$

$$p_n = \frac{\lambda q_{n-1}}{c\mu} \qquad (n \ge c).$$

6.5   OTHER QUEUE DISCIPLINES

As we mentioned early in Chapter 1, Section 1.2.3, there are many possible approaches to the selection from the queue of customers to be served, and certainly FCFS is not the only choice available. We have already considered some priority models in Chapter 3, and mentioned in Chapter 1 at least two other important possibilities, namely, random selection for service (RSS) and last come, first served (LCFS). We might even consider a third to be general discipline (GD), that is, no particular pattern specified. We shall obtain results in this section for some of our earlier FCFS models under the first two of these three possible variations. To do any more would be quite time consuming, and it would not be in the interest of the reader to be bogged down in such detail.

We should note that the system state probabilities do not change when the discipline is modified from FCFS to another. As observed before, the proof of Little's formula remains unchanged, and thus the average waiting time is the same. But there will indeed be changes in the waiting-time distribution, and, of course, this implies that any higher moment generalization of Little's formula is, in general, no longer applicable. For the sake of illustration we shall now derive the waiting-time distribution for the $M/M/c$ under the RSS discipline and for the $M/G/1$ under LCFS, two results that are thought to be quite typical.

We begin with a discussion of the waiting times for the $M/M/c$ when service is in random order. In the usual way let us define $W_q(t)$ as the CDF of the line delay, and then it may be written that

$$W_q(t) = 1 - \sum_{j=0}^{\infty} p_{c+j} \tilde{W}_q(t \mid j) \qquad (t \geq 0),$$

where $\tilde{W}_q(t \mid j)$ represents the probability that the delay undergone by an arbitrary arrival who joined when $c + j$ were in the system is more than t. But, from the results of Section 2.3, Equation (2.24), where we see that $p_{c+j} = p_c \rho^j$, we may rewrite the above equation as

$$W_q(t) = 1 - p_c \sum_{j=0}^{\infty} \rho^j \tilde{W}_q(t \mid j).$$

To calculate $\tilde{W}_q(t \mid j)$ we observe that the waiting times depend not only on the number of customers found in line by an arbitrary arrival, but also on the number who arrive afterward. We shall then derive a differential–difference equation for $\tilde{W}_q(t \mid j)$ by considering this Markov process over the time intervals $(0, \Delta t)$ and $(\Delta t, t + \Delta t)$, and evaluating the appropriate CK equation. There are three possible ways to have a waiting time greater than $t + \Delta t$, given that $c + j$ were in the system upon arrival: (1) the interval

$(\Delta t, t + \Delta t)$ passes without any change; (2) there is another arrival in $(0, \Delta t)$ [thus bringing the system size, not counting the first arrival, up to $c + j + 1$], and then the remaining waiting time is greater than t, given that $c + j + 1$ were in the system at the instant of arrival; and (3) there is a service completion in $(0, \Delta t)$ [thus leaving $c + j - 1$ of the originals], the subject customer is not the one selected for service, and the line wait now exceeds t, given that $c + j - 1$ were in the system. Thus

$$\tilde{W}_q(t + \Delta t \,|\, j) = [1 - (\lambda + c\mu) \,\Delta t] \tilde{W}_q(t \,|\, j) + \lambda \,\Delta t \, \tilde{W}_q(t \,|\, j + 1)$$

$$+ \frac{j}{j+1} c\mu \,\Delta t \, \tilde{W}_q(t \,|\, j-1) + o(\Delta t)$$

$$[j \geq 0, \quad \tilde{W}_q(t \,|\, -1) \equiv 0]. \quad (6.22)$$

The usual algebra on Equation (6.22) leads to

$$\frac{d\tilde{W}_q(t \,|\, j)}{dt} = -(\lambda + c\mu) \tilde{W}_q(t \,|\, j) + \lambda \tilde{W}_q(t \,|\, j + 1)$$

$$+ \frac{j}{j+1} c\mu \tilde{W}_q(t \,|\, j - 1) \qquad (j \geq 0) \qquad (6.23)$$

with $\tilde{W}_q(t \,|\, -1) \equiv 0$ and $\tilde{W}_q(0 \,|\, j) = 1$ for all j. Equation (6.23) is somewhat complicated, but can be examined using the following approach. Assume that $\tilde{W}_q(t \,|\, j)$ has the Maclaurin series representation

$$\tilde{W}_q(t \,|\, j) = \sum_{n=0}^{\infty} \frac{\tilde{W}_q^{(n)}(0 \,|\, j) t^n}{n!} \qquad (j = 0, 1, \ldots),$$

with $\tilde{W}_q^{(0)}(0 \,|\, j) \equiv 1$. We thus get

$$W_q(t) = 1 - p_c \sum_{j=0}^{\infty} \rho^j \sum_{n=0}^{\infty} \frac{\tilde{W}_q^{(n)}(0 \,|\, j) t^n}{n!}. \qquad (6.24)$$

Then the derivatives can be directly determined by the successive differentiation of the original recurrence relation given by Equation (6.23). For example,

$$\tilde{W}_q^{(1)}(0 \,|\, j) = -(\lambda + c\mu) \tilde{W}_q^{(0)}(0 \,|\, j) + \lambda \tilde{W}_q^{(0)}(0 \,|\, j + 1) + \frac{j}{j+1} c\mu \tilde{W}_q^{(0)}(0 \,|\, j - 1)$$

$$= -(\lambda + c\mu) + \lambda + \frac{jc\mu}{j+1} = \frac{-c\mu}{j+1},$$

$$\tilde{W}_q^{(2)}(0|j) = -(\lambda + c\mu)\tilde{W}_q^{(1)}(0|j) + \lambda\tilde{W}_q^{(1)}(0|j+1) + \frac{j}{j+1}c\mu\tilde{W}_q^{(1)}(0|j-1)$$

$$= \frac{(\lambda + c\mu)c\mu}{j+1} - \frac{\lambda c\mu}{j+2} - \frac{(c\mu)^2}{j+1},$$

and so on. Putting everything together into (6.24) gives a final series representation for $W_q(t)$. The ordinary moments of the line delay may be found by the appropriate manipulation of $W_q(t)$ and its complement (this kind of manipulation is well detailed by Parzen, 1960).

For the other model of this section, let us now consider $M/G/1/\infty/LCFS$. In this case, the waiting time of an arriving customer who comes when the server is busy is the sum of the duration of time from his instant of arrival, T_A, to the first subsequent service completion, at T_S, and the length of the total busy period generated by the $n \geq 0$ other customers who arrive in (T_A, T_S).

To find $W_q(t)$ let us first consider the joint probability that n customers arrive in (T_A, T_S) and $T_S - T_A \leq x$. Then, for $R(t)$ equal to the CDF of the remaining service time (see Section 5.1.5),

$$\text{Pr}\{n \text{ arrivals} \in (T_A, T_S) \text{ and } T_S - T_A \leq x\} = \int_0^x \frac{(\lambda t)^n e^{-\lambda t}}{n!} \, dR(t),$$

$$R(t) = \mu \int_0^t [1 - B(x)] \, dx.$$

Since $\pi_0 = 1 - \rho$, it is found by an argument similar to that used for the $M/G/1$ busy period that

$$W_q(t) = 1 - \rho + \text{Pr}\{\text{system busy}\}$$
$$\times \sum_n \text{Pr}\{n \text{ customers arrive} \in (T_A, T_S), T_S - T_A \leq t - x,$$
$$\text{and total busy period generated by these arrivals is } x\},$$

$$(6.25)$$

because under the LCFS discipline the most recent arrival in the system is the one chosen for service when the server has just had a completion. But the busy-period distribution is exactly the same as that derived in the FCFS case in Section 5.1.6, since the sum total of customer service times is unchanged by the discipline [it will be denoted henceforth by $G(x)$, with $G^{(n)}(x)$ used for its n-fold convolution]. Therefore (6.25) may be rewritten as

$$W_q(t) = 1 - \rho + \rho \sum_{n=0}^{\infty} \int_0^t \int_0^{t-x} \frac{(\lambda u)^n e^{-\lambda u}}{n!} \, dR(u) \, dG^{(n)}(x).$$

A change of the order of integration then gives

$$W_q(t) = 1 - \rho + \rho \sum_{n=0}^{\infty} \int_0^t \frac{(\lambda u)^n e^{-\lambda u}}{n!} \, dR(u) \int_0^{t-u} dG^{(n)}(x)$$

$$= 1 - \rho + \rho \sum_{n=0}^{\infty} \int_0^t \frac{(\lambda u)^n e^{-\lambda u}}{n!} G^{(n)}(t - u) \, dR(u). \qquad (6.26)$$

Though this form of $W_q(t)$ is adequate for some purposes, it turns out to be possible to refine the result further to make it free of $G(t)$. The final version as given in Cooper (1981) is

$$W_q(t) = 1 - \rho + \lambda \sum_{n=1}^{\infty} \int_0^t \frac{(\lambda u)^{n-1} e^{-\lambda u}}{n!} [1 - B^{(n)}(u)] \, du, \qquad (6.27)$$

where $B^{(n)}(t)$ is the n-fold convolution of the service-time CDF.

There are, of course, numerous other models with various disciplines, but, as mentioned earlier, it is felt that this discussion should suffice for this text. It is to be emphasized, however, that each discipline must be approached in a unique manner, and any interested reader is referred to the literature, one reference again being Cooper (1981).

6.5.1 Conservation

Little's formula and global and local stochastic balance are examples of conservation laws (see Sections 1.5.1 and 2.2.1). The general idea of conservation is that the expected change of a state function is zero over any finite (including infinitesimal) span of time picked at random in the steady state. These sorts of results play a particularly vital role in the modeling of systems with priorities (see Section 3.4).

For example, Little's result says that the expected change of aggregate waiting time of customers in line or system is zero during a randomly chosen finite time interval. We would similarly find that the time needed to clear out the system is unchanged, on average, over time intervals. As a further illustration, the fact that $\rho = 1 - p_0$ for all $G/G/1$ queues expresses the fact that the expected net change in the number in system over a random time interval is zero, since the equation is equivalent to $\lambda = \mu(1 - p_0)$.

Another previous result that made use of a conservation argument is the proof that $\pi_n = p_n$ for $M/G/1$ queues given in Section 5.1.3. Actually, contained within that proof is a more general result that $q_n = \pi_n$ for any stationary $G/G/c$ queue. To see this, reconsider Equation (5.32) in Section 5.1.3.

The right-hand side is q_n, while the left-hand side is π_n. Further, nothing in the proof up to this point assumed anything about the number of servers or distributions for arrival or service times, so that these results hold for $G/G/c$. The final step of the $p_n = \pi_n$ proof required only Poisson arrivals, so that $p_n = \pi_n$ thus holds for all $M/G/c$ systems.

We can redo this argument in a slightly different fashion following the logic used in Krakowski (1974), and similar logic used by us previously in Chapter 2 when deriving q_n for the finite queueing models (e.g., see Section 2.4):

$$q_n \equiv \Pr\{n \text{ in system} \mid \text{arrival about to occur}\}$$

$$= \lim_{\Delta t \to 0} \frac{\Pr\{n \text{ in system and arrival within } (t, t + \Delta t)\}}{\Pr\{\text{arrival within } (t, t + \Delta t)\}}.$$

Similarly,

$$\pi_n = \lim_{\Delta t \to 0} \frac{\Pr\{n + 1 \text{ in system and a departure within } (t, t + \Delta t)\}}{\Pr\{\text{a departure within } (t, t + \Delta t)\}}.$$

For systems in steady state, the numerators of q_n and π_n must be equal, since the limiting transition rate (as $\Delta t \to 0$) from n to $n + 1$ must be equal to the limiting transition rate from $n + 1$ to n (which is what the respective numerators represent). Furthermore, the denominators are also equal for steady-state systems, since the arrival rate must equal the departure rate; hence $q_n = \pi_n$ for all $G/G/c$ systems.

Now if input is Poisson, then the numerator of q_n is [ignoring $o(\Delta t)$ terms] $p_n \lambda \Delta t$ and the denominator is $\lambda \Delta t$, so that

$$q_n = \lim_{\Delta t \to 0} \frac{p_n \lambda \, \Delta t}{\lambda \, \Delta t} = p_n.$$

From the above argument, it is also easy to get the relationship between q_n and p_n for $G/M/1$. The limiting transition rate for going from state $n + 1$ to n is μp_{n+1} and since for steady state this must equal the limiting transition rate of going from n to $n + 1$, we see that the numerator of q_n for the above equation must equal $p_{n+1} \mu \, \Delta t$ [again ignoring $o(\Delta t)$ terms]. The arrival rate must equal the departure rate, which is $\mu(1 - p_0) = \mu \rho$, so that the denominator is $\mu \rho \, \Delta t$ and we have

$$q_n = \lim_{\Delta t \to 0} \frac{p_{n+1} \mu \, \Delta t}{\mu \rho \, \Delta t} = \frac{\mu p_{n+1}}{\lambda}.$$

This is the same result as obtained in Section 6.4 using semi-Markov pro-

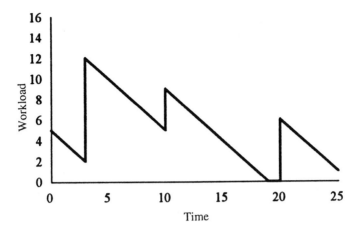

Fig. 6.1 Sample workload seen by a $G/G/1$ queue.

cesses, since $q_n = r_0^n(1 - r_0)$, r_0 being the root of the generating function equation $\beta(r) = r$ (see Section 5.3.1), and hence $p_{n+1} = \rho r_0^n(1 - r_0)$ as in Section 6.4. For further general discussion on the topic of conservation, we refer the reader to Krakowski (1973, 1974). See also the related work on level crossings by Brill and Posner (1977).

To see the role that this concept plays in priority queues, let us first define a work-conserving queue discipline to be one where the service need of each customer is unaltered by the queue discipline and where it is also true that the server is not forced to be idle with customers waiting. The relationship between virtual waits, backlog or workload (first introduced in Section 2.2.4, with a typical evolving pattern of workload shown in Fig. 6.1) in a priority system, and the customer service and waiting times is crucial to understanding how such systems operate. It is certainly fairly clear how the virtual wait relates to the more usual system parameters for nonpriority, FCFS $G/G/1$ queues, for example. If we separate the steady-state mean virtual delay V into the mean delay caused by those in queue (call it V_q) and the mean residual service time R, then it follows that

$$V_q = \frac{L_q}{\mu} = \frac{\lambda W_q}{\mu}.$$

When the input is Poisson, we see from Problem 5.5 that

$$V = V_q + R = \frac{\lambda}{\mu} W_q + \frac{\lambda}{2} E[S^2].$$

Hence, using the PK formula for this case and simplifying, we find that

$$V = \frac{\lambda}{\mu} \frac{\rho^2 + \lambda^2 \, \mathrm{Var}[S]}{2\lambda(1 - \lambda/\mu)} + \frac{\lambda(\mathrm{Var}[S] + 1/\mu^2)}{2}$$

$$= \frac{\rho^2 + \lambda^2 \, \mathrm{Var}[S]}{2\lambda(1 - \lambda/\mu)},$$

which is precisely W_q, as it should be.

But it is of special interest to see how these latter results may differ from those for systems with priorities, since then an arrival's delay is not simply the remaining time necessary to complete service of those in front. A major quantity for an arbitrary system is the product of a customer's service time and line delay, since this represents the total contribution to the remaining workload over the full extent of the customer's waiting time. Furthermore, the contribution of a given customer's service time to V is $S^2/2$ (as the average value of 0 through S). When we assume that the discipline is work-conserving and nonpreemptive (Poisson input or otherwise), it follows that $R = \lambda S^2/2$ and that $E[S \cdot T_q] = V_q/\lambda$.

We note a number of consequences of this argument for the $M/M/1$ and $M/G/1$ queues. For any stationary $M/M/1/GD$, we see that

$$E[S \cdot T_q] = \frac{V_q}{\lambda} = \frac{V - R}{\lambda} = \frac{W_q - \lambda E[S^2]/2}{\lambda} = \frac{\rho}{\mu^2(1 - \rho)}.$$

This is also true for the preemptive case in light of the lack of memory of the exponential. For the $M/G/1/GD$, we find that

$$E[S \cdot T_q] = \frac{V - R}{\lambda} = \frac{W_q - \lambda E[S^2]/2}{\lambda} = \frac{\rho E[S^2]}{2(1 - \rho)}.$$

6.6 DESIGN AND CONTROL OF QUEUES

Models have, on occasion, been classified into two general types—descriptive and prescriptive. Descriptive models are models which *describe* some current real-world situation, while prescriptive models (also often called normative) are models which *prescribe* what the real-world situation should be, that is, the optimal behavior at which to aim.

Most of the queueing models presented thus far are descriptive, in that for given types of arrival and service patterns, and specified queue discipline and configuration, the state probabilities and expected-value measures of effectiveness which describe the system are obtained. This type of model does not attempt to prescribe any action (such as to put on another server or to change from FCFS to priority), but merely represents the current state of affairs.

On the other hand, consider a resource allocation problem, for example, using linear programming to determine how much of each type of product to make for a certain sales period. This model indicates the best setting of the variables, that is, how much of each product *should* be made, within the limitations of the resources needed to produce the products. This, then, is a prescriptive model, for it prescribes the optimal course of action to follow. There has been much work on prescriptive queueing models, and this effort is generally referred to under the title of *design and control of queues*.

Design and control models are those dealing with what the optimal system parameters (in this context, they are actually variables) should be (for example, the optimal mean service rate, the optimal number of channels, etc.). Just which and how many parameters are to be optimized depends entirely on the system being modeled, that is, the particular set of parameters that are actually subject to control.

Generally, the controllable parameters are the service pattern (μ and/or its distribution), number of channels, and queue discipline, or some combination of these. Occasionally, one may even have control over arrivals in that certain arrivals can be shunted to certain servers, truncation can be instituted, the number of servers removed, increased or decreased, and so on. The arrival rate can even be influenced by the levying of tolls. In many cases, however, some "design" parameters are beyond control or perhaps limited to a few possibilities. For example, physical space may prevent increasing the number of channels, and the workers may prevent any proposed decreases. Similar types of effects could fix or limit other potentially controllable parameters.

It is not always easy to make a distinction between those models classified as design and those classified as control. Generally, design models are *static* in nature, that is, cost or profit functions are superimposed on classical descriptive models, so that λ, μ, c, or some combination of these parameters can be optimized. Control models are more dynamic in nature and usually deal with determining the optimal control policy. For example, one may be operating an $M/M/c$ queue and desire to find the optimal c that balances customer waits against server idleness. Since it has already been determined that the form of the policy is always to have c servers, it remains to determine the optimal c by using the descriptive waiting-time information and superimposing a cost function that is to be minimized with respect to c. This is a design problem.

On the other hand, perhaps we desire to know if a single-variable (c) policy is optimal, that is, perhaps we should have a (c_1, c_2) policy where when the queue size exceeds N, we shift from c_1 to c_2 servers, and when the queue size falls back to N, we shift back to c_1 servers. Or perhaps the best policy is (c_1, c_2, N_1, N_2), where, when the queue builds up to N_1, a shift from c_1 to c_2 servers is effected, but we shift back to c_1 only when the queue falls to a different level N_2 $(N_2 < N_1)$. This type of problem is a control problem.

The methodology is vastly different between design and control problems, and perhaps this is the easiest way to classify a problem as to whether it is design or control. For design problems, basic probabilistic queueing results are used to build objective (cost, profit, time, etc.) functions to be minimized or maximized, often subject to constraints (which are also functions of the probabilistic queueing measures). Classical optimization techniques (differential calculus; linear, nonlinear, or integer programming) are then applied. For control problems, techniques such as dynamic programming (or variations such as value or policy iteration) are used. These latter techniques all fall under the category of Markov decision problems (MDPs).

As a simple illustration, consider an $M/M/c/K$ problem where a fixed fee per customer is charged but the cost of serving a customer depends on the total time the customer spends in the system. We desire to find the optimal truncation value K ($1 \le K < \infty$) which maximizes the expected profit rate. Letting R be the fee charged per customer served and C be the cost incurred per customer per unit time, we have the classic optimization problem

$$\max_{K} Z = R\lambda(1 - p_k) - C\lambda(1 - p_k)W, \qquad (6.28)$$

where p_K and W are given by Equations (2.32) and (2.35), respectively. We have an integer programming problem to solve, which can be done by stepping K from 1 and finding the K that yields the maximum profit, since it can be shown that Z is concave in K (Rue and Rosenshine, 1981).

In the above design problem, we assumed the "optimal" policy was a single-value K. It might have been that the policy should have been "turn customers away when queue size reaches K_1, but do not let customers enter again until queue size drops to K_2 $(K_1 > K_2)$." Actually, a single K value is optimal for this cost structure (Rue and Rosenshine, 1981), which can be proven by value iteration. We would classify this latter type of analysis as a control problem, as opposed to the design problem, where it was predecided that the policy form was a single-value K and the objective function Z was then maximized using classical optimization techniques.

In this section, we shall concentrate mostly on design problems (without getting too much into the details of the necessary optimization procedures), for which much of the appropriate probabilistic analysis has already been done in prior sections of this book.

6.6.1 Queueing Design Problems

Although it was not explicitly pointed out, we have already introduced the ideas of prescriptive models. The reader is specifically referred to Examples

2.6, 3.1, and 5.4 and to Problems 2.12, 2.13, 2.24, 2.25, 2.33, 2.34, 2.35, 2.38, 2.47, and 2.48 for some illustrations of prescriptive analyses. Because of the importance of the subject, we feel it warrants the added visibility provided by a subsection, and thus devote some further space to a discussion of the details of the topic.

One of the earliest uses of queueing design models in the literature is described in Brigham (1955). He was concerned with the optimum number of clerks to place behind tool-crib service counters in plants belonging to the Boeing Airplane Company. From observed data, Brigham inferred Poisson arrivals and exponential service, so that an $M/M/c$ model could be used. Costing both clerk idle time and customer waiting time, he presented curves showing the optimal number of clerks (optimal c) as a function of λ/μ and the ratio of customer waiting cost to clerk idle cost.

Morse (1958) considered optimizing the mean service rate μ for an $M/M/1$ problem framed in terms of ships arriving at a harbor with a single dock. He desired the value of μ that would minimize costs proportional to it and the mean customer wait W. This is easily done by taking the derivative of the total-cost expression with respect to μ and equating it to zero. Morse also treated an $M/M/1/K$ problem with a service cost proportional to μ and a cost due to lost customers. The problem was actually framed as a profit model (a lost customer detracts from the profit), and again differential calculus was employed to find the optimal μ.

Yet another model considered by Morse dealt with finding an optimal μ when there is customer impatience. The greater μ, the less impatience on the part of a customer. Again, a profit per customer and a service cost proportional to μ were considered (see Problem 6.24). Morse also considered a model dealing with optimizing the number of channels in an $M/M/c/c$ (no queue allowed to form). The cost of service was now assumed proportional to c rather than μ, and again lost customers were accounted for economically by a profit-per-customer term. His solution was in the form of a graph where, for a given λ, μ, and ratio of service cost to profit per customer, the optimal c could be found.

Finally, Morse considered several machine-repair problems, with a cost proportional to μ and a machine profitability dependent on the machine's uptime. For the latter model, he also built in a preventive maintenance function.

A second reference on economic models in queueing falling into the category of design discussed here is Hillier and Lieberman (1995, Chapter 16). They have considered three general classes of models, the first dealing with optimizing c, the second with optimizing μ and c, and the third with optimizing λ and c. Mention is also made of optimization with respect to λ, c, and μ. Many of the previously mentioned models of Morse turn out to be special cases of one of the classes above. We present here a few of the examples from Hillier and Lieberman, and refer the reader to this source for further cases.

We first consider an $M/M/c$ model with unknown c and μ. Assume that there is a cost per server per unit service of C_s and a cost of waiting per unit time for each customer of C_w. Then the expected cost rate (expected cost per unit time) is given by

$$E[C] = cC_s\mu + C_wL. \tag{6.29}$$

Considering first the optimal value of c, we see that the first term is independent of c for a fixed value of ρ. Furthermore, Morse showed that L is increasing in c for fixed ρ; thus it follows that $E[C]$ is minimized when $c = 1$. So we now consider finding the optimal value of μ for an $M/M/1$ queue.

From (6.29),

$$E[C] = C_s\mu + C_wL = C_s\mu + C_w\frac{\lambda}{\mu - \lambda}.$$

Taking the derivative $dE[C]/d\mu$ and setting it equal to zero yields

$$0 = C_s - \frac{\lambda C_w}{(\mu^* - \lambda)^2} \quad \Rightarrow \quad \mu^* = \lambda + \sqrt{\frac{\lambda C_w}{C_s}}.$$

Looking at the sign of the second derivative confirms that μ^* minimizes $E[C]$.

Hillier and Lieberman have pointed out an interesting interpretation of the results. If a service channel consists of a crew, such that the mean service rate is proportional to crew size, it is better to have one large crew ($M/M/1$) whose size corresponds to μ^* than several smaller crews (or $M/M/c$). Clearly, given an individual service rate μ, using μ^* to obtain the crew size (μ^*/μ) may not yield an integer value. However, the expected-cost function is convex, so that checking the integer values on either side of μ^*/μ will give the optimum crew size. All of this, of course, is true only as long as the assumptions of crew service rate being proportional to crew size and cost functions being linear are valid. Stidham (1970) extended this model by relaxing the exponential assumptions on interarrival and service times and also considered nonlinear cost functions. He showed that even for these relaxed assumptions, $c = 1$ is generally optimal.

Another interesting model treated by Hillier and Lieberman deals with finding optimal λ (optimal assignment of arrivals to servers) and c for a given μ. The situation considered is one in which a population (such as employees of a building) must be provided with a certain service facility (such as a rest room). The problem is to determine what proportion of the population to assign to each facility (or equivalently the number of facilities) and the optimal number of channels for each facility (in the rest-room example, a channel would correspond to a stall). To simplify the analysis, it is assumed that λ and c are the same for all facilities. Thus if λ_p is the mean arrival rate for the entire population, we need to find the optimal λ, say λ^* (optimal mean number of arrivals to assign to a facility). The optimal number of facilities then would be λ_p/λ^*.

Using C_s to denote the marginal cost of a server per unit time and C_f to denote a fixed cost per facility per unit time, we desire to minimize $E[C]$ with respect to c and λ (or c and n), where

$$E[C] = (C_f + cC_S)n + nC_WL$$

$$\text{subject to} \quad n = \lambda_p/\lambda.$$

Hillier and Lieberman have shown under quite general conditions ($C_f \geq 0$, $C_s \geq 0$, and $L = \lambda W$) that the optimal solution is to make $\lambda^* = \lambda_p$, or equivalently $n = 1$, that is, to provide only a single service facility. The problem then reduces to finding the optimal value for c which minimizes

$$E[C] = cC_S + C_WL,$$

a problem previously discussed by Hillier and Lieberman and by Brigham (1955).

It is interesting to consider further that $n = 1$ was shown to be optimal, regardless of the particular situation. For example, in the World Trade Center, this model would say to have only one (gigantic) rest room. This is obviously absurd, and the reason, as Hillier and Lieberman point out, is that travel time to the facility is not considered in the model. If travel time is insignificant (which might be the case in certain situations), then a single facility would be optimal. When travel time is explicitly considered, that is not always the case, depending on the amount of travel time involved. Hillier and Lieberman present a variety of travel-time models with examples, and again, it is recommended that the interested reader consult this source.

Most of the attention so far has been in designing optimal service facilities (c or μ), the above being one exception, since λ was also of concern. In the introduction to Section 6.6, we also considered a design model involving the arrival process with the objective function given by (6.28). It is interesting to return briefly to this model and reconsider the cost criteria of (6.28). It was desired there to find the K that maximized (6.28), namely, the average profit rate *to the system*. This is referred to in the design and control literature as *social optimization*, that is, it maximizes benefits to the entire system rather than the gain of an individual customer. We can reformulate this problem from an individual customer's standpoint by viewing R as a reward given directly to each customer served and C as a direct cost charged to the customer for each unit of time the customer spends in the system. This gives rise to the objective function

$$Z = R - \frac{C(N + 1)}{\mu},$$

where N is the number of customers an arrival finds in the system. The customer will serve its own self-interest (positive gain) by joining the queue as long as $C(N + 1)/\mu \leq R$ or equivalently, $N \leq R\mu/C - 1$. Hence, if every customer uses this strategy, an $M/M/1/K$ queue results, where $K = [R\mu/C]$, $[\cdot]$ indicating the greatest integer contained in the brackets. Rue and Rosenshine (1981) show that the optimal K for the system or *social* maximization criterion of (6.28) is less than or equal to the K coming from the individual optimal criterion. This difference between individual and social optimization comes up quite frequently in control problems involving the arrival-rate process.

To conclude this subsection on design models, we present an example dealing with design in a queueing network.

Example 6.3
This example is from Gross et al. (1983), and the area of application is repairable-item spares provisioning. We consider a three-node closed Jackson network. The first node (call it node U) represents the population of operating units. There are M servers (channels) at node 1, representing the desired M operating units. A queue at node 1 indicates spare units on hand, while an idle server(s) represents a population operating below the desired M level. The second node (node B) represents base (local) repair, and there are c_B repair channels. The final node (node D) represents depot (remote) repair, and this node has c_D servers.

Units fail according to a Poisson process with rate λ, and repair times are exponential with mean rates μ_B and μ_D for base and depot, respectively. When a unit fails, there is a probability α that it is field-repairable and $1 - \alpha$ that it has to go to the depot. Further, there is a probability β that it will still have to go to the depot after having been worked on in the field (probability $1 - \beta$ of returning from field to operational status). All depot units are fully repairable and return to operational status.

The problem is to find the optimal numbers of spares, repair channels at the base, and repair channels at the depot to satisfy a service-level constraint on operating population availability. Specifically, we desire to

$$\underset{y, c_b, c_D}{\text{minimize}} \quad Z = k_y y + k_B c_B + k_D c_D$$

$$\text{subject to} \quad \sum_{n=M}^{M+y} p_n \geq A,$$

where

p_n = steady-state probability that n units are operational,

M = number of components desired to be operating (operating population size),

A = minimum proportion of time all M components are to be operational (availability),

y = number of spare components to stock,

c_B = base repair capacity (number of channels),

c_D = depot repair capacity (number of channels), and

k_i = cost per unit ($i = y, D, B$) including annual operating costs and capital investment amortization of a spare or a repair channel.

The variables y, c_B, and c_D are the decision variables to be determined by an optimization algorithm, with the steady-state probabilities $\{p_n\}$ determined using the closed Jackson network theory from Chapter 4.

Holding times at all nodes are assumed to be independent exponentially distributed random variables. At node U, the holding time is the time to failure of a component, with the mean failure rate denoted by $\lambda \equiv \mu_U$. At nodes B and D, the holding times are repair times, and the mean repair rates are denoted by μ_B and μ_D, respectively. Thus from Equation (4.17) we have

$$p_{n_U, n_B, n_D} = \frac{1}{G(N)} \frac{\rho_U^{n_U}}{a_U(n_U)} \frac{\rho_B^{n_B}}{a_B(n_B)} \frac{\rho_D^{n_D}}{a_D(n_D)} \qquad (n_U + n_B + n_D = N = M + y),$$

with

$$a_i(n) = \begin{cases} n! & (n < c_i, \quad i = U, B, D), \\ c_i^{n - c_i} c_i! & (n \geq c_i, \quad i = U, B, D). \end{cases}$$

The matrix giving the routing probabilities, (r_{ij}), for this problem is

$$\mathbf{R} = \{r_{ij}\} = \begin{matrix} U \\ B \\ D \end{matrix} \begin{pmatrix} \overset{U}{0} & \overset{B}{\alpha} & \overset{D}{1 - \alpha} \\ 1 - \beta & 0 & \beta \\ 1 & 0 & 0 \end{pmatrix}.$$

Now, we substitute these $\{r_{ij}\}$ in Equation (4.14) and Buzen's algorithm to calculate the constant $G(N)$. Once the joint probabilities are obtained, we calculate the marginal probabilities $\{p_{n_U}\}$ required for the constraint, again using Buzen's algorithm.

The resultant probability distribution is a function of the decision variables y, c_B, and c_D. The distribution exhibits certain monotonicity properties in relation to these variables, which play a crucial role in developing an implicit enumeration optimization algorithm. We will not dwell on the optimization

Table 6.2   Sample Results

M	c_B^*	c_D^*	y^*	Z^*	A_1	A_2	No. of Calls to Buzen's Algorithm
5	2	2	3	96	.938	.982	25
10	3	3	5	154	.926	.988	38
20	4	6	8	252	.907	.989	66
30	6	8	11	348	.904	–	137

routine here (we refer the interested reader to the reference). However, in Table 6.2 we do present results for the specific problem given below:

$$\text{minimize}_{y,c_b,c_D} \quad Z = 20y + 8c_B + 10c_D$$

$$\text{subject to} \quad \sum_{n_U=M}^{M+y} p_{n_U} \geq .90 \quad (A_1),$$

$$\sum_{n_U=.9M}^{M+y} p_{n_U} \geq .98 \quad (A_2)$$

(6.30)

The parameters were set as follows: $\alpha = 0.5$, $\beta = 0.5$, $u = 1$, $\mu_B = \mu_D = 5$.

We note that the optimization algorithm was quite efficient for these small problems. The efficiency of the computations relies not only on the implicit enumeration scheme, but also on the efficiency of calculations of the probabilities, which in turn depends on the efficiency of Buzen's algorithm. Two constraints were used in all but the last set of parameters run for the problem formulated in (6.30). The second constraint requires 90% of the population to be operating 98% of the time, while the first constraint requires the entire (M) population to be operating 90% of the time. The more constraints that are imposed, the more efficient the implicit enumeration scheme is, so that the most demanding run of (6.30) was the last one, given in Table 6.2.

In any problem like this, because of the complexity of the probabilities and the requirement of integer values for the decision variables, a search technique such as implicit enumeration is usually required; that is, a set of decision variables must first be specified, and then the constraints and objective function are evaluated. Implicit enumeration schemes appear to be quite well suited to these types of problems.

6.6.2   Queueing Control Problems

As previously mentioned, the focus in control problems is on determining the form of the optimal policy. Often, researchers are interested in determining when stationary policies (such as those treated in the previous section) are truly optimal. Also, although it seems to a somewhat lesser extent, there is some concern about actually determining the optimal values of the control parameters.

The methodology used generally involves an analysis of the functional equation of dynamic programming arising from an objective function which minimizes (maximizes) total discounted cost (profit) streams or average cost (profit) rates over either a finite or an infinite planning horizon. Since this type of analysis is rather different from anything we have thus far encountered in the text, we merely present to the reader some general notions and point to appropriate references.

The earliest efforts involving queueing control in the literature appear to date back to Romani (1957) and Moder and Phillips (1962). The variable under control was the number of servers in an $M/M/c$ queue, and while the form of the policy was prechosen so that their analyses would be technically categorized under our previous heading of static design, they did form the nucleus for the sizable research effort on service-rate control which followed. Romani considered a policy where, if the queue builds up to a certain critical value, additional servers are added as new arrivals come, thus preventing the queue from ever exceeding the critical value. Servers are then removed when they finish service and no one is waiting in the queue.

Moder and Phillips (1962) modified the Romani model in that there are a certain number of servers, say c_1, always available. If the size of the queue exceeds a critical value M_1, additional servers are added as arrivals enter in a manner similar to the Romani model, except here there is a limit, c_2, to how many servers can be added. Furthermore, in the Moder–Phillips model, the added servers are removed when they finish serving and the queue has fallen below another critical value M_2.

The usual measures of effectiveness, such as idle time, mean queue lengths, and mean wait, are derived, as well as the mean number of channel starts. No explicit cost functions or optimizations are considered, nor is any attempt made to prove the chosen policy to be optimal, so that these models not only fall into the static design category, but also are in reality more descriptive than prescriptive, although the measures of effectiveness are compared for various values of certain parameters such as the switch points (critical values of queue size for which servers are added and removed).

Yadin and Naor (1967) further generalized the Moder–Phillips model by assuming that the service rate can be varied at any time and is under the control of the decisionmaker. The class of policies considered is of the following form. Denoting the feasible service capacities by $\mu_0, \mu_1, \ldots, \mu_k, \ldots$, where $\mu_{k+1} > \mu_k$ and $\mu_0 = 0$, the policy is stated as "whenever system size reaches a value R_k (from below) and service capacity equals μ_{k-1}, the latter is increased to μ_k; whenever system size drops to S_k (from above) and service capacity is μ_{k+1}, the latter is decreased to μ_k." The $\{R_k\} = \{R_1, R_2, \ldots, R_k, \ldots\}$ and $\{S_k\} = \{S_0, S_1, \ldots, S_k, \ldots\}$ are vectors of integers, ordered by $R_{k+1} > R_k$, $S_{k+1} > S_k$, $R_{k+1} > S_k$, $S_0 = 0$, and are the policy parameters; that is, a specific set of values for $\{R_k\}$ and $\{S_k\}$ yields a specific decision rule (policy) from the class of policies described above within the quotation marks. Input is assumed to be Poisson and service exponential, so the queueing model is essentially $M/M/1$ with state-dependent service. Note that this type of $M/M/1$ model with state-dependent service can also represent a situation where additional channels are added and removed as the state of the system changes.

Given the class of policies above, the authors derive the steady-state probabilities, the expected system size, and expected number of rate switches per unit time. While they do not specifically superimpose any cost functions in order to compare various policies of the class studied, they do discuss the problem in general terms. Given a feasible set $\{\mu_k\}$, a cost structure made up of costs proportional to customer wait, service rate, and number of rate switches, and sets $\{R_k\}$ and $\{S_k\}$ which represent a particular policy, the expected cost of the policy can be computed. Thus various policies (different sets $\{R_k\}$, $\{S_k\}$) can be compared. They further point out the extreme difficulty, however, of trying to find the optimal policy, that is, finding the particular sets $\{R_k\}$ and $\{S_k\}$ which minimize expected cost.

Gebhard (1967) considered two particular service-rate-switching policies for situations with two possible service rates, also under the assumption of Poisson input and exponential service (state-dependent $M/M/1$). He refers to the first as single level control, and the policy is "whenever the system size is $\leq N_1$, use rate μ_1; otherwise, use rate μ_2." The second policy considered, called bilevel hysteretic control, can be stated as "when the system size reaches a value N_2 from below, switch to rate μ_2; when the system size drops to a value N_1 from above, switch back to rate μ_1." The term *hysteretic* (which was also used by Yadin and Naor) stems from the control loop which can be seen in a plot of system size versus service rate. Gebhard actually compared his two policies for specific cost functions (including both service and queueing costs) after deriving the steady-state probabilities, expected system size, and expected rate of service switching.

As the Yadin–Naor and Gebhard papers appeared simultaneously, it was not specifically pointed out that the Gebhard policies were special cases of the Yadin–Naor policies. The single-level control policy is the case for which, in the Yadin–Naor notation, $R_1 = 1$, $R_2 = N_1 + 1$, and $S_1 = N_1$; and the bilevel hysteretic control is the case for which $R_1 = 1$, $R_2 = N_2$, and $S_1 = N_1$.

Heyman (1968) considered an $M/G/1$ state-dependent model where there are two possible service rates, one being zero (the server is turned off) and the other being μ (the server is turned on). He included a server startup cost, a server shutdown cost, a cost per unit time when the server is running, and a customer waiting cost. Heyman was able to *prove* that the form of the optimal policy is "turn the server on when there are n customers in the system, and turn the server off when the system is empty." Heyman examined various combinations of cases involving discounting or not discounting costs over time and a finite or infinite planning horizon.

This paper is the first of those mentioned so far in this subsection that properly fit the category of queueing control, since the emphasis is on the determination of what the optimal form of the class of policies should be. Heyman also considered the problem of determining the optimal n for the various combinations of cases mentioned above (infinite horizon with and without discounting, and finite horizon), although this is not the prime thrust of his work.

Sobel (1969) looked at the same problem as Heyman, namely, starting and stopping service but generalized the results to $G/G/1$, as well as assumed a more general cost structure. Considering only the criterion of average cost rate (undiscounted) over an infinite horizon, he showed that almost any type of stationary policy, is equivalent to one that is a slight generalization of Heyman's policy, and further, that it is optimal under a wide class of cost functions. The policy form is "provide no service (server turned off) if system size is m or less; when the system size increases to M ($M > m$), turn the server on and continue serving until the system size again drops to m." He refers to these as (M, m) policies, and one can readily see that Heyman's is a proper subset of Sobel's class of policies, namely, (M, m) policies where $m = 0$. Sobel's results are strictly qualitative in that he is interested in showing that almost any type of policy imaginable for this kind of problem falls into his class of (M, m) policies. He also shows conditions under which (M, m) policies are optimal, but does not deal with determining, for specific costs, the optimum values of M and m. (For a complete discussion of this and other related material, see Heyman and Sobel, 1984.)

The reader having some familiarity with inventory theory will notice the similarity of the (M, m) policies to the classical (S, s) inventory policies. Sobel does, in one section of his paper, show the applicability of these (M, m) policies to a production inventory system, namely, that the rule becomes, "if inventory is at least as high as M, do not produce until inventory drops to m; at that time, start to produce and continue until the inventory level reaches M once again."

Sobel (1974) gives a survey of work on control of service rates, and the reader desiring to delve further into the topic is referred to that source. Another reference on service control problems is Bengtsson (1983). Much of this material is again covered in Heyman and Sobel (1984).

While the early control work centered on the control of service parameter(s), and work continues in this area, much of the later work has involved controlling the arrival process. Forms of the optimal policies for both social and individual optimization and comparisons of the optimal policies for these cases have been a major focus. Stidham (1982) and Rue and Rosenshine (1981) are recommended references for readers interested in delving further into this area. General survey papers on design and control are Crabill et al. (1977) and Stidham and Prabhu (1974). More recently, Serfozo (1981) and Serfozo and Lu (1984) have studied Markovian queues with simultaneous control of arrival and service rates, and derived conditions for the existence of natural monotone optimal policies.

A somewhat different type of control problem involves determining the optimal stopping time for operating a transient queueing system, that is, determining when to shut down the queue ("close up shop"). The cost trade-offs here generally involve loss of revenue versus incurring overtime costs. The methodology for treating these problems is different than the dynamic programming techniques used for the arrival or service control problems, and we refer the interested reader to Prabhu (1974). An additional, useful, and more up-to-date reference on control problems in queueing, and more generally on Markov decision processes, is Puterman (1991).

6.7   STATISTICAL INFERENCE IN QUEUEING

The role of statistics (as contrasted with probability) in queueing analyses is focused on the estimation of arrival and service parameters and/or distributions from observed data. Since one must, in practice, often use observable data to decide on what arrival and service patterns of a queueing system actually are, it is extremely important to utilize the data to the fullest extent possible. Since there are also many statistical problems associated with the use of discrete-event simulation modeling in queueing analyses, we have divided the coverage of the broad topic of inference between this section and the later one on simulation in Section 7.4. In the current section, we examine only those statistical questions related to what might be called *a posteriori* analysis; that is, data are collected on the behavior of a queueing problem, and we wish to make inferences about the underlying interarrival and service structure giving rise to these observations. We contrast this with the *a priori* analysis necessary for selecting the appropriate interarrival and/or service distributions to input to a queueing model (analytical or simulation analyses) or for the isolated examination of data on interarrival and/or service times. These latter problems are to be addressed in Section 7.4; of course, we recognize the overall importance of distribution selection and estimation throughout all of queueing theory.

Known statistical procedures can help in making the best use of existing data, or in determining what and how much new data should be taken. That this is an important facet of real queueing studies is readily seen, since the output from a queueing model can be no better than its input. Furthermore, when statistical procedures are used to estimate input parameters (say from data), the output measures of effectiveness actually become random variables, and it is often of interest to obtain confidence statements concerning them. The primary references for the material to follow are a survey paper by Cox entitled "Some Problems of Statistical Analysis Connected with Congestion," which appeared in Smith and Wilkinson (1965), as well as Clarke (1957), Billingsley (1961), Wolff (1965), and Harris (1974).

As already stated, it is our intention in this section to describe the problems of statistical inference which arise when it is assumed that a specific model form is applicable. This in turn leads to a further subdivision into problems of estimation, hypothesis testing, and confidence statements, though all of course are clearly related.

The initial step in any statistical procedure is a determination of the availability of sample information. The method that is chosen for estimation and the form of the estimators depend very much on the completeness of the monitoring process. It is one thing if one can observe a system fully over a certain period and is thus able to record the instants of service for each customer, but quite a different problem to have such incomplete information as only the queue size at each departure.

The earliest work on the statistics of queues did, in fact, assume that the subject queue was fully observed over a period of time and therefore that complete information was available in the form of the arrival instants and the points of the beginning and end of the service of each customer. As would then be expected, the queue was assumed to be a Markov chain in continuous time. Clarke (1957) began a sequence of papers on this and related topics by obtaining the maximum-likelihood estimators for the arrival and service parameters of an $M/M/1$ queue, in addition to the variance–covariance matrix for the two statistics. This work was followed shortly thereafter by a similar exposition for $M/M/\infty$ by Beneš (1957). Clarke's work was fundamental, and therefore a brief discussion of it follows.

We recall that the class known as maximum-likelihood estimates (MLEs) can be obtained as follows. First we form the likelihood L as a function of the (one or more) model parameters equal to the joint density of the observed sample. Then the MLEs of the parameters are the values maximizing L. Remember that it is often mathematically easier to find the maximum of the natural logarithm of L, which is generally written as $\mathscr{L}$.

To describe Clarke's work, begin by assuming that a stationary $M/M/1$ queue is being observed, with unknown mean arrival rate λ and mean service rate μ ($\rho \equiv \lambda/\mu$). We suppose that the queue begins operation with no customers present and then consider both the conditional likelihood given $N(0) = n_0$ and the likelihood ignoring the initial condition. It is clear that times between transitions are exponential, with mean $1/(\lambda + \mu)$ when the zero state is not occupied and mean $1/\lambda$ when the zero state is occupied. All jumps are upward from zero, while jumps upward from nonzero states occur with probability $\lambda/(\lambda + \mu)$ and jumps downward with probability $\mu/(\lambda + \mu)$, all independent of previous queue history.

Let us further assume that we are observing the system for a fixed amount of time t, where t is sufficiently large to guarantee some appropriate number of observations and must be chosen independently of the arrival and service processes, so that the sampling interval is independent of λ and μ. Let us then use t_e and t_b to denote, respectively, the amounts of time the system is empty and busy ($t_b = t - t_e$). In addition, let n_c denote the number of customers who have been served, n_{ae} the number of arrivals to an empty system, and n_{ab} the number to a busy system, with the total number of arrivals $n_a = n_{ae} + n_{ab}$. This is essentially Clarke's notation, and it is most convenient to use in this situation.

The likelihood function is then made up of components which are formed from the following kinds of information:

1. intervals of length x_b spent in a nonzero state and ending in an arrival or departure;
2. intervals of length x_e spent in the zero state and ending in an arrival;
3. the very last (unended) interval (length x_l) of observation;
4. arrivals at a busy system;
5. departures;
6. the initial number of customers, n_0.

The contributions to the likelihood of each of these are

1. $(\lambda + \mu)e^{-(\lambda+\mu)x_b}$;
2. $\lambda e^{-\lambda x_e}$;
3. $e^{-\lambda x_l}$ or $e^{-(\lambda+\mu)x_l}$;
4. $\lambda/(\lambda + \mu)$;
5. $\mu/(\lambda + \mu)$;
6. $\Pr\{N(0) = 0\}$.

Since $\Sigma x_b = t_b$, $\Sigma x_e = t_e$, and $n_{ae} + n_{ab} = n_a$, the likelihood is found to be (with the unended time x_l properly included in either t_b or t_e)

$$L(\lambda, \mu) = (\lambda + \mu)^{n_c + n_{ab}} e^{-(\lambda+\mu)\Sigma x_b} \lambda^{n_{ae}} e^{-\lambda\Sigma x_e} \left(\frac{\lambda}{\lambda + \mu}\right)^{n_{ab}} \left(\frac{\mu}{\lambda + \mu}\right)^{n_c} \Pr\{n_0\}$$

$$= e^{-(\lambda+\mu)t_b} \lambda^{n_{ae}} e^{-\lambda t_e} \mu^{n_c} \Pr\{n_0\}.$$

The log-likelihood function $\mathcal{L}$ corresponding to the above L is clearly given by

$$\mathcal{L}(\lambda, \mu) = -\lambda t - \mu t_b + n_a \ln \lambda + n_c \ln \mu + \ln \Pr\{n_0\}. \tag{6.31}$$

In the event that the queue is in equilibrium, the initial size may be ignored, and then the MLEs $\hat{\lambda}$ and $\hat{\mu}$ ($\hat{\rho} = \hat{\lambda}/\hat{\mu}$) will be given by the solution to

$$\frac{\partial \mathcal{L}}{\partial \lambda} = 0, \qquad \frac{\partial \mathcal{L}}{\partial \mu} = 0.$$

In this case,

$$\frac{\partial \mathcal{L}}{\partial \lambda} = -t + \frac{n_a}{\lambda}, \qquad \frac{\partial \mathcal{L}}{\partial \mu} = -t_b + \frac{n_c}{\mu}.$$

Thus

$$\hat{\lambda} = \frac{n_a}{t}, \qquad \hat{\mu} = \frac{n_c}{t_b}. \tag{6.32}$$

Note that this is what one would obtain by observing individual interarrival and service times and taking their sample averages (ignoring the last unended interval); we later show in Section 7.4 that these are also the MLEs.

It must be that $\hat{\rho} = \hat{\lambda}/\hat{\mu} < 1$, since equilibrium has been assumed; but if this condition is violated, then we must assume that $\hat{\lambda}/\hat{\mu} \doteq 1$ and minimize $\mathcal{L}(\lambda, \mu) + \theta(\lambda, \mu)$, where θ is a Lagrange multiplier, and obtain as the common estimator $\hat{\lambda} = \hat{\mu} = (n_a + n_c)/(t + t_b)$.

If we now instead assume that $N(0)$ cannot be ignored, then in order to obtain any meaningful results, some assumption must be made regarding the distribution of this initial size. If ρ is known to be less than one, then by choosing $\Pr\{N(0) = n_0\}$ to be $\rho^{n_0}(1 - \rho)$ we would immediately place ourselves in the steady state. On the other hand, we may want to do the estimation under the assumption that ρ can indeed be greater than one. But then the suggested geometric distribution for system size would not be appropriate. In this case, an alternative approach must be tried, and the choice is somewhat arbitrary.

For illustrative purposes, let us now use $\Pr\{n_0\} = \rho^{n_0}(1 - \rho)$, in which case Equation (6.31) becomes

$$\mathcal{L}(\lambda, \mu) = -\lambda t - \mu t_b + n_a \ln \lambda + n_c \ln \mu + n_0(\ln \lambda - \ln \mu) + \ln\left(\frac{1 - \lambda}{\mu}\right).$$

Then

$$\frac{\partial \mathcal{L}}{\partial \lambda} = -t + \frac{n_a}{\lambda} + \frac{n_0}{\lambda} - \frac{1}{\mu - \lambda},$$

$$\frac{\partial \mathcal{L}}{\partial \mu} = -t_b + \frac{n_c}{\mu} - \frac{n_0}{\mu} + \frac{\lambda}{\mu(\mu - \lambda)}.$$

The estimators $\hat{\lambda}$ and $\hat{\mu}$ are thus the solution to

$$0 = -t + \frac{n_a + n_0}{\hat{\lambda}} - \frac{1}{\hat{\mu} - \hat{\lambda}},$$

$$0 = -t_b + \frac{n_c - n_0}{\hat{\mu}} + \frac{\hat{\lambda}}{\hat{\mu}(\hat{\mu} - \hat{\lambda})}. \tag{6.33}$$

The first of these equations simplifies to

$$\hat{\mu} - \hat{\lambda} = \frac{\hat{\lambda}}{n_a + n_0 - \hat{\lambda}t} .$$

Then eliminating $\hat{\mu}$ from the second equation gives a quadratic in $\hat{\lambda}$, which would then be used to obtain two values of $\hat{\lambda}$. Any negative value obtained is rejected, and for the remaining values of $\hat{\lambda}$, the corresponding value of $\hat{\mu}$ is obtained. In addition, one would reject any $(\hat{\lambda}, \hat{\mu})$ pair for which $\hat{\mu} \leq 0$ or $\hat{\lambda}/\hat{\mu} > 1$. If both solutions are valid, then the one that maximizes the likelihood function is kept. If neither solution is valid and it is positiveness that is violated, let the violating parameter be equal to a small positive ϵ; otherwise, let $\hat{\lambda} = \hat{\mu}$.

An alternative approach, incorporating the initial state, would be to adjust (6.32) by subtracting the initial system size minus the estimated mean equilibrium system size divided by the observing time from $\hat{\mu}$, since the effect of the difference of n_0 from the steady-state mean will then be removed. The reverse is true of $\hat{\lambda}$; that is, this quantity is added. This then gives the approximations

$$\hat{\lambda} \doteq \frac{n_a}{t} + \frac{n_0 - (n_a t_b/n_c t)/(1 - n_a t_b/n_c t)}{t},$$

$$\hat{\mu} \doteq \frac{n_c}{t_b} + \frac{n_0 - (n_a t_b/n_c t)/(1 - n_a t_b/n_c t)}{t_b}, \tag{6.34}$$

where $(n_a t_b/n_c t)/(1 - n_a t_b/n_0 t)$ is an estimate of L, namely, $\hat{L} = \hat{\rho}/(1 - \hat{\rho})$. It should be noted that all the estimators presented here for λ and μ are indeed consistent.

Cox (1965) and Lilliefors (1966) have considered the problem of finding confidence intervals for the actual $M/M/1$ traffic intensity from the MLEs given by Equation (6.32). Since the individual interarrival times are IID exponential random variables, the quantity t is Erlang type n_a with mean n_a/λ; hence λt is Erlang type n_a with mean n_a. Likewise, the quantity μt_b is Erlang type n_c with mean n_c. If the sampling stopping rule is carefully specified to guarantee independence of n_a and n_c, then the distribution for the ratio $t_b/t\rho$ is given as $F_{2n_c,2n_a}(t_b/t\rho)$, where $F_{a,b}(x)$ is the usual F distribution with degrees of freedom a and b. (The twos in the degrees of freedom enter when the Erlang distributions are converted to χ^2 distributions, the ratio of which then yields the F distribution.) But, from (6.32), $t_b/t\rho = (n_c/n_a)\hat{\rho}/\rho$, and thus confidence intervals can be readily found for ρ by the direct use of the F distribution, with the upper $1 - \alpha$ confidence limit, say ρ_u, found from the F table by the equation

$$\frac{n_c \hat{\rho}}{n_a \rho_u} = F_{2n_c, 2n_a}(\alpha/2),$$

and lower $1 - \alpha$ confidence limit, say ρ_l, from

$$\frac{n_c \hat{\rho}}{n_a \rho_l} = F_{2n_c, 2n_a}(1 - \alpha/2).$$

In addition, confidence intervals can be found for any of the usual measures of effectiveness which are functions of ρ.

Example 6.4
As a simple illustration, let us consider the following problem. Observations are made of an $M/M/1$ queue, and it is noted at time $t = 400$ hr that all of 60 arrivals have been duly served and departed. Of the 400 hr of observation, the server was actually busy for a total of 300 hr. Let us then find a 95% confidence interval for the traffic intensity ρ.

By the previous discussion, $\hat{\lambda} = 60/400 = \frac{3}{20}$ and $\hat{\mu} = 60/300 = \frac{1}{5}$, so that $\hat{\rho} = \frac{3}{4}$. Furthermore, we are interested in confidence intervals at a level $\alpha = 0.05$. Then the appropriate upper and lower limits for degrees of freedom 120 and 120 are found to be approximately

$$\frac{n_c \hat{\rho}}{n_a \rho_u} \doteq 0.70 \quad \Rightarrow \quad \rho_u \doteq \frac{\hat{\rho}}{0.70} \doteq 1.07$$

and

$$\frac{n_c \hat{\rho}}{n_a \rho_l} \doteq 1.43 \quad \Rightarrow \quad \rho_l \doteq \frac{\hat{\rho}}{1.43} \doteq 0.52.$$

We therefore conclude with 95% confidence that ρ will fall in the interval (0.52, 1.07).

These same kinds of ideas can be nicely extended to exponential queues with many servers, and to cases with Erlang input and/or service. In addition, Billingsley (1961) has made a detailed study of likelihood estimation for Markov chains in continuous time, including limit theory and hypothesis testing, and then these results were used and extended to obtain results for birth–death queueing models by Wolff (1965).

Suppose that we now wish to apply the likelihood procedure to an $M/G/1$ queue. The approach is similar except that the loss of memorylessness is going to alter the likelihood function, since no use can be made of data which distinguish between empty and busy intervals, and there are now four components of the likelihood, namely:

1. interarrival intervals of length x, which are exponential, with contribution $\lambda e^{-\lambda x}$;
2. service times of duration x for the n_c completed customers, with contribution $b(x)$;
3. time spent in service (say x_l) by the very last customer, with contribution $1 - B(x_l)$;
4. the initial number of customers.

Hence the likelihood may be written as

$$L(\lambda, \mu) = e^{-\lambda(t-x_l)} \lambda^{n_a} \left(\prod_{i=1}^{n_c} b(x_i) \right) [1 - B(x_l)] \, \Pr\{n_0\},$$

and the log likelihood as

$$\mathcal{L}(\lambda, \mu) = n_a \ln \lambda - \lambda(t - x_l) + \sum_{i=1}^{n_c} \ln b(x_i) + \ln[1 - B(x_l)] + \ln \Pr\{n_0\}.$$

Then derivatives are taken in the usual way, and the procedure follows that of the $M/M/1$ thereafter.

To illustrate this method, the following example is presented.

Example 6.5
Let us find the MLEs for an $M/E_2/1$ queue with mean arrival rate λ and mean service time $2/\mu$. From the above, since $b(t) = \mu^2 t e^{-\mu t}$, the log likelihood may be written as

$$\mathcal{L}(\lambda, \mu) = n_a \ln \lambda - \lambda(t - x_l) + \sum_{i=1}^{n_c} (2 \ln \mu + \ln x_i - \mu x_i)$$

$$+ \ln\left(1 - \int_0^{x_l} \mu^2 t e^{-\mu t} \, dt \right) + \ln \Pr\{n_0\}.$$

But the integral of an Erlang may be rewritten in terms of a Poisson sum [see Equation (1.15)] as

$$\int_0^{x_l} \mu^2 t e^{-\mu t} \, dt = 1 - e^{-\mu x_l} \sum_{i=0}^{1} \frac{(\mu x_l)^i}{i!};$$

hence

$$\mathscr{L}(\lambda, \mu) = n_a \ln \lambda - \lambda(t - x_l) + 2n_c \ln \mu + \sum_{i=1}^{n_c} \ln x_i - \mu \sum_{i=1}^{n_c} x_i$$

$$+ \ln(e^{-\mu x_l} + \mu x_l e^{-\mu x_l}) + \ln \Pr\{n_0\}.$$

The partial derivatives can be computed to be

$$\frac{\partial \mathscr{L}}{\partial \lambda} = \frac{n_a}{\lambda} - (t - x_l) + \frac{\partial \ln \Pr\{n_0\}}{\partial \lambda},$$

$$\frac{\partial \mathscr{L}}{\partial \mu} = \frac{2n_c}{\mu} - \sum_{i=1}^{n_c} x_i - x_l + \frac{x_l}{1 + \mu x_l} + \frac{\partial \ln \Pr\{n_0\}}{\partial \mu}.$$

If now the initial state is assumed to be chosen free of λ and μ, then the MLEs for λ and μ are found by equating the partial derivatives to zero as

$$\hat{\lambda} = \frac{n_a}{t - x_l},$$

and the appropriate solution $\hat{\mu}$ to the quadratic equation

$$x_l \left(x_l + \sum_{i=2}^{n_c} x_i \right) \hat{\mu}^2 + \left(\sum_{i=1}^{n_c} x_i - 2n_c x_l \right) \hat{\mu} - 2n_c = 0.$$

Thus far we have been in the position of assuming that we were dealing with simple processes subject to complete information. But suppose now that certain kinds of information are just not available. For example, suppose we observe only the stationary output of an $M/G/1$ (G known) queue and are then asked to estimate the mean service and interarrival times. Under the assumption that the stream is in equilibrium, the mean interarrival time $1/\lambda$ must equal the long-term arithmetic mean of the interdeparture times. If the mean of the departure process is denoted by $\bar{d}$, then the maximum-likelihood estimator of λ is $\hat{\lambda} = 1/\bar{d}$.

If service is exponential, then we know that the limiting distribution of output is the same as that of the input. Hence no inference is possible about the mean service time. But if service is assumed to be other than exponential, then the estimation is possible. The CDF of the interdeparture process of any $M/G/1$ queue is given by (recalling that $p_0 = 1 - \lambda/\mu$)

$$C(t) = \frac{\lambda}{\mu} B(t) + \left(1 - \frac{\lambda}{\mu}\right) \int_0^t B(t - x)\lambda e^{-\lambda x} \, dx, \qquad (6.35)$$

where the last term is the convolution of service and interarrival-time CDFs. In the special case where the service time is constant, (6.35) reduces to

$$C(t) = \begin{cases} 0 & (t < 1/\mu), \\ \lambda/\mu & (t = 1/\mu), \\ \lambda/\mu + (1 - \lambda/\mu)(1 - e^{-\lambda(t-1/\mu)}) & (t > 1/\mu). \end{cases}$$

Since there is nonzero probability associated with the point $t = 1/\mu$, we may directly obtain our estimate by equating $1/\mu$ with the minimum observed interdeparture time.

On the other hand, if the distribution is other than exponential or deterministic, an approach is to take LSTs of Equation (6.35), yielding

$$C^*(s) = \frac{(1 + s/\mu)B^*(s)}{1 + s/\lambda}, \tag{6.36}$$

where $B^*(s)$ is the LST of the service-time CDF, whose form is known, but not the values of its parameters. Then the moments of $C(t)$ and $B(t)$ may be directly related by the successive differentiation of Equation (6.36), using enough equations to determine all parameters of $B(t)$. However, there still exists a small problem: successive interdeparture times will be correlated, and this correlation must be considered when calculating moments from data. For example, if enough data are present, data spread sufficiently far apart may be considered to form an approximately random sample so that formulas based on uncorrelated observations can be used. It would be advisable, nevertheless, to test for lack of correlation by computing the sample correlation coefficient between successive observations before making any definitive statements. We have asked the reader to solve a problem of this type at the end of the chapter as Problem 6.32. There is also a discussion of this and related questions in Cox (1965).

Suppose we slightly modify the previous problem and instead now consider an *M/G/*1 with the form of *G* known and observations now made on both the input and output. The analysis might then proceed in a very similar manner, except that, since we are observing ordered instants of arrival and departure, the successive customer waiting times are now available. The relevant relationship between the transforms of $B(t)$ and the system-wait CDF $W(t)$ is [see Equation (5.37a)]

$$W^*(s) = \frac{(1 - \lambda/\mu)sB^*(s)}{s - \lambda + \lambda B^*(s)}. \tag{6.37}$$

Again, problems of autocorrelation surface. But assuming that these are taken into account as previously discussed, we know that in the event $B(t)$ is exponential,

$$\frac{dW^*(s)}{ds}\bigg|_{s=0} = -\frac{1}{\mu - \lambda}.$$

Hence

$$\hat{W} = \frac{1}{\hat{\mu} - \hat{\lambda}} \quad \Rightarrow \quad \hat{\mu} = \hat{\lambda} + \frac{1}{\hat{W}}.$$

In fact, for any one-parameter service distribution, we may directly appeal to the PK formula [found via the first derivatives of Equation (6.37) or from the results of Section 5.1], namely,

$$W = \frac{1}{\mu} + \frac{(\lambda/\mu)^2 + \lambda^2 \sigma_S^2}{2\lambda(1 - \lambda/\mu)}.$$

In the deterministic case, for example, we find that

$$W = \frac{1}{\mu} + \frac{\lambda/\mu^2}{2(1 - \lambda/\mu)} = \frac{2 - \lambda/\mu}{2\mu(1 - \lambda/\mu)}.$$

Therefore we may write that

$$\hat{W} = \frac{2 - \hat{\lambda}/\hat{\mu}}{2\hat{\mu}(1 - \hat{\lambda}/\hat{\mu})},$$

or that $\hat{\mu}$ is the appropriate solution to the quadratic

$$2\hat{W}\hat{\mu}^2 - 2(\hat{W}\hat{\lambda} + 1)\hat{\mu} + \hat{\lambda} = 0,$$

where $\hat{\lambda}$ is found as before from $\bar{d}$. Again, we leave the Erlang case as a problem (see Problem 6.33).

As for $M/G/\infty$, we have shown previously in Section 5.2.2 that both output and system size are nonhomogeneous Poisson streams. Hence we should be able to estimate parameters by converting all the observable processes to Poisson processes, from which it is easy to obtain appropriate estimators.

PROBLEMS

*Whenever a problem is best solved by use of this book's accompanying software, we have added a boldface **C** to the right of the problem number.*

6.1. Reformulate the $M/E_2/1$ rootfinding problem of Section 3.3.4 to the type required in Section 6.1, and then verify the answers from Section 3.3.4.

6.2. Find the general form of $W_q(t)$ for the $D/E_k/1$.

6.3. You have learned that the Laplace transform of the distribution function of the system waiting time in a $G/G/1$ system is

$$\overline{W}(s) = \frac{1}{s} - \frac{3s^2 + 22s + 36}{3(s^2 + 6s + 8)(s + 3)}.$$

Find $W(t)$ and its mean.

6.4. The Bearing Straight Corporation of Example 5.2 now finds that its machines do not break down as a Poisson process, but rather as the two-point distribution given below:

Interarrival Time	Probability	$A(t)$
9	$\frac{2}{3}$	$\frac{2}{3}$
18	$\frac{1}{3}$	1

Given that the service has the same two-point distribution as it did earlier, use Lindley's equation to find the line-delay CDF.

6.5. Use the approach indicated for $M/G/1$ waiting times in Section 5.1.5 to verify the waiting-time result of Example 6.1.

6.6. Consider a $G/G/1$ queue whose service times are uniformly distributed on $(0, 1)$, with $\lambda = 1$. A specific arrival in the steady state encountered 12 people in the system when it came in. What is its mean line wait?

6.7. Find the probability mass functions for the arrival counts associated with deterministic and Erlang input streams.

6.8. Show that the virtual idle time for a $G/M/1$ has CDF

$$F(u) = A(u) + \int_u^\infty e^{-\mu(1-r_0)(t-u)} \, dA(t).$$

6.9. Verify the derivation of Equation (6.17) from (6.16).

6.10. (a) Show that the poles of Equation (6.17) are distinct.
(b)C Fully solve an $M/D/2$ with $\lambda = \mu = 1$.

6.11. Use the generating function for the state probabilities of Problem 6.10(b) to find the variance of the line wait under the same assumptions.

6.12. Consider an $M^{[X]}/M/1$ queue in which if there are two or more customers waiting when the server is free, then the next two customers are served together. The service time is assumed exponential with mean $1/\mu$ regardless of the number being served. Find all the single-step transition probabilities of the imbedded Markov chain. Then use the results of Section 6.4 to obtain the chain's steady-state distribution.

6.13. Do as in Problem 6.12 for the two-channel exponential-exponential machine repair problem, and also find the general-time stationary probabilities.

6.14C. Consider the $G/M/3$ queue with the same interarrival distribution as in Problem 6.4, and mean service rate $\mu = 0.035$. Find the general-time probabilities.

6.15C. Find the general-time probability distribution associated with the queue in Example 5.6, as well as the mean system and queue sizes.

6.16C. Find the general-time probability distribution associated with the queue in Problem 5.28(b), as well as the mean system and queue sizes.

6.17C. For Problem 5.32, use the results of Section 6.4 to get general-time probabilities.

6.18C. Find the general-time probability distribution associated with the queue in Problem 5.33.

6.19. Find p_0, p_1, L, and W for a $G/G/1/1$ queue with $\lambda = \mu = 1$.

6.20. Consider an $M/M/1/\infty$ queue for which $\lambda = 1$ and $\mu = 3$. Every customer going through the system pays an amount $15, but costs the system $6 per unit time it spends in the system.
(a) What is the average profit rate of this system?
(b) A bright young OR analyst from a prestigious middle Atlantic state university tells management that the profit can be increased

by shutting off the queue at a certain point, that is, by preventing customers from entering whenever the queue gets to a certain size. Do you agree, and if so, what is the optimal point at which to prevent arrivals from joining?

6.21. Redo Problem 2.48 for $C_1 = \$24$, $C_2 = \$138$, and down-time cost = $\$10$/hr.

6.22. Consider a two-server system, where each server is capable of working at two speeds. The service times are exponential with mean rate μ_1 or μ_2. When there are k (>2) customers in the system, the mean rate of both servers switches from μ_1 to μ_2. Suppose that the cost per operating hour at low speed is $\$50$ and at high speed is $\$220$, and the cost of waiting time (time spent in the system) per customer is $\$10$/hr. The arrival rate is Poisson with mean 20/hr, while the service rates μ_1 and μ_2 are 7.5 and 15/hr, respectively. Use the results of Problem 2.49 to find the optimal k. Compare the solution with that of Problem 2.48.

6.23. Consider an $M/M/1$ queue with a three-state service rate as explained in Problem 2.50. Suppose that management policy sets the upper switch point k_2 at 5. Using the results of Problem 2.50, find the optimal k_1 (the lower switch point) for a system with mean arrival rate of 40, μ_1 of 20, μ_2 of 40, and μ of 50. Assume that the waiting cost per customer is $\$10$/hr and the service costs per hour are $\$100$, $\$150$, and $\$250$, respectively.

6.24. Suppose we have an $M/M/1$ queue with customer balking. Furthermore, suppose the balking function b_n is given as $e^{-n/\mu}$, where μ is the mean service rate (see Section 2.9.1). It is known that the salary of the server depends on his or her skill, so that the marginal cost of providing service is proportional to the rate μ and, in fact, is estimated as $\$1.50\mu$/hr. Arrivals are Poisson with a mean rate of 10/hr. The profit per customer served is estimated at $\$75$. Use the results of Problem 2.53 and find the optimal value of μ.

6.25C. Consider a tool crib where the manager believes that costs are really only associated with idle time, that is, the time that the clerks are idle (waiting for mechanics to come) and the time that the mechanics are idle (waiting in the queue for a clerk to become available). Using costs $\$30$ per hour idle per clerk and $\$70$ per hour idle per mechanic, with $1/\lambda = 50$ sec, $1/\mu = 60$ sec, find the optimal number of clerks. Comment.

6.26. Consider an $M/M/1$ queue where the mean service rate μ is under the control of management. There is a customer waiting cost per unit time spent in the system of C_W and a service cost which is proportional to the *square* of the mean service rate, the constant of proportionality being C_S. Find the optimal value of μ in terms of λ, C_S, and C_W. Solve when $\lambda = 10$, $C_S = 2$, and $C_W = 20$.

6.27. Find the quadratic in $\hat{\lambda}$ that arises in the simultaneous solution of Equation (6.33), and then solve.

6.28. Get maximum-likelihood estimators for λ and μ in an $M/M/1$, first ignoring the initial state and then with an initial state of 4 in the steady state, from the data

$$t = 150, \qquad t_b = 100, \qquad n_a = 16, \quad \text{and} \quad n_c = 12.$$

6.29. Under the assumption that n_0 is always fixed at 0, find the MLEs for λ, μ, and ρ in an $M/M/1$ from the data

$$t = 150, \qquad t_b = 100, \qquad n_a = 16, \quad \text{and} \quad n_c = 8.$$

6.30. Give a 95% confidence interval for the ρ in Problem 6.29.

6.31. Find the formulas for the MLEs of λ and μ in an $M/E_3/1$.

6.32. Use Equation (6.36) to find the estimators of the service parameter from the output of an $M/E_2/1$ queue.

6.33. Use Equation (6.37) to find the estimators of the service parameter from the system waiting times of the $M/E_2/1$ queue.

6.34. The observed output (interdeparture times) of an $M/G/1$ queue is as follows:

0.6, 2.4, 1.0, 1.1, 0.2, 0.2, 0.2, 2.7, 1.5, 0.3, 0.6, 0.9, 0.5, 0.2, 0.5, 3.8, 0.4, 0.1, 1.5, 1.4, 0.7, 0.5, 0.1, 0.6, 1.6, 1.5, 0.5, 0.8, 1.3, 0.4, 0.7, 2.4, 2.4, 0.3, 0.8, 0.9, 1.5, 0.3, 1.2, 1.0, 0.6, 0.1, 0.4, 0.3, 2.5, 3.5, 0.8, 0.6, 9.5, 1.6.

Test to determine whether this stream is exponential and therefore whether the queue is $M/M/1$.

CHAPTER 7

Bounds, Approximations, Numerical Techniques, and Simulation

The use of approximating and/or numerical procedures in queueing theory has become most important. Once we leave the arena of steady-state Markovian queues, nice closed-form analytical results are quite elusive. In order to obtain useful solutions in these cases, one must resort to simulation, approximations, or numerical techniques. Typical examples of problems where concise closed-form solutions are particularly difficult to find are the $G/G/c$, non-phase-type $G/G/1$, transient problems, and non-Markovian networks.

The technical presentation in this chapter is divided into four parts. First, we offer a derivation of some popular upper and lower bounds for expected line waits (and therefore also for the mean queue sizes) of both steady-state $G/G/1$ and $G/G/c$ queues. Then we move on to the derivation of some commonly used approximations. A number of these are developed directly from the bounds presented in the first part, with the remainder derived in an independent fashion. Included here is a section on approximations and limit theorems for heavy-traffic and nonstationary queues. One highlight is a brief discussion of the use of diffusion models. Following this, we offer an overview of the use of numerical computer-based techniques in the solution of complex queueing problems for which closed-form, easy-to-use solutions are not available. Finally, we conclude this chapter with a brief overview of discrete event simulation methodology for queueing modeling.

7.1 BOUNDS AND INEQUALITIES

In the following section, we first present upper and lower bounds for the mean line delay of a steady-state $G/G/1$ queue as a function only of the first and second moments of its interarrival and service times. This is succeeded by the derivation of a further lower bound when the full forms of the interarrival and service-time distribution functions are known—but possibly too complicated to analyze in one of the usual ways. In Section 7.1.2 we then present some comparable results for the general multiserver queue.

7.1.1   General Single-Server Queues

This section is concerned with a discussion of the use of bounds and inequalities for arbitrary stationary $G/G/1$ queues. Much of the following development parallels a paper by Kingman (1962) and one by Marshall (1968).

We begin by finding some fairly simple relationships between the moments of the interarrival times, service times, idle periods, and waiting times for an arbitrary $G/G/1$ queue with $\rho < 1$. The starting point again is the iterative equation for the line delays (see Section 6.2),

$$W_q^{(n+1)} = \max(0, W_q^{(n)} + U^{(n)}) \qquad (U^{(n)} \equiv S^{(n)} - T^{(n)}).$$

If we now let $X^{(n)}$ be the slack in the event of idle time, that is,

$$X^{(n)} = -\min(0, W_q^{(n)} + U^{(n)}),$$

then

$$W_q^{(n+1)} = W_q^{(n)} + U^{(n)} + X^{(n)}. \tag{7.1}$$

note that X_n can be interpreted as a virtual idle time. Since the queue is stationary and $E[W_q^{(n+1)}] = E[W_q^{(n)}]$, when expectations are taken of both sides of (7.1) it is found that $0 = E[U] + E[X]$ or $E[U] = -E[X]$. But $E[U] = 1/\mu - 1/\lambda$ and $E[X] = \Pr\{$ system found empty by an arrival$\} \cdot$ E[length of idle period] $= q_0 E[I]$, where, as usual, we denote the arrival point probabilities as $\{q_n\}$ and the length of the idle period by I. This then gives

$$E[I] = \frac{E[X]}{q_0} = -\frac{E[U]}{q_0} = \frac{1/\lambda - 1/\mu}{q_0}. \tag{7.2}$$

The next result we find is a formula for the expected wait for a stable $G/G/1$ in terms of the first and second moments of U and I. First, square both sides of (7.1) to get

$$(W_q^{(n+1)})^2 + 2W_q^{(n+1)}X^{(n)} + (X^{(n)})^2 = (W_q^{(n)})^2 + 2W_q^{(n)}U^{(n)} + (U^{(n)})^2. \tag{7.3}$$

We see from the definitions of $W_q^{(n+1)}$ and $X^{(n)}$ that $2W_q^{(n+1)}X^{(n)} = 0$, since one or the other of the two factors must always be zero. Since $W_q^{(n)}$ and $U^{(n)}$ are independent, when expectations are taken of both sides of (7.3) and n goes to ∞, we find that

$$0 = E[U^2] - E[X^2] + 2E[U]W_q,$$

or

$$W_q = \frac{E[X^2] - E[U^2]}{2E[U]}. \tag{7.4}$$

But $E[X^2] = \Pr\{\text{system found empty by an arrival}\} \cdot E[(\text{length of idle period})^2]$. Hence

$$W_q = \frac{q_0 E[I^2] - E[U^2]}{2E[U]},$$

and, from Equation (7.2),

$$\boxed{W_q = \frac{-E[I^2]}{2E[I]} - \frac{E[U^2]}{2E[U]}.} \tag{7.5}$$

For Poisson arrivals, $E[I] = 1/\lambda$ and $E[I^2] = 2/\lambda^2$, and remembering that $E[U^2] = \text{Var}[U] + E^2[U]$, Equation (7.5) checks out correctly to give the PK expression

$$W_q = \frac{\rho^2 + \lambda^2 \sigma_B^2}{2\lambda(1 - \rho)}.$$

A similar expression can then be found for the variance by cubing Equation (7.1). The final result is given by Marshall (1968).

The foregoing results now lead to some bounds valid for all $G/G/1$ queues. The first is a lower bound on the mean idle time. From Equation (7.2), we see that since $q_0 \le 1$, we have $E[I] \ge 1/\lambda - 1/\mu$, with the inequality binding for $D/D/1$. We can use this inequality to derive an upper bound for W_q. Begin by rewriting (7.5) as

$$W_q = \frac{-(\text{Var}[I] + E^2[I])}{2E[I]} - \frac{E[U^2]}{2E[U]}.$$

Since the variance must be positive, it follows from (7.2) that

$$W_q \leq \frac{-E^2[I]}{2E[I]} - \frac{E[U^2]}{2E[U]} = \frac{1}{2}\left(-E[I] - \frac{E[U^2]}{E[U]}\right) = \frac{1}{2}\left(\frac{E[U]}{q_0} - \frac{E[U^2]}{E[U]}\right).$$

Recognizing that $E[U] = 1/\mu - 1/\lambda < 0$ and $q_0 < 1$, which implies that $E[U]/q_0 < E[U]$, we see that

$$W_q \leq \frac{1}{2}\left(E[U] - \frac{E[U^2]}{E[U]}\right) = \frac{1}{2}\left(\frac{E^2[U] - E[U^2]}{E[U]}\right) = \frac{1}{2}\left(\frac{-\text{Var}[U]}{E[U]}\right)$$

$$= \frac{1}{2}\left(\frac{\text{Var}[S] + \text{Var}[T]}{1/\lambda - 1/\mu}\right),$$

which can be rewritten as

$$W_q \leq \frac{\lambda(\sigma_A^2 + \sigma_B^2)}{2(1 - \rho)}. \tag{7.6}$$

For a lower bound similarly dependent upon interarrival and service-time moments only, we refer to the work of Marchal (1978). To begin, go back to Equation (7.4), where we see that we can bound W_q from below if we can find a lower bound for $E[X^2]$. To do so, we recognize from (7.1) that the random variable X is stochastically smaller than the interarrival time variable T, since X is either 0 or $T - (W_q + S)$. [By stochastically smaller we mean that $\Pr\{X \leq x\} \geq \Pr\{T \leq x)$ for all x, and we write it as $X \leq_{\text{st}} T$.] It then follows that $E[X^2] \leq E[T^2]$.

Thus we have

$$W_q \geq \frac{E[T^2] - E[U^2]}{2E[U]} = \frac{\lambda^2(\sigma_B^2 + 1/\mu^2 - 2/\mu\lambda)}{2\lambda(1 - \rho)},$$

or

$$W_q \geq \frac{\lambda^2 \sigma_B^2 + \rho(\rho - 2)}{2\lambda(1 - \rho)}.$$

Because of the term -2ρ in the numerator, this lower bound is positive if and only if $\sigma_B^2 > (2 - \rho)/\lambda\mu$, and is thus not always of value. Nevertheless, it can be quite useful in a variety of important situations.

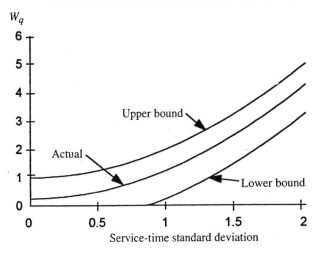

Fig. 7.1   The performance of the bounds in an $M/G/1$ situation.

Marchal (1974) has provided an interesting picture of the performance of the upper and lower bounds for the $M/G/1$. In Figure 7.1 we offer his graph of the bounds plotted with the service-time standard deviation as abscissa, assuming $\lambda = 1$ and $\rho = \frac{1}{2}$. We note the roughly parallel, parabolic nature of the bounds around the actual value computed from the PK formula as $\sigma_B^2 + \frac{1}{4}$.

Suppose it is now assumed that the interarrival and service distributions are both known and given by $A(t)$ and $B(t)$, respectively. Then another lower bound on the stationary line wait may be found to be $W_q \geq r_0$, where r_0 is the unique nonnegative root when $\rho < 1$ of

$$f(z) = z - \int_{-z}^{\infty} [1 - U(t)] \, dt = 0,$$

where, as before (in Chapter 6), $U(t)$ is the CDF of $U = S - T$.

To prove this assertion, we begin by observing that $f(z)$ does indeed have a unique nonnegative root. It is easily seen that $f(z)$ is monotonically increasing for $z \geq 0$, since

$$f'(z) = 1 - [1 - U(-z)] = U(-z) \geq 0.$$

When $z = 0$,

$$f(0) = -\int_0^\infty [1 - U(t)]\, dt < 0,$$

since the integrand is always positive; when z is large, say M,

$$f(z) = M - \int_{-M}^\infty [1 - U(t)]\, dt = M - \int_{-M}^\infty \int_t^\infty dU(x)\, dt$$

$$= M - \int_{-M}^\infty \int_{-M}^x dt\, dU(x) = M - \int_{-M}^\infty (x + M)\, dU(x)$$

$$\geq M - \int_{-\infty}^\infty (x + M)\, dU(x) = M - (E[U] + M)$$

$$= M - \left(\frac{1}{\mu} - \frac{1}{\lambda}\right) - M = -\left(\frac{1}{\mu} - \frac{1}{\lambda}\right) = \frac{1 - \rho}{\lambda} > 0.$$

Since $f(z)$ is monotonic and goes from $f(0) < 0$ to $f(M) > 0$, we have thus shown there is a unique nonnegative root when $\rho < 1$, which will be henceforth known as r_0. It therefore remains for us to show that $W_q \geq r_0$. Now rewrite $f(z)$ as $f(z) = z - f_1(z)$, where $f_1(z) = \int_{-z}^\infty [1 - U(t)]\, dt$. Then we have that

$$f_1(z) = \int_{-z}^\infty [1 - U(t)]\, dt \quad \begin{cases} > z & (z < r_0), \\ \leq z & (z \geq r_0). \end{cases} \tag{7.7}$$

The function $f(z)$ is, in fact, continuous and convex, and we are going to apply Jensen's inequality for the expected value of a convex function of a nonnegative random variable (see Parzen, 1960, for example) to get a relationship between W_q and f_1. Given that $W_q^{(n)}$ is (say) x, we see that

$$E[W_q^{(n+1)} | W_q^{(n)} = x] = E[\max(0, x + U^{(n)}]$$

$$= \int_{-x}^\infty (x + t)\, dU^{(n)}(t) = \int_{-x}^\infty t\, dU^{(n)}(t) + x[1 - U^{(n)}(-x)]$$

$$= \int_0^\infty \int_0^t dv\, dU^{(n)}(t) - \int_{-x}^0 \int_t^0 dv\, dU^{(n)} + x[1 - U^{(n)}(-x)]$$

$$= \int_0^\infty \int_v^\infty dU^{(n)}(t)\, dv - \int_{-x}^0 \int_{-x}^v dU^{(n)}(t)\, dv + x[1 - U^{(n)}(-x)].$$

Integrating over v gives

$$E[W_q^{(n+1)} \mid W_q^{(n)} = x]$$

$$= \int_0^\infty [1 - U^{(n)}(v)]\, dv - \int_{-x}^0 [U^{(n)}(v) - U^{(n)}(-x)]\, dv + x[1 - U^{(n)}(-x)]$$

$$= \int_0^\infty [1 - U^{(n)}(v)]\, dv$$

$$\quad - \int_{-x}^0 \{[1 - U^{(n)}(-x)] - [1 - U^{(n)}(v)]\}\, dv + x[1 - U^{(n)}(-x)]$$

$$= \int_0^\infty [1 - U^{(n)}(v)]\, dv - x[1 - U^{(n)}(-x)]$$

$$\quad - \int_{-x}^0 [1 - U^{(n)}(v)]\, dv + x[1 - U^{(n)}(-x)]$$

$$= \int_{-x}^\infty [1 - U^{(n)}(v)]\, dv = f_1(x).$$

Hence by the law of total probability, $E[W_q^{(n+1)}] = \int_0^\infty f_1(x)\, dW_q^{(n)}(x)$. But Jensen's inequality then tells us that for a convex function f, $E[f(x)] \geq f(E[x])$, so that $\int_0^\infty f_1(x)\, dW_q^{(n)}(x) \geq f_1(E[W_q^{(n)}])$, and hence $E[W_q^{(n+1)}] \geq f_1(E[W_q^{(n)}])$, or, in the steady state,

$$W_q \geq f_1(W_q) = \int_{-W_q}^\infty [1 - U(t)]\, dt. \qquad (7.8)$$

We now assume that $W_q < r_0$ and proceed to prove the result by contradiction.

Equation (7.7) says that

$$\int_{-W_q}^\infty [1 - U(t)]\, dt > W_q$$

for $W_q < r_0$. But this is a contradiction of (7.8); hence the result is shown and $r_0 \leq W_q$.

Putting the upper and lower bounds together gives

$$\boxed{\max\left(0, r_0, \frac{\lambda^2 \sigma_B^2 + \rho(\rho - 2)}{2\lambda(1 - \rho)}\right) \leq W_q \leq \frac{\lambda(\sigma_A^2 + \sigma_B^2)}{2(1 - \rho)}.}$$

To illustrate, let us look at $M/M/1$ to see what the bounds look like. Going back to Section 6.2, Equation (6.12), we have

$$U(t) = \begin{cases} \dfrac{\mu e^{\lambda t}}{\lambda + \mu} & (t < 0), \\[3mm] 1 - \dfrac{\lambda e^{-\mu t}}{\lambda + \mu} & (t \geq 0). \end{cases}$$

The lower bound r_0 is then found by solving $f(z) = 0$; that is,

$$0 = r_0 - \int_{-r_0}^{\infty} [1 - U(t)]\, dt = r_0 - \int_{r_0}^{0} \left(1 - \frac{\mu e^{\lambda t}}{\lambda + \mu}\right) dt - \int_{0}^{\infty} \frac{\lambda e^{-\mu t}}{\lambda + \mu}\, dt$$

$$= r_0 - r_0 + \frac{\mu(1 - e^{-\lambda r_0})}{\lambda(\lambda + \mu)} - \frac{\lambda}{\mu(\lambda + \mu)} = \frac{\mu^2 - \lambda^2 - \mu^2 e^{-\lambda r_0}}{\lambda \mu(\lambda + \mu)}.$$

So $\mu^2 - \lambda^2 = \mu^2 e^{-\lambda r_0}$, or $1 - \rho^2 = e^{-\lambda r_0}$. Hence finally,

$$r_0 = -\frac{1}{\lambda} \ln(1 - \rho^2).$$

The upper bound here is

$$\frac{\lambda(1/\lambda^2 + 1/\mu^2)}{2(1 - \lambda/\mu)} = \frac{1}{\lambda} \frac{1 + \rho^2}{2(1 - \rho)}.$$

Note that both r_0 and the upper bound go to ∞ as ρ goes to 1, as expected. To get a better idea of the behavior of the bounds, we supply a graph in Figure 7.2 of the bounds and the actual mean for all ρ between zero and one. Indeed, such asymptotic behavior is true of the upper bound for all $G/G/1$ in the sense that the bound always gets asymptotically sharper. In turn, the lower bound gets sharper as ρ goes to zero.

In his paper, Marshall goes on to get some more results for special classes of possible arrival distributions, but it is not necessary to go into them here. The reader can pursue them in the cited reference. Suffice it to say that the bounds become tighter as more information is used on the interarrival and service times.

Example 7.1
To see how these results might apply to a real problem, let us examine their application to a $G/G/1$ queue in which both the interarrival times and service times are empirical. Assume that the service times are as given by Table 5.2,

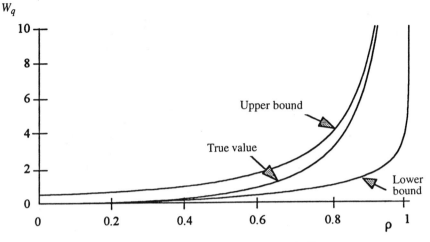

Fig. 7.2 Bounds on the expected wait in the $M/M/1$ queue.

namely, $b(9) = \frac{2}{3}$ and $b(12) = \frac{1}{3}$, and that the interarrival-time probabilities are given by $a(10) = \frac{2}{5}$ and $a(15) = \frac{3}{5}$.

A little bit of calculation gives $\sigma_B^2 = 2 \text{ min}^2$ and $\sigma_A^2 = 6 \text{ min}^2$, with $\mu = \frac{1}{10}$, $\lambda = \frac{1}{13}$, and $\rho = \frac{10}{13}$. So we can immediately calculate the upper bound from (7.6) as

$$W_q \le \frac{\frac{1}{13}(8)}{2(\frac{3}{13})} = \frac{4}{3} \text{ min.}$$

The obtaining of the lower bound is slightly more difficult, since we must find the nonnegative root of the nonlinear equation

$$f(r_0) = 0 = r_0 - \int_{-r_0}^{\infty} [1 - U(t)] \, dt.$$

(We use the second type of lower bound in view of the availability of so much information.) We begin by first calculating $U(t)$ directly from the empirical forms of the interarrival and service distributions. It is thus found that the only possible values of the random variable $U = S - T$ are $-6 = 9 - 15$ (with probability $(\frac{2}{3})(\frac{3}{5}) = \frac{2}{5}$), $-3 = 12 - 15$ (with probability $\frac{1}{5}$), $-1 = 9 - 10$ (with probability $\frac{4}{15}$), and 2 (with probability $\frac{2}{15}$). Therefore,

$$U(t) = \begin{cases} 0 & (t < -6), \\ \frac{2}{5} & (-6 \le t < -3), \\ \frac{3}{5} & (-3 \le t < -1), \\ \frac{13}{15} & (-1 \le t < 2), \\ 1 & (t \ge 2), \end{cases} \Rightarrow 1 - U(t) = \begin{cases} 1 & (t < -6), \\ \frac{3}{5} & (-6 \le t < -3), \\ \frac{2}{5} & (-3 \le t < -1), \\ \frac{2}{15} & (-1 \le t < 2), \\ 0 & (t \ge 2). \end{cases}$$

So

$$\int_{-r_0}^{\infty} [1 - U(t)]dt = \int_{-r_0}^{2} [1 - U(t)]\, dt$$

$$= \begin{cases} \frac{2}{15}r_0 + \frac{4}{15} & (0 \le r_0 \le 1), \\ \frac{2}{5}r_0 & (1 < r_0 \le 3), \\ \frac{3}{5}r_0 - \frac{9}{15} & (3 < r_0 \le 6), \\ r_0 - 3 & (r_0 > 6). \end{cases}$$

Since the upper bound is just over one, we surmise that the lower bound should probably be less than one; that is to say, it is the solution to $r_0 = \frac{2}{15}r_0 + \frac{4}{15} \Rightarrow r_0 = \frac{4}{13}$, which must be correct, since r_0 is the unique nonnegative solution.

Therefore,

$$\tfrac{4}{13}\min \le W_q \le \tfrac{4}{3}\min.$$

But, of course, we should realize that an exact answer is indeed available for this problem. Recall that we did such a discrete example in Section 6.2.2, where we went directly to a discrete version of the $G/G/1$'s stationary delay equation given in (6.7) and solved for the CDF $W_q(t)$ as a regular Markov chain. The same is possible here and the outline of the solution is provided in the following.

The feasible values of the random variable U here are $(-6, -3, -1, 2)$, with respective probabilities $(\frac{2}{5}, \frac{1}{5}, \frac{4}{15}, \frac{2}{15})$. The stationary equations for the steady-state waiting probabilities $\{w_j, j \ge 0\}$ are found from (6.7) as follows:

$$w_0 = w_0 p_{00} + w_1 p_{10} + w_2 p_{20} + w_3 p_{30} + w_4 p_{40} + w_5 p_{50} + w_6 p_{60},$$

$$w_1 = w_2 p_{21} + w_4 p_{41} + w_7 p_{71},$$

$$w_j = \sum_i w_i p_{ij} \quad (j \ge 2).$$

The nonzero transition probabilities $\{p_{ij}\}$ are seen to be

$$p_{00} = \Pr\{U < 0\} = \tfrac{13}{15}, \quad p_{02} = \Pr\{U = 2\} = \tfrac{2}{15}, \quad p_{10} = \Pr\{U \le -1\} = \tfrac{13}{15},$$

$$p_{13} = \Pr\{U = 2\} = \tfrac{2}{15}, \quad p_{20} = \Pr\{U \le -2\} = \tfrac{3}{5}, \quad p_{21} = \Pr\{U = -1\} = \tfrac{4}{15},$$

$$p_{24} = \tfrac{2}{15}, \quad p_{30} = \tfrac{3}{5}, \quad p_{32} = \tfrac{4}{15}, \quad p_{35} = \tfrac{2}{15}, \quad p_{40} = \tfrac{2}{5}, \quad p_{41} = \tfrac{1}{5},$$

$$p_{43} = \tfrac{4}{15}, \quad p_{46} = \tfrac{2}{15}, \quad p_{50} = \tfrac{2}{5}, \quad p_{52} = \tfrac{1}{5}, \quad p_{64} = \tfrac{4}{15}, \quad p_{57} = \tfrac{2}{15},$$

$$p_{i,i-6} = \tfrac{2}{5}, \quad p_{i,i-3} = \tfrac{1}{5}, \quad p_{i,i-1} = \tfrac{4}{15}, \quad p_{i,i+2} = \tfrac{2}{15} \quad (i \ge 6).$$

Thus the $\{w_j\}$ are found as the solution to

$$w_j = \tfrac{2}{15}w_{j-2} + \tfrac{4}{15}w_{j+1} + \tfrac{1}{5}w_{j+3} + \tfrac{2}{5}w_{j+6} \qquad (j \geq 2),$$

$$w_1 = \tfrac{4}{15}w_2 + \tfrac{1}{5}w_4 + \tfrac{2}{5}w_7, \qquad\qquad\qquad (7.9)$$

$$w_0 = \tfrac{13}{15}(w_0 + w_1) + \tfrac{3}{5}(w_2 + w_3) + \tfrac{2}{5}(w_4 + w_5 + w_6).$$

The next step then is to obtain the roots of the eighth-degree operator polynomial equation formed from the top line of (7.9), namely,

$$6D^8 + 3D^5 + 4D^3 - 15D^2 + 2 = 0.$$

When a root-finding algorithm is applied, we find that there are four complex roots and three real roots in addition to the usual 1. They are (approximately)

$$0.3885, \quad -0.3481, \quad -1.2681, \quad 0.6353 \pm 1.0274i, \quad -0.5215 \pm 1.0296i.$$

Only the roots 0.3885 and -0.3481 have absolute values less than one, and hence the general solution is

$$w_j = C_1(0.3885)^j + C_2(-0.3481)^j. \qquad\qquad (7.10)$$

To determine C_1 and C_2, we set up a pair of simultaneous linear equations derived from the second and third lines of (7.9) and the fact that the $\{w_j\}$ must sum to one. As a result, we find that $C_1 = 0.4349$ and $C_2 = 0.3894$.

There is an important theory which connects the solution of arbitrary $G/G/1$ queues to the sort of discrete forms just discussed. This is the concept of the continuity of queues (e.g., see Kennedy, 1972, and Whitt, 1974). The key idea is that if the interarrival and service-time distributions can be expressed respectively, as the limits of sequences of distributions (say $\{A_n\} \to A$ and $\{B_n\} \to B$) then the measures of effectiveness for the $G/G/1$ formed from A and B can be found as the limits of those obtained from the sequence of queues formed from A_n and B_n. If we now permit the sequences $\{A_n\}$ and $\{B_n\}$ to be constructed as increasingly accurate discrete approximations, then the previously discussed method can be used to measure the individual queues formed from A_n and B_n, and the limits estimated accordingly. This works theoretically, but unfortunately the method gets computationally overpowering rather quickly if convergence does not occur early. If that happens, we can simply use a small number of interarrival and service-time values as approximations to keep the computations manageable and then proceed as in Example 7.1.

We should also note that the concept of stochastic dominance can be used to bound and/or approximate. The key idea is that if the interarrival random variable of one queue (say T_1) is stochastically smaller than that of a second queue (T_2) and the service-time variable of the second (S_2) is stochastically

Table 7.1 Delay Probabilities for the First Three Customers

Delay	Customer 1	Probability 2	3
0	$\frac{13}{15} = .8667$	$\frac{187}{225} = .8311$	$\frac{2803}{3375} = .8305$
1	0	$\frac{8}{225} = .0356$	$\frac{104}{3375} = .0308$
2	$\frac{2}{15} = .1333$	$\frac{26}{225} = .1156$	$\frac{374}{3375} = .1108$
3	0	0	$\frac{32}{3375} = .0095$
4	0	$\frac{4}{225} = .0178$	$\frac{52}{3375} = .0154$
5	0	0	0
6	0	0	$\frac{8}{3375} = .0024$
Mean	$\frac{4}{15} = .2667$	$\frac{76}{225} = .3378$	$\frac{1204}{3375} = .3567$

smaller than that of the first (S_1), then it follows that the waiting times of the first queue will be stochastically larger than those of the second. It will then result that W and W_q are larger for queue 1 than for queue 2. Thus we have a potential tool for bounding complicated $G/G/1$ systems by finding a solvable combination of interarrival and service-time CDFs which obey the ordering.

It is quite valuable to apply the concept of stochastic ordering directly to the calculations of Example 7.1. The point is that each successive probability of a wait j, w_j, gets closer to its stationary limit, with the kth customer's waiting time stochastically larger than its predecessor. Thus we may take the calculations as far as seems necessary for convergence. In a good many cases, there is rapid convergence to a good lower bound. For our example, the first three customers (after the initial customer) turn out to have the following delay probabilities in Table 7.1.

Using (7.10), we can calculate the true $\{w_j\}$, and calculate W_q as $\Sigma j w_j$. This gives

$$w_0 = 0.8244, \quad w_1 = 0.0334, \quad w_2 = 0.1128, \quad w_3 = 0.0091,$$
$$w_4 = 0.0156, \quad w_5 = 0.0019, \quad w_6 = 0.0022,$$

and using 50 terms in $\Sigma j w_j$ yields a W_q of 0.37724. Notice that even with only three customers, the values in the table for customer 3 are not too far off.

7.1.2 Bounds for Multiple-Server General Queues

It should be pretty clear, from the limited number of results we have presented thus far for queues with more than one server, that bounds and inequalities could be particularly useful for analyzing the $G/G/c$. To get some bounds for the $G/G/c$, we recall some previous results in Chapters 2–5 on the relative merit of single-server queues as opposed to comparable multichannel queues. For example, we cite Problems 2.25 and 2.26. It is in this spirit that we proceed here.

To begin, assume that the $G/G/c$ queue we wish to bound has mean arrival rate λ and mean service rate μ at each channel. Let us denote the mean line wait in the original system by W_q. Now suppose that the discipline is altered by requiring that the customers be assigned in cyclic order to the c servers (that is, the first to server 1, the second to server 2, . . ., the $(c + 1)$st to server 1, . . .), with no jockeying allowed. Then each server faces a single-server $G/G/1$, in which the interarrival time is the c-fold convolution of the original interarrival distribution, with no change in the service-time process. The mean waiting time in queue in the single-server system (call it W_{q_1}) is an upper bound on the mean waiting time in the original multiserver system, since the condition which was added constitutes a restriction on the original system (see Brumelle, 1971). In fact, we can make the stronger observation that the waiting times are stochastically ordered.

Now in order to find a lower bound on the mean queue wait in the multichannel system, consider a second $G/G/1$ queue in which the interarrival distribution is identical to the original but where all c servers work on every customer simultaneously at the combined rate μ to satisfy a mean customer service requirement of $1/\mu$. We denote the average remaining work in this system as ω_2 and note that Brumelle has shown ω_2 to be less than or equal to the mean remaining work for the multichannel system (call it ω). To compare them, we use an argument like that presented for the virtual-wait derivation of Section 6.5.1. Virtual wait (in queue) and remaining work are the same for single-channel queues but different for multichannel systems, since there it is not necessary for all the work ahead of an arbitrary arrival to empty for the customer to get into service—only enough to free up one of the servers.

For multiple-server systems the mean unfinished work can be written as

$$\omega = \frac{L_q}{\mu} + r\,\mathrm{E}[\text{remaining service time per server}],$$

since $r = \lambda/\mu$ is the expected number of busy servers. Therefore,

$$\omega = \frac{\lambda W_q}{\mu} + \frac{\lambda}{\mu}\frac{\mathrm{E}[S^2]}{2/\mu} = \frac{\lambda W_q}{\mu} + \frac{\lambda}{\mu}\frac{\sigma_B^2 + 1/\mu^2}{2/\mu}.$$

Now, for the combined-service $G/G/1$, we first note that $p_0 = 1 - \lambda/c\mu$, since the total service rate is $c\mu$. The c remaining service times (one for each server), given that service is ongoing, are each now going to be

$$\frac{\sigma_B^2/c^2 + 1/c^2\mu^2}{2(1/c\mu)} = \frac{\sigma_B^2 + 1/\mu^2}{2c/\mu},$$

since all servers work simultaneously and we thus must scale down the original service time by a factor of c. Therefore we see that the remaining work for the $G/G/1$ is

$$\omega_2 = \frac{\lambda W_{q2}}{\mu} + c\frac{\sigma_B^2 + 1/\mu^2}{2c/\mu}\frac{\lambda}{cu} = \frac{\lambda W_{q2}}{\mu} + \lambda\frac{\sigma_B^2 + 1/\mu^2}{2c}.$$

When ω and ω_2 are compared, it turns out that using the Brumelle result $\omega \geq \omega_2$ yields

$$W_q \geq W_{q2} - \frac{\mu(c-1)(\sigma_B^2 + 1/\mu^2)}{2c}.$$

If now the single-server systems we have used to obtain W_{q1} and W_{q2} are in fact, solvable in themselves, then the bounds follow nicely as

$$W_{q2} - \frac{\mu(c-1)(\sigma_B^2 + 1/\mu^2)}{2c} \leq W_q \leq W_{q1}. \tag{7.11}$$

Otherwise, we would use the earlier $G/G/1$ bounds to bound W_{q1} and W_{q2} in turn.

For example, the above arguments may be translated into the following upper and lower bounds for the $G/G/c$ queue. For the upper bound, the mean interarrival time for the $G/G/1$ now becomes c/λ and arrival-time variance $c\sigma_A^2$. It then follows that the inequality of (7.6) must be altered to be

$$W_{q1} \leq \frac{\lambda(c\sigma_A^2 + \sigma_B^2)}{2c(1 - \lambda/c\mu)} = \frac{\lambda(c\sigma_A^2 + \sigma_B^2)}{2c(1 - \rho)} \qquad (\rho = \lambda/c\mu).$$

A tighter upper bound was proposed by Kingman and proven by De Smit (1973), where the variance of the service time can be divided by c, yielding

$$W_{q1} \leq \frac{\lambda(c\sigma_A^2 + \sigma_B^2/c)}{2c(1 - \rho)} \qquad (\rho = \lambda/c\mu).$$

To get the lower bound, we use the Marchal single-server inequality but with the mean service time $1/c\mu$ and service-time variance σ_B^2/c. We see that this $G/G/1$ bound becomes

$$W_{q2} \geq \frac{\lambda \sigma_B^2/c + \rho(\rho - 2)}{2\lambda(1 - \rho)} = \frac{\lambda \sigma_B^2 + c\rho(\rho - 2)}{2\lambda c(1 - \rho)} \qquad (\rho = \lambda/c\mu),$$

and thus from (7.11) the $G/G/c$ W_q is bounded, for $\rho = \lambda/c\mu$, by

$$\max\left(0, \frac{\lambda \sigma_B^2 + c\rho(\rho - 2)}{2\lambda c(1 - \rho)} - \frac{\mu(c - 1)(\sigma_B^2 + 1/\mu^2)}{2c}\right) \leq W_q \leq \frac{\lambda(c\sigma_A^2 + \sigma_B^2/c)}{2c(1 - \rho)}.$$

7.2 APPROXIMATIONS

In this section, we use a classification scheme based on (but modified from) Bhat et al. (1979). We categorize approximations into three types. The first makes use of *bounds and inequalities*. For example, one such scheme is a weighted average of the upper and lower bounds for a $G/G/1$ queue with the weighting factor depending on the traffic intensity. The second type of approximation deals with using a known queueing system to approximate one for which results are not readily obtainable. We refer to this type of approximation as a *system approximation*. An example is the use of $M/E_k/c$ or $M/H_k/c$ models to approximate an $M/G/c$. The third type of approximation involves approximating the queueing process itself by a process which is easier to deal with. Examples of such *process approximations* are replacement of the discrete queueing process by a continuous diffusion or fluid process, and using asymptotic or limiting results. The following subsections present these ideas in more detail.

7.2.1 Using Bounds and Inequalities to Approximate

As typical of what might be done, let us again look at the work of Marchal (1978). Based on the fact that the upper bound of (7.6) gets better as $\rho \to 1$, it might make sense to try to multiply the bound by a fractional function in ρ which itself approaches one with ρ. Marchal proposed the quotient

$$\frac{1 + \mu^2\sigma_B^2}{1/\rho^2 + \mu^2\sigma_B^2} = \frac{\rho^2 + \lambda^2\sigma_B^2}{1 + \lambda^2\sigma_B^2}$$

as such a factor to scale down the bound for use as an approximation. This was chosen to make the answer binding for the $M/G/1$ queue. The resulting approximation for W_q is therefore given by

$$\hat{W}_q = \frac{\lambda(\sigma_A^2 + \sigma_B^2)}{2(1 - \rho)} \frac{\rho^2 + \lambda^2 \sigma_B^2}{1 + \lambda^2 \sigma_B^2}.$$

To see that the answer is truly binding for $M/G/1$, we let $\sigma_A^2 = 1/\lambda^2$. Then we note that the approximation simplifies to

$$\frac{\lambda(1/\lambda^2 + \sigma_B^2)}{2(1 - \rho)} \frac{\rho^2 + \lambda^2 \sigma_B^2}{1 + \lambda^2 \sigma_B^2} = \frac{\rho^2 + \lambda^2 \sigma_B^2}{2\lambda(1 - \rho)},$$

which is precisely the PK formula. Marchal has shown that this formula also works well for $G/M/1$ queues. The performance of the approximation for arbitrary $G/G/1$'s deteriorates as the service times or interarrival times deviate further from exponentiality. Its ability to estimate, however, does go up with increasing values of the traffic intensity, in light of the asymptotic sharpness of the upper bound.

Marchal also suggests the possibility of using a different weighting factor, namely,

$$\frac{\rho^2 \sigma_A^2 + \sigma_B^2}{\sigma_A^2 + \sigma_B^2}.$$

Then a new approximation can be derived as

$$\hat{W}_q = \frac{\rho(\lambda^2 \sigma_A^2 + \mu^2 \sigma_B^2)}{2\mu(1 - \rho)}.$$

Besides being exact for $M/G/1$, it is also exact for $D/D/1$. The formula has a nice "product form" of three terms: (1) a traffic-intensity factor, (2) a variability factor, and (3) a time-scale factor, and thus can be written as

$$\hat{W}_q = \left(\frac{\rho}{1 - \rho}\right)\left(\frac{C_A^2 + C_B^2}{2}\right)\left(\frac{1}{\mu}\right),$$

where C represents the coefficient of variation (standard deviation/mean). Marchal (1978) has also done similarly for the multiserver bounds we presented earlier.

A simple approximation for any $G/G/c$ queue is the Allen–Cunneen (AC) approximation formula (see Allen, 1990). If, for an $M/M/c$, we denote the probability that all servers are busy by p_{cb}, the AC approximation is

$$\hat{W}_q = \frac{p_{cb}}{c(1 - \rho)}\left(\frac{C_A^2 + C_B^2}{2}\right)\left(\frac{1}{\mu}\right).$$

Note that when $c = 1$, we have $p_{cb} = \rho$, and the AC approximation reduces to the one given previously by Marchal. From Equation (2.30), we have that $p_{cb} = r^c p_0 / c! (1 - \rho)$, and the AC approximation can be written as

$$\hat{W}_q = \frac{r^c p_0}{c \cdot c! (1 - \rho)^2} \left(\frac{C_A^2 + C_B^2}{2} \right) \left(\frac{1}{\mu} \right).$$

7.2.2 System Approximations

As we noted earlier, examples of system approximations are the use of either an $M/E_k/c$ or an $M/H_k/c$ model to approximate an $M/G/c$. Recall our earlier comments that the Erlang family provides great modeling flexibility, particularly when it is generalized to the more global phase-type distributions first mentioned in Section 3.3.2. Recall also that both the usual Erlang and mixture of exponentials are easily expressed in phase form and that their use as service distributions leads to totally solvable systems.

Also in this spirit, we note some results of Section 6.2.1, where we showed that the $G/G/1$ problem can be greatly simplified if it may be assumed that both interarrival and service distributions can be expressed as convolutions of independent and not necessarily identical exponential random variables (usually called generalized Erlangs, GE). When the means of such exponentials are allowed to come in conjugate pairs (so that their Stieltjes transforms are inverse polynomials), Smith (1953) calls the family K_n (with n the degree of the defining polynomial). Other authors (for example, Cohen, 1982) define K_n as the class of distributions whose transforms are rational functions (clearly including the inverse polynomials); but we shall instead call these R_n (with n the degree of the denominator's polynomial) or Coxian, after the early work of Cox (1955a). The class K_n includes all regular Erlangs, but not mixed exponentials and mixed Erlangs, which are, however, members of R_n. When the denominator polynomial of a K_n transform has real and distinct roots, the associated distribution is called a GH, for generalized hyperexponential (e.g., see Botta and Harris, 1986). We also mention that the phase-type distributions (PH) have rational transforms as well, though not necessarily of the inverse polynomial form. Thus we may symbolically represent the relationship of those respective families as (see Harris, 1985)

$$GE \subset K_n \subset R_n, \quad GH \subset K_n \subset R_n, \quad \text{and} \quad PH \subset R_n.$$

Now, under a double K_n assumption (i.e., that the queue is $K_m/K_n/1$) it turns out (just as for $GE/GE/1$) that

$$W_q^*(s) = \frac{\prod_{i=1}^{n} (-z_i/\mu_i)(s + \mu_i)}{s \prod_{i=1}^{n} (s - z_i)},$$

where the $\{\mu_i\}$ are the individual parameters of the exponential decomposition of the service-time CDF and the $\{z_i\}$ are roots of the polynomial equation $A^*(-s)B^*(s) - 1 = 0$ with negative real parts. This result for the waiting-time transform is a nice simplification, since it now puts $W_q^*(s)$ into an invertible form. A partial-fraction expansion is then performed to give

$$W_q^*(s) = \frac{1}{s} + \sum_{i=1}^{n} \frac{k_i}{s - z_i} \qquad \text{(all } z_i \text{ assumed distinct)}.$$

Thus

$$W_q(t) = 1 + \sum_{i=1}^{n} k_i e^{z_i t},$$

where the $\{k_i\}$ would be determined in the usual way and are arbitrary in sign, with complex conjugates of the $\{k_i\}$ paired with the complex conjugates of the $\{z_i\}$. A completely parallel result exists for distributions in R_n.

Now, most importantly, Smith (1953) has also shown that a comparable result follows even for arbitrary interarrival times. That is, the $G/K_n/1$ and $G/R_n/1$ queues have mixed exponential waits independent of the form of G, where now the $\{z_i\}$ are the roots with negative real parts to the possibly transcendental equation $A^*(-s)B^*(s) - 1 = 0$. Thus we are now in an excellent position to well approximate the most awkward $G/G/1$ queues.

In closing here, we should point out that the matching of any approximant to an original queue structure is likely to involve the use of procedures for parameter estimation. For this, the interested reader is referred to any basic statistical text and our later discussion in Section 7.4.2.

7.2.3 Process Approximations

As noted earlier, we define a process approximation as one where the actual problem is replaced by a nonqueueing one which is simpler to work with. The primary examples of interest are the use of creative probability arguments on random walks, stochastic convergence, and the like to solve heavy-traffic and nonstationary queueing problems, and the use of continuous-time diffusion models to solve queueing problems in heavy traffic.

Heavy-Traffic and Nonstationary Queues

It is our intent in this subsection to indicate some interesting limit results and approximations for $G/G/1$ models in which the traffic intensity is just barely less than one $(1 - \epsilon < \rho < 1)$ or is equal to or greater than one $(\rho \geq 1)$. The former will be said to be saturated or in heavy traffic, while the latter will be described as nonstationary, divergent, or unstable. Of course, any transient results derived earlier in the text are valid for heavy traffic and divergent queues, since no assumptions are made on the size of ρ.

For limit results, the virtual wait $V(t)$ (the wait a customer would undergo if that customer arrives at time t) and the actual wait $W^{(n)}$ (the actual waiting time of the nth customer) will be used to construct random sequences, which are then shown to stochastically converge. These convergence theorems will then be applied to obtain approximate distributions of the actual and virtual waits in queue of a customer and their averages. In addition, bounds on the average waiting times will be obtained, some of these directly from the limit results. While we shall present some nice theoretical results, the practicality of these is subject to question, since the results depend on allowing time to go to ∞. This is unrealistic, as it is unlikely that systems with congestion would be permitted to run for very long before remedial action was taken.

If we were to go back to Lindley's approach to waiting times for the $G/G/1$, or even to the $M/M/1$ case, and try to get results for $\rho \geq 1$, we would fail. The convergence theory which exists for stationary queues cannot be directly extended in either case, generally because the relevant quantities get excessively large as time increases, so that a completely new approach is required. This new approach is to appropriately scale and shift the random variable to permit some form of convergence. For example, when we study $W^{(n)}$, the appropriate quantity to consider in the event that $\rho \geq 1$ is $(W^{(n)} - an)/(b\sqrt{n})$, where a and b will be suitably chosen constants. It will be shown that the distribution of this new random variable will converge to a normal or a truncated normal distribution for $\rho > 1$ and $\rho = 1$, respectively. Two references for the following discussion are contained in Smith and Wilkinson (1965)—specifically, the paper on heavy traffic queues (the $\rho < 1$ case) by Kingman, and the paper on divergent queues ($\rho \geq 1$) by Heathcote.

We shall begin by first discussing in some detail the heavy-traffic problem in the context of $G/G/1$. Here the system is completely specified by the sequences of interarrival and service times. If each of these sequences contains IID random variables which are independent of each other, then the main result that will be shown is that the line waiting times in heavy traffic are approximately exponentially distributed with mean

$$W_q^{(H)} = \frac{1}{2} \frac{\text{Var}[T] + \text{Var}[S]}{1/\lambda - 1/\mu} = \frac{\lambda(\sigma_A^2 + \sigma_B^2)}{2(1 - \rho)}, \tag{7.12}$$

which is indeed the upper bound for W_q presented in Section 7.1.1, Equation (7.6).

Recall from Section 6.2 that we defined the difference between the nth service and interval times to be $U^{(n)} \equiv S^{(n)} - T^{(n)}$. Let us also now define the nth partial sum of the $\{U^{(n)}\}$ to be $P^{(n)} = \sum_{i=0}^{n-1} U^{(i)}$, and let the mean and variance of the IID $\{U^{(n)}\}$ be denoted by $-\alpha$ and β^2, respectively. Then the expectation of $P^{(n)}$ is $-n\alpha$, while its variance is $n\beta^2$. Note that since we are in heavy traffic, α should be small. Now when the central limit theorem for IID random variables is applied to $\{P^{(n)}\}$, it is found that

$$Y_n \equiv \frac{P^{(n)} + n\alpha}{\sqrt{n}\beta} \xrightarrow{df} N(0, 1),$$

where *df* means "in distribution function." Without going through the details of the proof, suffice it to say that by using the asymptotic normality of Y_n, it can be shown that the LST of $W_q^{(n)}$ for large n is given approximately by

$$W_q^*(s) \doteq \left(1 + \frac{\beta^2 s}{2\alpha}\right)^{-1}$$

when α/β is small. From this we therefore deduce that the line delay is negative exponential with mean

$$\frac{\beta^2}{2\alpha} = -\frac{1}{2}\frac{\mathrm{Var}[U]}{\mathrm{E}[U]} = \frac{1}{2}\frac{\mathrm{Var}[S] + \mathrm{Var}[T]}{1/\lambda - 1/\mu}, \tag{7.13}$$

which is equivalent to both Equation (7.12) and the bound of (7.6).

The foregoing result is rather loose, so it becomes important to see how fast this limit may be approached. We have exact answers for the $M/G/1$ and can thus, for example, calculate the rate at which its W_q goes to the heavy-traffic result. For $M/G/1$, the PK formula gives the exact result as

$$W_q = \frac{\rho^2 + \lambda^2\sigma^2}{2\lambda(1 - \rho)} = \frac{\lambda(1/\mu^2 + \sigma^2)}{2(1 - \lambda/\mu)} = \frac{\mathrm{E}[S^2]}{2(1/\lambda - 1/\mu)} = \frac{\mathrm{E}[S^2]}{-2\mathrm{E}[U]}.$$

Comparing the above with Equation (7.12), we see that the rate of conver-gence of W_q to $W_q^{(H)}$ for the $M/G/1$ depends on the rate that $\mathrm{E}[S^2]$ goes to $\mathrm{Var}[U]$. Since for $M/G/1$, we have $\mathrm{Var}[U] = 1/\lambda^2 + \sigma_B^2$ and $\mathrm{E}[S^2] = \sigma_B^2 + 1/\mu^2$, it follows that W_q goes to $W_q^{(H)}$ as fast as ρ^2 goes to 1. We can also extend the $G/G/1$ bound given by (7.12) to $G/G/c$ by noting that the system is essentially always operating as a single-server queue with service rate $c\mu$. Hence it would be fair to expect exponential delays in heavy traffic with

$$W_q^{(H)} \doteq \frac{1}{2}\frac{\mathrm{Var}[T] + (1/c^2)\,\mathrm{Var}[S]}{1/\lambda - 1/c\mu}.$$

We now move to $\rho \geq 1$ and a brief study of divergent queues. Once again we make use of

$$W_q^{(n+1)} = \max(0, W_q^{(n)} + U^{(n)}) \tag{7.14}$$

where $U^{(n)} = S^{(n)} - T^{(n)}$, but we shall now operate under the assumption that $E[U] \geq 0$.

First, let $1/\mu - 1/\lambda = E[U] > 0$, that is, $\rho > 1$. We would then expect the difference $\Delta W_q^{(n)} \equiv W_q^{(n+1)} - W_q^{(n)}$ to behave like $U^{(n)}$ when n gets large, since by observation of (7.14) we see that it is unlikely that $W_q^{(n+1)}$ will ever be zero again when $\rho > 1$. But we have already stated earlier in this section that $\Sigma U^{(i)}$ properly normalized as

$$Y_n = \frac{\Sigma U^{(i)} + n\alpha}{\sqrt{n}\beta} \qquad (\alpha < 0)$$

converges in distribution function to the unit normal. Since $W_q^{(0)} = 0$,

$$\sum_{i=0}^{n-1} \Delta W_q^{(i)} = W_q^{(n)} = \sum_{i=0}^{n-1} U^{(i)} = P^{(n)},$$

and hence

$$\Pr\left\{\frac{W_q^{(n)} + n\alpha}{\sqrt{n}\beta} \leq x\right\} \rightarrow \int_{-\infty}^{x} \frac{e^{-t^2/2}}{\sqrt{2\pi}} \, dt. \tag{7.15}$$

One immediate result of (7.15) is the determination of an estimate for the probability of no wait for n large. This is clearly given by

$$\Pr\{W_q^{(n)} = 0\} = \Pr\left\{\frac{W_q^{(n)} + n\alpha}{\sqrt{n}\beta} = \frac{n\alpha}{\sqrt{n}\beta}\right\} = \int_{-\infty}^{n\alpha/(\sqrt{n}\beta)} \frac{e^{-t^2/2}}{\sqrt{2\pi}} \, dt = \Phi(n\alpha/\sqrt{n}\beta),$$

where $\Phi(\cdot)$ is the CDF of the standard normal, $N(0, 1)$.

We can get an interesting alternative approximation to this probability which does not use the normal distribution by using Chebyshev's inequality, which says that

$$\Pr\{|X - \mu| \geq k\sigma\} \leq \frac{1}{k^2}$$

for any variable X possessing two moments. Applying this result here, it is found that

$$\Pr\{W_q^{(n)} = 0\} = \Pr\{X + n\alpha \leq n\alpha\} = \frac{\Pr\{|X + n\alpha| \geq n\alpha\}}{2} \leq \frac{\beta^2}{2n\alpha^2},$$

which provides a fairly reasonable upper bound for the probability that an arbitrary arrival encounters an idle system. Notice that this number does go to zero as n approaches ∞, as required.

There are some additional limit theorems for the special $G/G/1$ queue with $G = M$ and $\rho = 1$. These results will be stated without proof, and the interested reader is referred to Cohen (1982).

For $V^m(t)$ defined as the maximum of the virtual wait on the interval $(0, t)$, $W_q^{(k)m}$ as the largest value of $W_q^{(n)}$ between $n = 1$ and k, and $X^m(t)$ as the maximum system size on $(0, t)$, we have that $\mu V^m(t)/t$, $\mu W_q^{(k)m}/k$, and $X^m(t)/t$ all converge in distribution, as t (or k) goes to ∞, to a random variable with CDF

$$G(x) = \begin{cases} e^{-1/x} & (x > 0), \\ 0 & (x \le 0). \end{cases} \tag{7.16}$$

Similar results were obtained by Iglehart (1972) for arbitrary $G/G/1$ queues with both $\rho = 1$ and $\rho > 1$.

Comparable results are available for $G/M/1$ queues which say that for $\rho = 1$ and T_2 the second moment of the interarrival time, the sequences $2V^m(t)/(\lambda T_2)$, $2W_q^{(k)m}/(\lambda n T_2)$, and $2X^m(t)/(\lambda^2 t T_2)$ all converge in distribution function to the $G(x)$ of (7.16).

The final result we wish to state is that the limit of the time-average virtual wait at t almost surely (with probability one, often abbreviated by a.s.) goes to $\lambda - 1$ as t goes to ∞, that is,

$$\lim_{t \to \infty} \frac{V(t)}{t} \overset{\text{a.s.}}{=} \lambda - 1.$$

Diffusion Approximations

Since we have already seen that heavy-traffic $G/G/1$ queues have exponential waiting times, it might therefore seem quite logical that these queues should be birth–death and of the $M/M/1$ type when $\rho = 1 - \epsilon$, independent of the form of the arrival and service distributions (remember that $M/G/1$ queues have exponential waits if $G = M$, and that $G/M/1$ queues also have exponential waits if $G = M$—see Problem 5.35). However, this is not correct, since we know from Equation (7.13) that the interarrival- and service-time variances must be involved. But we can obtain a fairly simple approximate result for the transient system-size distribution of the $M/M/1$ in heavy traffic by using a continuous approximation of the diffusion type. After that we shall use another, but similar, approach to get a diffusion approximation for the transient distribution of the line delay for the $M/G/1$ in heavy traffic.

The $M/M/1$ diffusion result for the system-size distribution is found by first considering the elementary random-walk approximation of a birth–death process which changes every Δt units and has transition probabilities given by

Pr{state goes up by one over unit interval} $= \lambda \, \Delta t,$

Pr{state goes down by one over unit interval} $= \mu \, \Delta t,$

Pr{no change} $= 1 - (\lambda + \mu) \, \Delta t,$

assuming for the moment that since the queue is in heavy traffic, the walk is unrestricted with no impenetrable barrier at 0.

Now the elementary random walk is clearly a Markov chain, since its future behavior is only a probabilistic function of its current position and is independent of past history. More specifically, we may write that the probability $p_k(n)$ of finding the walk in position k after n steps is given by the CK equation

$$p_k(n + 1) = p_k(n)[1 - (\lambda + \mu) \, \Delta t] + p_{k+1}(n)\mu \, \Delta t + p_{k-1}(n)\lambda \, \Delta t,$$

which may be rewritten after a bit of algebra as

$$\frac{p_k(n + 1) - p_k(n)}{\Delta t} = \frac{\mu + \lambda}{2} [p_{k+1}(n) - 2p_k(n) + p_{k-1}(n)]$$

$$+ \frac{\mu - \lambda}{2} [p_{k+1}(n) - p_k(n)]$$

$$+ \frac{\mu - \lambda}{2} [p_k(n) - p_{k-1}(n)]. \tag{7.17}$$

We then observe that Equation (7.17) has been written in terms of the discrete version of derivatives: the left-hand side with respect to step (time) and the right-hand side with respect to the state variable. If we then appropriately take limits of (7.17) so that the time between transitions shrinks to zero, while simultaneously the size of the state steps goes to zero, we find ourselves ending up with a partial differential equation of the diffusion type.

To be more exact, let the length of a unit state change be denoted by θ and the step time by Δt; then (7.17) leads to

$$\frac{p_{k\theta}(t + \Delta t) - p_{k\theta}(t)}{\Delta t} = \frac{\mu + \lambda}{2} [p_{(k+1)\theta}(t) - 2p_{k\theta}(t) + p_{(k-1)\theta}(t)]$$

$$+ \frac{\mu - \lambda}{2} [p_{(k+1)\theta}(t) - p_{k\theta}(t)]$$

$$+ \frac{\mu - \lambda}{2} [p_{k\theta}(t) - p_{(k-1)\theta}(t)]. \tag{7.18}$$

Now let both Δt and θ go to 0, preserving the relationship that $\Delta t = \theta^2$ (which guarantees that the state variance is meaningful) and at the same time letting k increase to ∞ in such a way that $k\theta \rightarrow x$. Then $p_{k\theta} \rightarrow p(x, t \mid X_0 = x_0)$, which is now the probability density for the system state X, given that the queueing system began operation at a size of x_0. Utilizing the definitions of first and second derivatives, Equation (7.18) becomes

$$\frac{\partial p(x, t \mid x_0)}{\partial t} = \left(\frac{\mu + \lambda}{2}\right) \frac{\partial^2 p(x, t \mid x_0)}{\partial x^2} + (\mu - \lambda) \frac{\partial p(x, t \mid x_0)}{\partial x}, \qquad (7.19)$$

one form of the well-known diffusion equation, which among other things describes the movement of a particle under Brownian motion (e.g., see Prabhu, 1965a, or Heyman and Sobel, 1982). This particular form of the diffusion equation often goes under the name of Fokker–Planck and, in addition, turns out to be the version of the forward Kolmogorov equation that is found for the continuous-state Markov process.

So now we would like to solve (7.19) under the boundary conditions that

$$p(x, t \mid x_0) \geq 0,$$

$$\int_{-\infty}^{\infty} p(x, t \mid x_0) \, dx = 1,$$

$$\lim_{t \rightarrow 0} p(x, t \mid x_0) = 0 \qquad (x \neq x_0),$$

where the first two are usual properties of densities, while the third is essentially a continuity requirement at time 0. It can then be shown (see Prabhu, 1965a) that the solution to (7.19) is given by

$$p(x, t \mid x_0) = \frac{e^{-[x - x_0 + (\mu - \lambda)t]^2 / 2(\mu + \lambda)t}}{\sqrt{2\pi t(\mu + \lambda)}},$$

which is a Wiener process starting from x_0 with drift $-(\mu - \lambda)t$ and variance $(\mu + \lambda)t$. That is to say, the random process $X(t)$ is normal with mean $x_0 - (\mu - \lambda)t$ and variance $(\mu + \lambda)t$ so that

$$\Pr\{n - \tfrac{1}{2} < X(t) < n + \tfrac{1}{2} \mid X_0 = x_0\} \doteq \int_{n-1/2}^{n+1/2} N(x_0 - [\mu - \lambda]t, [\mu + \lambda]t) \, dt.$$

But we observe that this result is not very meaningful, since $\lambda < \mu$ and the drift is therefore negative, thus bringing the process eventually to one with a negative mean (this is due to neglecting the impenetrable barrier at 0; however, the approximation would be valid if x_0 were large, since it would then be unlikely for the queue to empty). In order to counter this possibility we must impose an impenetrable barrier upon the walk at $x = 0$. This is added to the problem in the form of the additional boundary condition that

$$\lim_{x \to 0} \frac{\partial p(x, t \mid x_0)}{\partial t} = 0 \qquad \text{(for all } t\text{).}$$

This is used because the process cannot move beyond zero to the negatives and therefore $p(x, t \mid x_0) = 0$. Thus $\Delta p(x, t \mid x_0) = 0$ for $x < 0$ and $\lim_{x \to 0} \Delta p(x, t \mid x_0) = 0$. The new solution is then given by

$$p(x, t \mid x_0) = \frac{1}{\sqrt{2\pi(\lambda + \mu)t}} \left[e^{-[x - x_0 - (\lambda - \mu)t]^2 / 2(\mu + \lambda)t} \right.$$

$$+ e^{-2x(\mu - \lambda)/(\mu + \lambda)} \left(e^{-[x + x_0 - (\lambda - \mu)t]^2 / 2(\mu + \lambda)t} \right.$$

$$\left. \left. + \frac{2(\mu - \lambda)}{\mu + \lambda} \int_x^\infty e^{-[y + x_0 - (\lambda - \mu)t]^2 / 2(\mu + \lambda)t} \, dy \right) \right]. \qquad (7.20)$$

This solution is then valid as an approximation for any $M/M/1$ provided $\rho = 1 - \epsilon$.

It is particularly interesting to make two additional computations. The first is to allow $\lambda = \mu$ throughout the derivation, and the second is to look at the limiting behavior of $p(x, t \mid x_0)$ as t goes to ∞. When $\lambda = \mu$ Equation (7.19) is quite simplified and becomes

$$\frac{\partial p(x, t \mid x_0)}{\partial t} = \lambda \frac{\partial^2 p(x, t \mid x_0)}{\partial x^2}.$$

Under the same augmented boundary conditions as lead to (7.20), this differential equation has the solution

$$p(x, t \mid x_0) = \frac{1}{\sqrt{4\pi\lambda t}} (e^{-(x - x_0)^2 / 4\lambda t} + e^{-(x + x_0)^2 / 4\lambda t}),$$

again a Wiener process, but one with no drift. This one may be used to approximate the transient solution for any $M/M/1$ with $\rho = 1$.

On the other hand, when we let $t \to \infty$ in (7.19), it is found that

$$0 = \left(\frac{\mu + \lambda}{2} \right) \frac{d^2 p(x)}{dx^2} + (\mu - \lambda) \frac{dp(x)}{dx},$$

which is a homogeneous, second-order linear differential equation with solution

$$p(x) = C_1 + C_2 e^{-2[(\mu - \lambda)/(\mu + \lambda)]x}.$$

Since $p(x)$ must integrate to one, we see that $C_1 = 0$ and $C_2 = (\mu + \lambda)/[2(\mu - \lambda)]$. Thus $p(x)$ is an exponential density, as might have been expected, since $M/M/1$ lengths are geometric in distribution, which is just the discrete analog of the exponential. That this mean of $(\mu + \lambda)/[2(\mu - \lambda)] = (1 + \rho)/[2(1 - \rho)]$ makes sense can be seen by noting that for ρ nearly 1, $(1 + \rho)/2 \doteq \rho$; hence $(1 + \rho)/[2(1 - \rho)] \doteq \rho/(1 - \rho)$, the usual $M/M/1$ result.

Another approach to this approximation which leads to the same result is the use of the central limit theorem on the IID random variables making up the random walk. One then uses the same kind of limiting argument to get the results in terms of the same variables as before.

The $M/G/1$ heavy-traffic diffusion approximation for line delay is due to Gaver (1968). In order to approximate the conditional "density function," say $w_q(x, t \mid x_0)$ (recall from Chapter 2 that this is not a true density, since there is a nonzero probability of a zero wait), of the virtual line delay $V(t)$ (which is easier to use than the actual for the derivation here—keep in mind that both are the same for $M/G/1$), its mean μ and variance σ^2 are approximated by

$$\Delta t = E[V(t + \Delta t) - V(t) \mid V(t)] = (\lambda E[S] - 1) \Delta t + o(\Delta t),$$

$$\sigma^2 \Delta t = \text{Var}[V(t + \Delta t) - V(t) \mid V(t)] = \lambda E[S^2] \Delta t + o(\Delta t),$$

since the change in the virtual wait over the time increment Δt assuming a loaded system is the total service time needed to serve all arrivals over Δt, minus Δt. But it is known that $V(t)$ is a continuous-parameter continuous-state Markov process and hence will satisfy the Fokker–Planck equation for its conditional "density" $w_q(x, t \mid x_0)$ given by (see Newell, 1972)

$$\frac{\partial w_q(x, t \mid x_0)}{\partial t} = -\mu \frac{\partial w_q(x, t \mid x_0)}{\partial x} + \frac{\sigma^2}{2} \frac{\partial^2 w_q(x, t \mid x_0)}{\partial x^2},$$

subject to the boundary conditions

$$w_q(x, t \mid x_0) \geq 0,$$

$$\int_0^\infty w_q(x, t \mid x_0) \, dx = \frac{\lambda}{\mu},$$

$$\lim_{t \to 0} w_q(x, t \mid x_0) = 0 \qquad (x \neq 0).$$

From the earlier discussion of heavy traffic, it is to be expected that if

$$\frac{-\mu}{\sigma^2} = \frac{1 - \lambda E[S]}{\lambda E[S^2]}$$

is positive and small, then the diffusion solution should provide a good approximation. The final expression for $w_q(x, t \mid x_0)$ is found to be

$$w_q(x, t \mid x_0) = \frac{1}{\sqrt{2\pi t \sigma^2}} \left[e^{-(x-x_0-\mu t)^2/\sigma^2 t} \right.$$
$$\left. + e^{2x\mu/\sigma^2} \left(e^{-(x-x_0-\mu t)^2/\sigma^2 t} + \frac{2\mu}{\sigma^2} \int_x^\infty e^{-(y-x_0-\mu t)^2/\sigma^2 t} \, dy \right) \right].$$

This subsection was by no means meant to exhaust the subject of diffusion approximations in queueing, but rather to give the reader a brief introduction to the area. The interested reader will find a limited amount of literature on the subject, including in addition to Heyman and Sobel (1982), Prabhu (1965a), and Newell (1972) such references as Feller (1971) and Karlin and Taylor (1975).

7.3 NUMERICAL TECHNIQUES

In contrast to the approximation techniques presented in the previous section, we now return our concentration to exact analysis of the situation of interest. In many cases where we are unable to find neat, closed-solution analytical formulas, we can still analyze the system and obtain numerical answers, often to a desired, prespecified error tolerance.

For many steady-state situations, the analysis involves solving simultaneous linear-algebraic equations, perhaps large numbers of them; of interest are efficient procedures for solving such systems. In the case of transient solutions, we are often faced with solving sets of linear first-order differential equations, and a variety of numerical techniques for doing so exist.

The disadvantage of numerical solution procedures is that all parameters must be specified numerically before answers can be obtained, so that when parameters are changed, the calculations must be redone. However, where no other methods can suffice, this is a small price to pay for obtaining actual answers.

7.3.1   Steady-State Solutions

For queues in which a Markov analysis is possible, the steady-state solution is found by solving the stationary equations for a discrete-parameter Markov chain (DPMC)

$$\pi = \pi P,$$

$$\pi e = 1,$$

or for a continuous-parameter Markov chain (CPMC)

$$0 = pQ,$$

$$1 = pe,$$

where π (or p) is the steady-state probability vector, P the discrete-parameter Markov-chain transition probability matrix, Q the infinitesimal generator of the continuous-time Markov chain, and e a vector of ones (see Section 1.9.5). For situations such as finite-source queues, queues with limited waiting capacity, and so on, P and Q are finite-dimensional matrices, and numerically, the problem reduces to solving a system of simultaneous linear equations. The equation $\pi = \pi P$ can always be written as $0 = \pi Q$, where $Q = P - I$ for the DPMC case.

As an example, consider an $M/G/1/3$ queue. We desire to find the departure-point steady-state system-size probabilities. For this case, the are three possible states of the system at a departure point, namely, a departing customer can see an empty system, a system with one remaining customer, or a system with two remaining customers. Thus, from Section 5.1.7, the P matrix for the imbedded Markov chain is

$$P = \begin{pmatrix} k_0 & k_1 & 1 - k_0 - k_1 \\ k_0 & k_1 & 1 - k_0 - k_1 \\ 0 & k_0 & 1 - k_0 \end{pmatrix},$$

where, as usual,

$$k_n = \Pr\{n \text{ arrivals during a service period}\} = \frac{1}{n!} \int_0^\infty (\lambda t)^n e^{-\lambda t} \, dB(t).$$

Thus if $B(t)$ and λ are specified, the values k_0 and k_1 can be calculated and the problem reduces to solving a 3×3 set of linear equations (one equation of $\pi = \pi P$ is always redundant):

$$(\pi_0, \pi_1, \pi_2) = (\pi_0, \pi_1, \pi_2) \begin{pmatrix} k_0 & k_1 & 1 - k_0 - k_1 \\ k_0 & k_1 & 1 - k_0 - k_1 \\ 0 & k_0 & 1 - k_0 \end{pmatrix},$$

$$\pi_0 + \pi_1 + \pi_2 = 1.$$

Let us now reconsider Example 5.2. Suppose that whenever three or more machines are down (a situation that is intolerable for production purposes), an outside repairman is called in, so that as far as the internal repair process is concerned, we have an $M/G/1/3$ system. For the two-point service distribution,

$$k_n = \frac{2}{3n!} e^{-3/4} (\tfrac{3}{4})^n + \frac{1}{3n!} e^{-1},$$

and thus

$$k_0 = \tfrac{2}{3} e^{-3/4} + \tfrac{1}{3} e^{-1} = .43 \qquad k_1 = \tfrac{1}{2} e^{-3/4} + \tfrac{1}{3} e^{-1} = .36.$$

To find π we must thus solve

$$(\pi_0, \pi_1, \pi_2) = (\pi_0, \pi_1, \pi_2) \begin{pmatrix} .43 & .36 & .21 \\ .43 & .36 & .21 \\ 0 & .43 & .57 \end{pmatrix},$$

$$\pi_0 + \pi_1 + \pi_2 = 1.$$

Rewriting in the $0 = \pi Q$ form, we have

$$(0, 0, 0) = (\pi_0, \pi_1, \pi_2) \begin{pmatrix} -.57 & .36 & .21 \\ .43 & -.64 & .21 \\ 0 & .43 & -.43 \end{pmatrix},$$

$$1 = \pi_0 + \pi_1 + \pi_2.$$

Since one of the equations in $0 = \pi Q$ is always redundant, we can replace the last column of the Q matrix by a column of ones and the last 0 element of the left-hand-side 0 vector by a one, thereby incorporating the summability to one condition. Thus we must solve

$$(0, 0, 1) = (\pi_0, \pi_1, \pi_2) \begin{pmatrix} -.57 & .36 & 1 \\ .43 & -.64 & 1 \\ 0 & .43 & 1 \end{pmatrix}, \qquad (7.21)$$

a system of the form

$$b = \pi A,$$

where b is the "modified" zero vector and A is the "modified" Q matrix. It suffices then to find A^{-1}, since the solution is bA^{-1}. In fact, since b is a vector of all zeros except the last element, we need only find the last row A^{-1}. This last row then contains the $\{\pi_i\}$. It is easy to find the last row of A^{-1} to be $(.29, .38, .33)$, so that $\pi_0 = .29$, $\pi_1 = .38$, and $\pi_2 = .33$. Hence, solving these types of problems for large state spaces often boils down to obtaining or generating an efficient method for matrix inversion.

There is an alternative way of solving these problems for handling redundancy in $0 = \pi Q$. Instead of replacing one of the equations with the sum of the probabilities equal to 1, we can arbitrarily set one of the π_i equal to one, solve an $(n-1) \times (n-1)$ nonsingular system (i.e., solve for $n-1$ variables in terms of the remaining one), and then renormalize the resulting π_i. In the above example, suppose we set π_2 equal to one. Then the resulting $(n-1) \times (n-1)$ system of equations becomes

$$(0, -.43) = (\pi_0, \pi_1) \begin{pmatrix} -.57 & .36 \\ .43 & -.64 \end{pmatrix},$$

which still has the form $b = \pi A$, but now b and π are two-component vectors and A is a 2×2 matrix. The solution is then $\pi = bA^{-1}$, which turns out to be $(0.88, 1.17)$. Now we must include $\pi_2 = 1$, so that renormalizing on the sum $0.88 + 1.17 + 1 = 3.05$ yields $\pi_0 = 0.88/3.05 = 0.29$, $\pi_1 = 1.17/3.05 = 0.38$, and $\pi_2 = 1/3.05 = 0.33$, the same answers as before.

In real systems, we may well end up with matrices with thousands or tens of thousands of rows and columns; one can easily conceive of queueing network problems with a million states. Thus efficient procedures for solving *large* systems of equations are crucial.

In many queueing applications, the matrix to be inverted is sparse; that is, most of the elements are zero. For example, in birth–death processes, only the elements on the main, super-, and subdiagonals can be nonzero; the other elements are zero [see Section 1.9.3, just prior to Equation (1.30)]. Good sparse-matrix computer packages are available that can handle fairly large systems. Also, one can utilize a large-scale linear programming package, since the Markov queueing equations are linear. By formulating a linear program with equality constraints and a suitable objective function, such a package will yield solutions to queueing types of equations.

Iterative solution techniques have also been shown to be efficient for many large-scale Markovian queueing systems. For example, we could use the basic Markov-chain recursive equation (1.22), $\pi^{(m)} = \pi^{(m-1)}P$, stopping at a suitably large m. Using this procedure for the above example with

$$\boldsymbol{\pi}^{(0)} = (1, 0, 0), \quad \boldsymbol{P} = \begin{pmatrix} .43 & .36 & .21 \\ .43 & .36 & .21 \\ 0 & .43 & .57 \end{pmatrix}$$

yields

$$\boldsymbol{\pi}^{(1)} = \boldsymbol{\pi}^{(0)}\boldsymbol{P} = (.43, .36, .21), \qquad \boldsymbol{\pi}^{(2)} = \boldsymbol{\pi}^{(1)}\boldsymbol{P} = (.34, .37, .29),$$

$$\boldsymbol{\pi}^{(3)} = \boldsymbol{\pi}^{(2)}\boldsymbol{P} = (.31, .38, .31). \qquad \boldsymbol{\pi}^{(4)} = \boldsymbol{\pi}^{(3)}\boldsymbol{P} = (.30, .38, .33),$$

$$\boldsymbol{\pi}^{(5)} = \boldsymbol{\pi}^{(4)}\boldsymbol{P} = (.29, .38, .33), \qquad \boldsymbol{\pi}^{(6)} = \boldsymbol{\pi}^{(5)}\boldsymbol{P} = (.29, .38, .33),$$

and we see that after only five iterations, the vector $\boldsymbol{\pi}$ is, to two decimal places, equal to the steady-state solution previously obtained. This type of procedure, as we show a little later, can be used for general systems of linear equations [e.g., those given by (7.1)], and as applied above is called Jacobi stepping.

A variation of this procedure, which is called Gauss–Seidel stepping, uses each new $\pi_j^{(m)}$ as it is calculated for calculating $\pi_{j+1}^{(m)}$, rather than using only the $\boldsymbol{\pi}^{(m-1)}$ elements. For example, designating $\boldsymbol{P}_i$ as the ith column of the $\boldsymbol{P}$ matrix,

$$\pi_0^{(1)} = \boldsymbol{\pi}^{(0)}\boldsymbol{P}_1 = (1, 0, 0) \begin{pmatrix} .43 \\ .43 \\ 0 \end{pmatrix} = .43.$$

Now adjust $\boldsymbol{\pi}^{(0)}$ from $(1, 0, 0)$ to $(.43, 0, 0)$, so that we now calculate

$$\pi_1^{(1)} = (.43, 0, 0)\boldsymbol{P}_2 = (.43, 0, 0) \begin{pmatrix} .36 \\ .36 \\ .43 \end{pmatrix} = .15.$$

Continuing,

$$\pi_2^{(1)} = (.43, .15, 0)\boldsymbol{P}_3 = (.43, .15, 0) \begin{pmatrix} .21 \\ .21 \\ .57 \end{pmatrix} = .12.$$

Now

$$\pi_0^{(2)} = (.43, .15, .12)\boldsymbol{P}_1 = .25, \qquad \pi_1^{(2)} = (.25, .15, .12)\boldsymbol{P}_2 = .20,$$

$$\pi_2^{(2)} = (.25, .20, .12)\boldsymbol{P}_3 = .16.$$

Iterating again yields

$$\pi_0^{(3)} = (.25, .20, .16)P_1 = .19, \qquad \pi_1^{(3)} = (.19, .20, .16)P_2 = .21,$$

$$\pi_2^{(3)} = (.19, .21, .16)P_3 = .18,$$

and twice more gives

$$\pi_0^{(4)} = .17, \qquad \pi_1^{(4)} = .21, \qquad \pi_2^{(4)} = .18,$$

$$\pi_0^{(5)} = .16, \qquad \pi_1^{(5)} = .21, \qquad \pi_2^{(5)} = .18.$$

Normalizing after five iterations, we get an estimate of the steady-state π of

$$(.16/.56, .21/.56, .18/.56) = (.29, .38, .32),$$

which agrees, quite well, to two decimal places with what we obtained earlier by matrix inversion and Jacobi stepping.

These same types of techniques (Jacobi and Gauss–Seidel) can be used to solve any finite system of linear equations, such as the stationary equations $0 = pQ$ for continuous-parameter Markov chains (see, e.g., Maron, 1982, or Cooper, 1981). In fact, the Jacobi and Gauss–Seidel procedures really refer to solving sets of equations and not to stepping through finite discrete-parameter Markov chains (we took the liberty of using these names to describe the aforementioned Markov-chain stepping procedure for convenience).

Consider, for example, the finite system of equations for a continuous-parameter Markov chain,

$$0 = pQ,$$
$$1 = pe.$$

Since, as mentioned earlier, one equation of the set $0 = pQ$ is redundant, we can replace the last column of the Q matrix by ones and the last element of the 0 vector by a one. This incorporates the equation $1 = pe$ so that we have the linear system of equations

$$b = pA,$$

where $b = (0, 1)$ and

$$A = \begin{pmatrix} -q_0 & q_{0,1} & \cdots & q_{0,N-2} & 1 \\ q_{10} & -q_1 & \cdots & q_{1,N-2} & 1 \\ \vdots & \vdots & & \vdots & \vdots \\ q_{N-1,0} & q_{N-1,1} & \cdots & q_{N-1,N-2} & 1 \end{pmatrix}. \qquad (7.22)$$

Using straightforward solution techniques, as illustrated in the beginning of this section (and, in fact, as we mentioned before in the $M/G/1/3$ example), all we need is the last row of A^{-1}, since only the last element of b is nonzero. Further, the last element of b is a one, so we have

$$\mathbf{p} = (a_{N-1,0}, a_{N-1,1}, \ldots, a_{N-1,N}),$$

where $a_{N-1,j}$ is the jth element of the last row of A^{-1}. We could, of course, have used the alternative formulation of setting one p_i equal to one, solving a nonsingular $(N-1) \times (N-1)$ system $b = pA$, and then renormalizing the $\{p_i\}$. We will mention more about this alternative problem setup a little later.

Standard inversion techniques such as Gauss–Jordan pivoting are quite adequate for moderately sized systems; however, to model large-state-space systems (for networks, N's of 10,000, 50,000, or even 100,000 are not unrealistic), Jacobi and Gauss–Seidel iterative techniques appear to be more efficient.

The Jacobi technique for solving sets of simultaneous equations is "mechanically" similar to stepping through a discrete-parameter Markov chain. If we break up the matrix A into a lower-triagular matrix L, a diagonal matrix D, and an upper-triagular U matrix, so that $L + D + U = A$, then the equations $b = pA$ can be written $b = p(L + D + U)$ or $pD = b - p(L + U)$. We start with some arbitrary $p^{(0)}$ and iterate using the equation

$$p^{(n+1)}D = b - p^{(n)}(L + U)$$

until we satisfy some stopping criterion. Convergence is not, in general, guaranteed; it often depends on the order in which the equations are written and on the particular problem generating the equations. For equations emanating from continuous-time Markov chains as illustrated above, Cooper and Gross (1991) prove that convergence is guaranteed when using the nonsingular $(N-1) \times (N-1)$ set generated by setting one of the p_i to one, although in our example, we shall see that we also obtain convergence for the $N \times N$ system given by (7.22).

The Gauss–Seidel procedure is a modification of the Jacobi technique in that as each new element of $p^{(n+1)}$ is calculated, it is used to replace the old element of $p^{(n)}$ in calculating the next element of $p^{(n+1)}$, as was done when stepping through a Markov chain.

To illustrate the Jacobi and Gauss–Seidel methods for solving equations, we use the previous *discrete-time* Markov-chain example for the $N \times N$ system of equations given in (7.21). The matrix of that example

$$A = \begin{pmatrix} -.57, & .36 & 1 \\ .43 & -.64 & 1 \\ 0 & .43 & 1 \end{pmatrix}$$

can be written as $A = L + D + U$, where

$$L = \begin{pmatrix} 0 & 0 & 0 \\ .43 & 0 & 0 \\ 0 & .43 & 0 \end{pmatrix}, \quad D = \begin{pmatrix} -.57, & 0 & 0 \\ 0 & -.64 & 0 \\ 0 & 0 & 1 \end{pmatrix}, \quad U = \begin{pmatrix} 0 & .36 & 1 \\ 0 & 0 & 1 \\ 0 & 0 & 0 \end{pmatrix},$$

so that $\boldsymbol{\pi}^{(n+1)}D = \boldsymbol{b} - \boldsymbol{\pi}^{(n)}(L + U)$ yields

$$(\pi_0^{(n+1)}, \pi_1^{(n+1)}, \pi_2^{(n+1)}) \begin{pmatrix} -.57, & 0 & 0 \\ 0 & -.64 & 0 \\ 0 & 0 & 1 \end{pmatrix}$$

$$= (0, 0, 1) - (\pi_0^{(n)}, \pi_1^{(n)}, \pi_2^{(n)}) \begin{pmatrix} 0 & .36 & 1 \\ .43 & 0 & 1 \\ 0 & .43 & 0 \end{pmatrix}.$$

Hence,

$$-.57\pi_0^{(n+1)} = 0 - .43\pi_1^{(n)},$$
$$-.64\pi_1^{(n+1)} = 0 - .36\pi_0^{(n)} - .43\pi_2^{(n)},$$
$$\pi_2^{(n+1)} = 1 - \pi_0^{(n)} - \pi_1^{(n)},$$

or equivalently,

$$\pi_0^{(n+1)} = .75\pi_1^{(n)},$$
$$\pi_1^{(n+1)} = .56\pi_0^{(n)} + .67\pi_2^{(n)}, \tag{7.23}$$
$$\pi_2^{(n+1)} = 1 - \pi_0^{(n)} - \pi_1^{(n)}.$$

Thus, for the Jacobi method, we pick a starting vector $\boldsymbol{\pi}^{(0)}$ [say $(1, 0, 0)$] and we find

$$\pi_0^{(1)} = .75(0) = 0, \quad \pi_1^{(1)} = .56(1) + .67(0) = .56, \quad \pi_2^{(1)} = 1 - 1 - 0 = 0,$$
$$\pi_0^{(2)} = .75(.56) = .42, \quad \pi_1^{(2)} = .56(0) + .67(0) = 0, \quad \pi_2^{(2)} = 1 - 0 - .56 = .44,$$

and so on.

For the Gauss–Seidel procedure (7.23) is modified as follows:

$$\pi_0^{(n+1)} = 0.75\pi_1^{(n)},$$

$$\pi_1^{(n+1)} = 0.56\pi_0^{(n+1)} + 0.67\pi_2^{(n)},$$

$$\pi_2^{(n+1)} = 1 - \pi_0^{(n+1)} - \pi_1^{(n+1)}.$$

This modification, in matrix notation, is $(U^T + D)\boldsymbol{\pi}^{(n+1)} = \boldsymbol{b} - L^T\boldsymbol{\pi}^{(n)}$, $\boldsymbol{\pi}^{(n+1)}$ and $\boldsymbol{\pi}^{(n)}$ now being column vectors, and T indicating transpose. The calculations proceeds as

$$\pi_0^{(1)} = .75(0) = 0, \quad \pi_1^{(1)} = .56(0) + .67(0) = 0, \quad \pi_2^{(1)} = 1 - 0 - 0 = 1,$$

$$\pi_0^{(2)} = .75(0) = 0, \quad \pi_1^{(2)} = .56(0) + .67(0) = .67, \quad \pi_2^{(2)} = 1 - 0 - .67 = .33,$$

and so on. Table 7.2 shows 21 iterations, rounded to two decimal places.

It appears that, to two decimal places, Gauss–Seidel converged in about 14 iterations, while it is not clear that Jacobi has as yet converged after 21

Table 7.2   Jacobi and Gauss–Siedel Calculations

Iteration Number	Jacobi			Gauss–Seidel		
	π_0	π_1	π_2	π_0	π_1	π_2
0	1	0	0	1	0	0
1	0	.56	0	0	0	1
2	.42	0	.44	0	.67	.33
3	0	.53	.58	.50	.50	0
4	.40	.39	.47	.38	.21	.42
5	.29	.54	.21	.16	.37	.48
6	.40	.31	.17	.27	.47	.25
7	.23	.34	.29	.36	.37	.28
8	.25	.32	.42	.28	.34	.38
9	.24	.43	.42	.25	.40	.34
10	.32	.42	.32	.30	.40	.30
11	.31	.40	.26	.30	.37	.33
12	.30	.35	.29	.28	.38	.35
13	.26	.36	.35	.28	.39	.33
14	.27	.38	.38	.29	.38	.32
15	.29	.40	.35	.29	.38	.33
16	.30	.39	.31	.28	.38	.33
17	.30	.38	.30	.29	.38	.33
18	.29	.37	.33	.29	.38	.33
19	.28	.38	.35	.28	.38	.33
20	.28	.39	.35	.29	.38	.33
21	.29	.39	.32	.29	.38	.33

iterations. As mentioned previously, convergence is often a problem with these techniques. There are many examples wherein these iterative techniques do not converge for some ordering of the equations but do for others. Also, the starting guess can affect convergence. However, experience has shown that if a limiting distribution exists for a set of equations generated from a Markov process, convergence appears to result (see Cooper, 1981). Cooper and Gross (1991), as stated above, have proven that convergence is guaranteed for a *continuous-time* Markov process if one uses the $(N-1) \times (N-1)$ formulation of the stationary equations (setting one of the $p_i = 1$, solving for the reduced set of $N-1$ $\{p_i\}$, and then renormalizing), but not necessarily for the $N \times N$ formulation where one of the equations is replaced by the $\Sigma p_i = 1$ as shown in (7.22).

In comparing the Markov-chain iterative procedure $\boldsymbol{\pi}^{(n+1)} = \boldsymbol{\pi}^{(n)}\boldsymbol{P}$ which we know from Markov theory *will* converge for appropriate $\boldsymbol{P}$, we see that after the five iterations presented previously, we are closer to the solution than with either the Jacobi or the Gauss–Seidel method applied to the stationary equations. (No general conclusions should be drawn as to the relative merits of these procedures, however; the interested reader is referred to Gross et al. 1984).

For continuous-parameter Markov chains, if we wish to utilize the Markov-chain stepping theory instead of solving the stationary equations, an appropriate imbedded discrete-parameter chain must be found which yields the same steady-state probabilities. For example, in birth–death processes one might think of using the imbedded jump (transition point) chain, but since this is periodic, no steady state exists. We will, however, present in the next section on transient solution techniques an imbedded chain which does have a steady-state solution and for which the imbedded discrete-parameter process probabilities are identical to the general-time (continuous-parameter process) steady-state probabilities.

Another major problem with these iterative procedures is choosing a stopping criterion. Generally, the Cauchy criterion is used, namely, the calculations stop when $\max_i |p_i^{(n+1)} - p_i^{(n)}| < \epsilon$. This is not always a good choice, depending on the type and rate of convergence, and in certain cases can lead to significant errors (see Gross et al., 1984).

There are ways of speeding up Gauss–Seidel convergence by using a weighting scheme (called overrelaxation; again, see Maron, 1982). Rather than dwell any further on solving steady-state equations at this time, we turn our attention toward numerical techniques for obtaining transient solutions, and will mention a way to utilize these also for steady state.

7.3.2 Transient Solutions

For Markovian queues, transient solutions, conceptually, can be obtained by solving the Kolmogorov differential equations,

$$p'(t) = p(t)Q.$$

Again, as long as the Q matrix is of finite dimension, numerical techniques for solving these linear, first-order differential equations can be employed. Numerical integration methods such as the Euler, Taylor, Runge–Kutta (RK), or predictor–corrector (pc) methods have long been employed in solving systems of differential equations.

Another method that is particularly well suited for queueing models is referred to as the *randomization technique*. It also has the advantage of having a probabilistic interpretation, and we shall derive it by a direct probabilistic analysis. We illustrate some of these techniques in the following sections.

Numerical Integration Methods

Numerical integration methods can be employed to solve a general system of ordinary differential equations described by

$$p'(t) \equiv \begin{pmatrix} p_1'(t) \\ p_2'(t) \\ \vdots \\ p_k'(t) \end{pmatrix} = \begin{pmatrix} f_1(p_1, \ldots, p_k; t) \\ f_2(p_1, \ldots, p_k; t) \\ \vdots \\ f_k(p_1, \ldots, p_k; t) \end{pmatrix} \equiv f(p, t)$$

with known initial value $p(t_0)$. The standard techniques are generally variations of Euler, Taylor, RK, or pc methods.

Taylor and RK methods are based on formulas that approximate the Taylor series solutions

$$p_i(t + h) = p_i(t) + hp'(t) + \frac{h^2}{2}p''(t) + \cdots + \frac{h^n}{n!}p_i^{(n)}(t) + R_n,$$

$i = 1, \ldots, k$. RK methods use approximations for the second and higher order derivatives, rather than doing the exact differentiation as prescribed for the Taylor methods. Euler's method is a special RK method, with $k = 1$. These methods have been used by several authors (e.g. Bookbinder and Martell, 1979; Grassmann, 1977; Liittschwager and Ames, 1975; and Neuts, 1973) to find transient solutions in queueing systems.

Predictor–corrector methods require information about several previous points in order to evaluate the next point. These methods involve using one formula to predict the next $p(t)$ value, followed by the application of a more accurate corrector formula. Unlike the Taylor and RK methods, pc methods are not self-starting; hence, they must use the RK or Taylor methods to obtain the first $p(t)$ value. Predictor–corrector methods can provide an estimate of the local truncation error at each step in the calculations, in contrast to the Taylor and RK methods, which cannot obtain such an estimate. Predictor–corrector methods have been used by Ashour and Jha (1973) for queueing problems.

We illustrate numerical integration methodology by considering some simple cases. For further details we refer the interested reader to Maron (1982). Consider an $M/M/1/1$ queue for which direct analytical methods do yield the transient solution (see Section 2.10.1). The differential equation to be solved, Equation (2.53), is

$$\frac{dp_1(t)}{dt} = -\mu p_1(t) + \lambda p_0(t) = -\mu p_1(t) + \lambda[1 - p_1(t)] = -(\mu + \lambda)p_1(t) + \lambda.$$
$$(7.24)$$

Letting $\lambda = 1$, $\mu = 2$, and assuming $p_1(0) = 0$, we have the solution from Equation (2.54) as $p_1(t) = (1 - e^{-3t})/3$.

Let us consider Euler's method of solving (7.24) numerically (this is an illustration only—one would never use numerical methods when an analytical solution exists). For small Δt,

$$\frac{dp_1(t)}{dt} \doteq \frac{p_1(t + \Delta t) - p_1(t)}{\Delta t},$$

so that (7.24) is approximately

$$p_1(t + \Delta t) \doteq p_1(t) - \Delta t(\mu + \lambda)p_1(t) + \lambda \Delta t \doteq p_1(t) - 3 \Delta t\, p_1(t) + \Delta t.$$
$$(7.25)$$

We can solve this recursively for $t = 0$, Δt, $2\Delta t$, and so on, and obtain the recursive relationship

$$p_1([n + 1]\Delta t) = (1 - 3\Delta t)p_1(n\Delta t) + \Delta t \quad (n \geq 1),$$
$$p_1(\Delta t) = (1 - 3\Delta t)p_1(0) + \Delta t = \Delta t.$$

Table 7.3 shows the solutions for $t = 0$, Δt, $2\Delta t$, and $3\Delta t$, where $\Delta t = 0.01$,

Table 7.3 Euler's Method versus Analytical Solution

n	$t = n\,\Delta t$	$p_1(n\,\Delta t)$	Exact $p_1(t)$	Error
0	0	0	0	
1	.01	.0100	.0099	.0001
2	.02	.0197	.0194	.0003
3	.03	.0291	.0287	.0004

and is compared to the analytical solution. For greater accuracy, Δt can be made smaller.

Euler's method is called a first-order method, for the following reason. Consider a Taylor series expansion of $p_1(t)$ about Δt, namely,

$$p_1(t + \Delta t) = p_1(t) + \Delta t\, p_1'(t) + \frac{\Delta t^2}{2!} p_1''(t) + \cdots + \frac{(\Delta t)^{n-1}}{(n-1)!} p_1^{(n-1)}(t) + R_n.$$

A first-order expansion, or approximation, is merely $p_1(t + \Delta t) \doteq p_1(t) + \Delta t\, p_1'(t)$, which, substituting for $p_1'(t)$ from (7.24), gives (7.25). We could get greater accuracy in a second-order approximation, say

$$p_1(t + \Delta t) \doteq p_1(t) + \Delta t\, p_1'(t) + \frac{\Delta t^2}{2} p_1''(t).$$

We have $p_1'(t)$ from our original equation, (7.24). To get $p_1''(t)$, we differentiate the right-hand side of (7.24), which yields $p_1''(t) = -(\mu + \lambda)p_1'(t)$. Now the second-order approximation becomes

$$p_1(t + \Delta t) \doteq p_1(t) + \Delta t[-(\mu + \lambda)p_1(t) + \lambda] + \frac{\Delta t^2}{2}[-(\mu + \lambda)p_1'(t)]$$

$$= [1 - (\mu + \lambda)\,\Delta t]p_1(t) + \lambda\,\Delta t - \frac{\Delta t^2}{2}(\mu + \lambda)[-(\mu + \lambda)p_1(t) + \lambda]$$

$$= \left(1 - (\mu + \lambda)\,\Delta t + (\mu + \lambda)^2 \frac{\Delta t^2}{2}\right)p_1(t) + \lambda\,\Delta t - (\mu + \lambda)\lambda \frac{\Delta t^2}{2}$$

$$= \left(1 - 3\,\Delta t + \frac{9\,\Delta t^2}{2}\right)p_1(t) + \Delta t - \frac{3\,\Delta t^2}{2}.$$

Table 7.4 shows the second-order approximation versus the first-order and exact solutions. We can see, of course, the increased accuracy of the higher-order approximation at the expense of more effort in setting up the approximating formula.

The numerical integration procedures for more complex problems are similar, except one must work with sets of equations, instead of just one as

Table 7.4 Comparison of First- and Second-Order Approximations

t	Exact	First Order	Error	Second Order	Error
0	0	0	0	0	0
.01	.0099	.0100	.0001	.0099	.0000
.02	.0194	.0197	.0003	.0195	.0001
.03	.0287	.0291	.0004	.0288	.0001

in the preceding example (see Problem 7.12). Nevertheless, if the set of equations is fully determined at some time t, it can be determined at time $t + \Delta t$, since $p_0(t + \Delta t), p_1(t + \Delta t), \ldots, p_N(t + \Delta t)$ become functions of $p_0(t)$, $p_1(t), \ldots, p_N(t)$, that is,

$$p_n(t + \Delta t) = f(p_0(t), p_1(t), \ldots, p_N(t)) \qquad (n = 0, 1, \ldots, N).$$

Numerical integration procedures then calculate the set of equations recursively in steps of Δt, starting with the known values $p_0(0), p_1(0), \ldots, p_N(0)$, as was done for the special case of $N = 1$ above.

A final comment on numerical integration methods: why are they called *numerical integration* methods? The answer lies in the fact that the formulas can be obtained from the fundamental theorem of calculus, namely,

$$p(t + \Delta t) = p(t) + \int_t^{t + \Delta t} p'(t) \, dt.$$

Since we do not know $p(t)$, we cannot use the above equation directly. What these methods do, then, is equivalent to computing the integral numerically; for example, a first-order approximation to the integral would be $\Delta t \, p'(t)$.

Randomization Technique

Although the randomization procedure is a computational technique for solving the set of differential equations, $p'(t) = p(t)Q$, arising from Markovian queueing systems, we develop the procedure by analyzing the stochastic process, rather than looking at the numerical solution of differential equations. Consider a finite birth–death process with a Q matrix given as follows:

$$Q = \begin{pmatrix} -\lambda_0 & \lambda_0 & 0 & 0 & 0 & \cdots & \cdots & & \cdots & 0 \\ \mu_1 & -\lambda_1 - \mu_1 & \lambda_1 & 0 & 0 & \cdots & \cdots & & \cdots & 0 \\ 0 & \mu_2 & -\lambda_2 - \mu_2 & \lambda_2 & 0 & \cdots & \cdots & & \cdots & 0 \\ \vdots & \vdots & \vdots & \vdots & \vdots & & & & & \vdots \\ \vdots & \vdots & \vdots & \vdots & \vdots & & & & & \dot{0} \\ 0 & 0 & 0 & 0 & 0 & \cdots & \mu_{N-1} & -\lambda_{N-1} - \mu_{N-1} & \lambda_{N-1} \\ 0 & 0 & 0 & 0 & 0 & \cdots & 0 & \mu_N & -\mu_N \end{pmatrix}$$

(7.26)

Suppose the process is in state n (n customers in the system). It remains in state n until either a birth or a death occurs. The time until a birth (arrival) is exponential with mean $1/\lambda_n$, and the time until a death (service completion) is exponential with mean $1/\mu_n$. The process leaves state n when the first of these two possible events occurs. Thus the time it spends in state n is the minimum of these two exponential random variables, which is itself an exponential random variable with mean $1/(\lambda_n + \mu_n)$.

When a transition occurs, the probability that it is a birth (see Section 1.9.4) is $\lambda_n/(\lambda_n + \mu_n)$, and the probability that it is a death is $\mu_n/(\lambda_n + \mu_n)$. Thus we can view the system as a Markov process with exponential holding times [mean of $(\lambda_n + \mu_n)^{-1}$ for holding in state n] and transition probabilities of $\lambda_n/(\lambda_n + \mu_n)$ and $\mu_n/(\lambda_n + \mu_n)$ for going up one or down one, respectively, from state n ($\mu_0 = \lambda_N = 0$). Furthermore, it can be shown that the occurrence of birth or death is independent of the holding time.

Suppose we wish to generate this process, for example, by simulating it in a Monte Carlo way. We could first create the transition-time occurrence and then generate, using simple Bernoulli probabilities, the change of state — that is, whether the process increases or decreases its state by one. The generation of the transition times is somewhat complicated in that we would have to sample from an exponential distribution with a state-dependent parameter $\lambda_n + \mu_n$.

To avoid this, we can reproduce this process in the following way. Consider the *minimum* mean holding time (time until the next transition occurs) of this process. This will correspond to the *maximum* value of $\lambda_n + \mu_n$, or equivalently, the minimum diagonal element of Q, the generator of the process. Call this value Λ. Let us denote the diagonal elements of the Q matrix by $-q_n$, that is,

$$q_n = \begin{cases} \lambda_0 & (n = 0), \\ \lambda_n + \mu_n & (n = 1, 2, \ldots, N - 1), \\ \mu_N & (n = N). \end{cases}$$

To reproduce our desired transition occurrence process, we could generate a true Poisson process with rate Λ (exponential holding times with constant mean $1/\Lambda$) and then *thin* the process to get the desired state-dependent transition rate. By thinning, we mean that whenever an occurrence is generated by the Poisson (Λ) process, if we are in state n, we draw from a Bernoulli probability distribution with success probability q_n/Λ, where a *success* indicates counting the occurrence as a transition and a *failure* (probability $1 - q_n/\Lambda$) indicates ignoring the occurrence. This thinning procedure on the Poisson (Λ) process generates our desired underlying state-dependent transition process.

Thus, to reproduce our process we can:

1. generate a Poisson process with rate Λ;
2. thin the Poisson (Λ) process by a Bernoulli *switch* with acceptance probability q_n/Λ and rejection probability $1 - q_n/\Lambda$;

3. generate the state change (up or down one unit) by another Bernoulli switch with an up probability of $\lambda_n/(\lambda_n + \mu_n)$ and a down probability of $\mu_n/(\lambda_n + \mu_n)$.

Denote the Poisson process as $\{N(t), t \geq 0\}$, that is, $N(t)$ equals the number of occurrences in $[0, t]$, and let Y_k equal the state of the system after the kth occurrence of the Poisson process, that is, Y_k equals $X(T_k)$, where $T_k = \min\{t : N(t) \geq k\}$. Viewing the process in this manner will allow us to derive the randomization computing algorithm that will yield $p(t)$.

Consider an element of the transient state probability vector $p(t)$, say $p_n(t) = \Pr\{X(t) = n\}$. Letting $p_{in}(t) = \Pr\{X(t) = n \mid X(0) = i\}$ gives

$$p_n(t) = \sum_{i=0}^{N} p_i(0)p_{in}(t). \tag{7.27}$$

Now considering our process as described above, we have, using the law of total probability,

$$p_{in}(t) = \sum_{k=0}^{\infty} \Pr\{Y_k = n \mid Y_0 = i\} \Pr\{N(t) = k\} = \sum_{k=0}^{\infty} \tilde{p}_{in}^{(k)} \frac{e^{-\Lambda t}(\Lambda t)^k}{k!}, \tag{7.28}$$

where $\tilde{p}_{in}^{(k)}$ represents the probability of going from state i to state n in k occurrences of the Poisson process. But

$$\tilde{p}_{in} = \Pr\{\text{going from } i \text{ to } n \text{ in one occurrence}\}$$

$$= \Pr\{\text{accepting the occurrence as a transition}\}$$

$$\times \Pr\{\text{transiting from } i \text{ to } n \mid \text{a transition takes place}\},$$

which is

$$\tilde{p}_{in}^{(1)} = \begin{cases} \dfrac{q_i}{\Lambda} \dfrac{\lambda_i}{\lambda_i + \mu_i} = \dfrac{\lambda_i}{\Lambda} & (n = i + 1), \\[2ex] \dfrac{q_i}{\Lambda} \dfrac{\mu_i}{\lambda_i + \mu_i} = \dfrac{\mu_i}{\Lambda} & (n = i - 1), \\[2ex] 1 - \dfrac{\lambda_i + \mu_i}{\Lambda} & (n = i), \\[2ex] 0 & (\text{elsewhere}). \end{cases}$$

Denote the matrix with elements $\tilde{p}_{in}^{(1)}$ by $\tilde{P}$; note that $\tilde{P} = Q/\Lambda + I$. We can obtain $\tilde{p}_{in}^{(k)}$ as elements of a matrix formed by multiplying $\tilde{P}$ by itself k times, that is,

$$\tilde{\boldsymbol{P}}^{(k)} \equiv \{\tilde{p}_{in}^{(k)}\} = \tilde{\boldsymbol{P}}^k.$$

Substituting this into Equations (7.27) and (7.28) yields

$$p(t) = \sum_{k=0}^{\infty} p(0)\tilde{\boldsymbol{P}}^{(k)} \frac{e^{-\Lambda t}(\Lambda t)^k}{k!}. \tag{7.29}$$

For computing purposes, one problem remains, namely, the infinite summation in (7.29). Thus we must truncate the sum at some value, say $T(t, \epsilon)$. We can set this value to guarantee a truncation error of less than a prespecified amount, say ϵ, since we are throwing away the tail of a Poisson distribution. From Equation (7.29) we have

$$p_n(t) = \sum_{k=0}^{\infty} \sum_{i=0}^{N} p_i(0)\tilde{p}_{in}^{(k)} \frac{e^{-\Lambda t}(\Lambda t)^k}{k!}$$

$$= \sum_{k=0}^{T(t,\epsilon)} \sum_{i=0}^{N} p_i(0)\tilde{p}_{in}^{(k)} \frac{e^{-\Lambda t}(\Lambda t)^k}{k!} + \sum_{T(t,\epsilon)+1}^{\infty} \sum_{i=0}^{N} p_i(0)\tilde{p}_{in}^{(k)} \frac{e^{-\Lambda t}(\Lambda t)^k}{k!}.$$

We desire

$$\sum_{T(t,\epsilon)+1}^{\infty} \sum_{i=0}^{N} p_i(0)\tilde{p}_{in}^{(k)} \frac{e^{-\Lambda t}(\Lambda t)^k}{k!} \equiv R_T < \epsilon.$$

But

$$R_T < \sum_{T(t,\epsilon)+1}^{\infty} \frac{e^{-\Lambda t}(\Lambda t)^k}{k!},$$

so that by finding $T(t, \epsilon)$ such that

$$\sum_{T(t,\epsilon)+1}^{\infty} < \epsilon \quad \Rightarrow \quad \sum_{k=0}^{T(t,\epsilon)} \frac{e^{-\Lambda t}(\Lambda t)^k}{k!} > 1 - \epsilon, \tag{7.30}$$

we have an error bound on $p_n(t)$ of ϵ.

While we have developed this procedure for a birth–death model, the computing formula (in vector–matrix form)

$$p(t) = \sum_{k=0}^{T(t,\epsilon)} p(0)\tilde{\boldsymbol{P}}^{(k)} \frac{e^{-\Lambda t}(\Lambda t)^k}{k!} \tag{7.31}$$

holds for any Markov process with infinitesimal generator

$$Q = \begin{pmatrix} -q_0 & q_{01} & q_{02} & \cdots & q_{0N} \\ q_{10} & -q_1 & q_{12} & \cdots & q_{1N} \\ \vdots & \vdots & \vdots & & \vdots \\ q_{N0} & q_{N1} & q_{N2} & \cdots & -q_N \end{pmatrix},$$

where $q_i = \sum_{j \neq i} q_{ij}$, $i = 0, 1, 2, \ldots, N$. The general expression for $\tilde{p}_{in}^{(1)}$ is

$$\tilde{p}_{in}^{(1)} = \begin{cases} q_{ij}/\Lambda & (i \neq n), \\ 1 - q_i/\Lambda & (i = n). \end{cases} \tag{7.32}$$

With respect to computational difficulties in problems with large state spaces, the major computing effort is raising the matrix $\tilde{P}$ to the kth power, since $T(t, \epsilon)$ can be sizable. This can be avoided by a recursive computing scheme as follows.

In (7.31), consider the product $p(0)\tilde{P}^{(k)}$, and call this $\phi^{(k)}$. This vector is the system state probability vector after k occurrences of the underlying Poisson (Λ) process, that is, the probability distribution of the state of the system of a discrete-parameter Markov chain after k transitions, namely, the Markov chain Y_k with transition probability matrix $\tilde{P}$. From Markov-chain theory we know that

$$\phi^{(k)} = \phi^{(k-1)}\tilde{P} \qquad (\tilde{P} = Q/\Lambda + I), \tag{7.33}$$

so that we can write (7.31) as

$$p(t) = \sum_{k=0}^{T(t,\epsilon)} \phi^{(k)} \frac{e^{-\Lambda t}(\Lambda t)^k}{k!} \tag{7.34}$$

and calculate $\phi^{(k)}$ recursively using (7.33).

What we have essentially done is to reduce the complex calculations of the continuous-parameter Markov-chain $X(t)$ with infinitesimal generator Q to calculations on the discrete-parameter Markov chain Y_k with transition probability matrix $\tilde{P}$ relating k to t by the Poisson (Λ) process. That is, for the Y_k process (often referred to as the uniformized, imbedded Markov chain), we measure time in number of occurrences of the Poisson (Λ) process and relate it to clock time through the Poisson probabilities.

It is very often the case for queueing problems that $\tilde{P}$ is a *sparse* matrix; that is, most of the elements are zero. For example, consider the Q matrix of (7.26), keeping in mind that $\tilde{P} = Q/\Lambda + I$. Of the $(N + 1)^2$ elements, fewer than $3(N + 1)$ are nonzero. Thus, in doing the matrix multiplication of (7.33), many zero multiplications transpire. There are ways to avoid this, and we refer the reader to Gross and Miller (1984) for one such procedure.

The randomization procedure can also be used as a vehicle for obtaining steady-state solutions. One way to do this is simply to make t large, or keep trying successive t values until $p(t)$ does not change appreciably with t. Another route is to consider only the uniformized imbedded discrete-parameter Markov chain Y_k with transition probability matrix $\tilde{P}$ and use (7.33) recursively until $\phi^{(k)}$ appears independent of time. It can easily be shown that this uniformized imbedded discrete-parameter Markov chain with transition probability matrix $\tilde{P}$ has the same steady-state probability distribution as the original continuous-parameter Markov chain, that is, $\lim_{k \to \infty} \phi^{(k)} = \lim_{t \to \infty} p(t)$, because

$$\phi = \phi\tilde{P} \;\Rightarrow\; \phi = \phi\left(\frac{Q}{\Lambda} - I\right) \;\Rightarrow\; 0 = \phi\frac{Q}{\Lambda} \;\Rightarrow\; 0 = \phi Q.$$

It would seem that using (7.33) should be a more efficient means of obtaining a steady-state distribution than using (7.34), although the mixing with Poisson probabilities done in (7.34) might tend to act as a smoothing procedure and in some cases might conceivably converge faster. One could also, of course, use Gauss–Seidel stepping on the discrete-parameter Markov chain of (7.33). Which of these alternatives is the best is an open question and obviously will depend, at least in part, on the specific problem being considered.

Our interest in this section is not to present a treatise on numerical analysis, but to point out that this can be a very important contribution in applying queueing theory—an area that unfortunately has not gotten the attention in the past that is its due. It is certainly nice to obtain closed-form solutions where that is possible, but in those many cases where it is not, numerical methods can provide a way to obtain meaningful answers.

7.4   DISCRETE-EVENT STOCHASTIC SIMULATION

It often turns out that it is not possible to develop analytical models for queueing systems. This can be due to the characteristics of the input or service mechanisms, the complexity of the system design, the nature of the queue discipline, or combinations of the above. For example, a multistation multiserver system with some recycling, where service times are (truncated) normally distributed and a complex priority system is in effect, is impossible to model analytically. Furthermore, even some of the models treated previously in the text provided only steady-state results, and if one were interested in transient effects or if the probability distributions were to change with time, it might not be possible to develop analytical solutions or efficient numerical schemes in these cases. For such problems as these, it may be necessary to resort to analyses by simulation. It should be emphasized, however, that if analytical models are achievable, they should be used, and that simulation should be resorted to only in cases where either analytical models are not achievable and approximations not acceptable or they are so complex that the solution time is prohibitive.

While simulation may offer a way out for many analytically intractable models, it is not in itself a panacea. There are a considerable number of pitfalls one may encounter in using simulation. Since simulation is comparable to analysis by experimentation, one has all the usual problems associated with running experiments in order to make inferences concerning the real world, and must be concerned with such things as run length, number of replications, and statistical significance. However, the theory of statistics (including, of course, experimental design) can be of help here.

Another drawback to simulation analyses occurs if one is interested in optimal design of a queueing system. Suppose that it is desired to determine the optimal number of channels or the optimal service rate for a particular system where conflicting system costs are known. If an analytical model can be developed, the mathematics of optimization (differential calculus, mathematical programming, etc.) can be utilized. However, if it is necessary to study the system using simulation, then one must rely on techniques for searching experimental output. These search techniques are often not as neat as the mathematics of optimization for analytical functions. Frequently the experimenter will merely try a few alternatives and simply choose the best among them. It might well be that none of the alternatives tried is optimal nor even near optimal. How close one gets to optimality in a simulation study often depends on how clever the analyst is in considering the alternatives to be investigated. Because of these potential drawbacks, simulation analysis has often been referred to as an "art". Nevertheless, with the advances in simulation methodology in these areas (see, e.g., Rubenstein, 1986) and the many fine researchers in the field chipping away at these problems, it is becoming increasingly competitive with analytical modeling in some situations, and in many situations, simulation is the only way to proceed. Simulation has found major uses in modeling transportation, manufacturing, and communication systems. Such systems are usually stochastic in nature, with a variety of random processes interacting in complex ways. Without simplifying assumptions about the nature of the randomness, routing probabilities, etc., analytical modeling is usually not an option.

The following presents an overview of what we believe to be the key points in discrete event stochastic simulation; for more details, the interested reader is referred to basic simulation texts such as Banks et al. (1996), Fishman (1978), Hoover and Perry (1989), Law and Kelton (1991), and Schriber (1991).

7.4.1   Elements of a Simulation Model

One can look at simulation modeling as being composed of three major elements: (1) input distribution selection (sometimes called input modeling) and generation, (2) bookkeeping, and (3) output analysis. Since we are interested in modeling stochastic systems, it is necessary to select and then generate the appropriate stochastic phenomena in the computer. For example, a manufacturing system may consist of a network of queues with a variety of different interarrival-time and service-time distributions. We must decide on which probability distributions we wish to use to represent these arrival and service mechanisms (sometimes we may use an empirical distribution made up from actual collected data). Then, random variates from these different distributions must be generated so that the system can be observed in action. Once these distributions are chosen and random variates generated, the bookkeeping phase keeps track of transactions moving around the system and keeps counters on ongoing processes in order to calculate appropriate performance measures. Output analysis has to do with computing measures of system effectiveness and employing the appropriate statistical techniques required to make valid statements concerning system performance.

The following very simple hypothetical example illustrates the three basic elements described above.

Example 7.2
A small manufacturer of specialty items has signed a contract with a prestigious customer for 20 orders of its premiere product. Management is concerned with current capacity and wishes to analyze the situation using discrete-event simulation. The customer will place orders at random times and of course would like them filled as soon as possible. Orders are placed only at the beginning of a month and could come as frequently as 2 months apart or as infrequently as 7 months apart, or anything in between, all with equal probability. Currently, the production capability for this product is such that orders are shipped only at the end of a month and the order filling time is equally likely between 1 and 6 months, inclusive. Only one order at a time can be processed, so that if a second order comes in while one is being prepared, it must wait until the order ahead of it is completed. For this capability, management would like to get an idea of the average number of orders in the system, the average time an order spends in the system, the maximum time an order spends in the system, and the percentage of time the system is idle. The date of the first order is known, and the production line will be set up just in time to receive the first order. The production line will be taken down after the last (20th) order is completed.

Input modeling has been simplified here, since we have decided that the probability distributions are discrete uniform distributions. This also makes generating the random input data easy, since it can be done by (the equivalent of a) roll of a fair die. Times between placement of orders are discrete-uniform (2,7) and service times are discrete-uniform (1,6). Thus, for generating the interarrival times, we simply roll the die 19 times and add one to each value to get the times between successive orders after the first one.

Table 7.5 Input Data

Time between orders:	–, 7, 2, 6, 7, 6, 7, 2, 5, 4, 5, 3, 2, 6, 2, 4, 2, 6, 5, 5
Service times:	1, 3, 2, 3, 6, 5, 4, 5, 1, 1, 3, 1, 3, 2, 2, 6, 5, 1, 3, 5

For the service times, the value of the roll itself suffices and we simply need to roll the die 20 more times. Table 7.5 gives the results of using a fair die to generate the input data.

Using the input data in Table 7.5, we can construct an abbreviated book-keeping table as shown in Table 7.6 (note that Tables 7.5 and 7.6 are in the same spirit as Tables 1.3 and 1.4 respectively, of Section 1.6). At clock 0, the first order comes into the system, has a service time of 1 month, and is due to depart at clock 1. At clock 1, the next arrival, order 2, is due in at clock $0 + 7 = 7$, and since no order is in the system, will depart at its arrival time plus service time, that is, $7 + 3 = 10$. The clock is advanced to time 7, and the next arrival (order 3) scheduled at $7 + 2 = 9$. Since order 3 arrives before order 2 leaves, the clock is advanced to time 9, the arriving order 3 enters the queue, and order 2 is still in service, but due to depart at time 10. Order 4 is due in at $9 + 6 = 15$. The clock is then advanced to time 10, where order 2 leaves the system, and order 3 enters service and is scheduled to depart at $10 + 2 = 12$. Order 4 is next to arrive, and it is due in at 15, so the clock advances to 12. The bookkeeping continues on in this fashion until the 20th order is processed.

Such a bookkeeping table (which can get very complicated for realistic, complex systems) allows one to make the performance-measure calculations. It is straightforward to use Table 7.6 and obtain the queue wait and total

Table 7.6 Bookkeeping[a]

Master Clock Time	Next Events		Transaction in Queue	Transaction in Service
	Arrival	Departure		
0	[2], 7	[1], 1		→[1]
1	[2], 7	[2], 10		[1]→
7	[3], 9	[2], 10		→[2]
9	[4], 15	[2], 10	→[3]	[2]
10	[4], 15	[3], 12	[3]→	→[3] [2]→
⋮	⋮	⋮	⋮	⋮
81	[20], 86	[19], 84		→[19]
84	[20], 86			[19]→
86		[20], 91		→[20]
91				[20]→

[a]Key: [n], t = [transation number], time of occurrence.

time in system for each order. For example, order 1 came into the system at time 0, went right into service, and left at time 1, spending zero time in queue and 1 month in the system. Order 2 arrived at time 7, also went directly into processing, and left at time 10, spending 3 months in the system. Order 3, however, arriving at time 9, had to enter the queue, since 2 was still in process. It left the queue for processing at time 10 and exited the system at time 12 (not shown in the abbreviated table), spending 1 month in queue waiting for processing and 3 months total time in the system. Average waiting times and maximum waiting times can then be easily calculated. Queue and system size values, as well as idle periods, can also be obtained from Table 7.6, although it is a little more work getting average figures, since the sizes must be weighted by the amount of time the queue and system stayed at their various sizes (see Section 1.6).

For the above example, the maximum number of orders in the queue was 1, the maximum number of orders in the system was 2, the maximum time an order spent in the system waiting to be processed was 4 months (order 17), and the maximum time an order spent in the system in toto was 9 months (also order 17). The average queue size was 0.13, the average system size was 0.81, the average percentage of the time the system was empty and idle was 32%, and the average waiting times in queue and system respectively were 0.6 and 3.7 months.

7.4.2 Input Modeling and Random-Variate Generation

In this subsection, we treat the topics of choosing the appropriate distributions to represent the stochastic mechanisms of the system (input modeling) and the generation of random variates from the chosen distributions, and also include a brief treatment of pseudorandom-number generators.

Input Modeling

Input modeling is not only appropriate for simulation, but is necessary for any probabilitstic modeling, including analytical and numerical treatments as well. It is most important, since the output of any model can be only as good as its input. It is also a topic which has received relatively little attention in the applied probability community, although it is currently generating quite a bit of activity in the simulation community. The two major problems in input modeling are the selection of a family of distributions (e.g., exponential, Erlang, normal, etc.), and, once the family is selected, estimating its parameters. We start with the easier of the two problems first, namely, parameter estimation once a distribution family is selected. Because of the importance of the exponential and its relatives (e.g., Erlang) in analytical modeling, we use these as our primary illustrations. However, the reader should keep in mind that with modern simulation packages, almost any known statistical distribution can be easily utilized, and this is one of the major advantages of going to simulation modeling.

Parameter Estimation

We assume that we have chosen the distribution family and have data available on actual transactions (interarrival times or service times). Let us assume we have a random sample of size n, say, $t_1, t_2, \ldots, t_n$. The two classical methods of parameter estimation are the method of maximum likelihood, with its resulting estimators referred to as MLEs (we have encountered these before in Section 6.7) and the method of moments (MOM) estimators.

Maximum-likelihood estimators have some nice statistical properties and can be obtained as follows. First consider the case where we believe our underlying distribution is exponential with parameter θ and we wish to estimate θ from the sample data. We form the likelihood function (joint density function of the sample), assuming the observations are *independent*, as

$$L(\theta) = \prod_{i=1}^{n} \theta e^{-\theta t_i} = \theta^n e^{-\theta \sum_{i=1}^{n} t_i},$$

where t_i is the ith sample observation (say interarrival or service time). The MLE of θ is the value which maximzes L. It is often more convenient to find the maximum of $\ln L = \mathcal{L}$. Thus $\hat{\theta}$, the MLE of θ, is the value for which

$$\max_{\theta} \mathcal{L}(\theta) = \max_{\theta}\left(n \ln \theta - \theta \sum_{i=1}^{n} t_i\right)$$

is attained. Taking $d\mathcal{L}/d\theta$ and setting it equal to zero yields the maximizing value as

$$\frac{n}{\hat{\theta}} - \sum_{i=1}^{n} t_i = 0 \quad \Rightarrow \quad \hat{\theta} = \frac{n}{\sum_{i=1}^{n} t_i} = \frac{1}{\bar{t}},$$

where $\bar{t} \equiv (\sum_{i=1}^{n} t_i)/n$ is the sample mean.

This is the same estimator that results when the empirical and theoretical means are equated and then solved for θ. This procedure (equating theoretical and sample moments) is called the MOM and is usually a very fast way to get estimates, though often giving quite different answers from the MLE. However, the MOM does not in general lead to statistics with all the good properties we can ascribe to the MLE. In our context, one particularly important concern is whether an estimator is consistent in the sense that $\hat{\theta} \to \theta$ in probability as the sample size goes to ∞. It does indeed turn out that the likelihood equation is guaranteed to have a consistent solution under fairly general conditions; but the MOM is not. On the other hand, we know how important moments are in queueing theory. For example, the PK formula for the $M/G/1$ depends only on the mean and the variance of the service time distribution. The Kingman–Marshall upper bound and the heavy-traffic approximation depend only on the first two moments (mean and variance) of the interarrival and service-time distributions. Thus, for estimating parameters for queueing models, even though MOM estimators are less desirable than MLEs in a statistical sense, one might well be better off using them.

As another example, let us consider finding MLEs and MOM estimates for the two parameters of the Erlang density. We will first derive the MLEs. For the Erlang density,

$$f(t) = \frac{\phi(\phi t)^{k-1} e^{-\phi t}}{(k-1)!},$$

so the likelihood function may be written as

$$L(\phi, k) = \prod_{i=1}^{n} \frac{\phi(\phi t_i)^{k-1} e^{-\phi t_i}}{(k-1)!} = \frac{\phi^{nk} e^{-\phi \sum_{i=1}^{n} t_i} (\prod_{i=1}^{n} t_i)^{k-1}}{[(k-1)!]^n}.$$

Thus the log likelihood is

$$\mathcal{L}(\phi, k) = nk \ln \phi - \phi \sum_{i=1}^{n} t_i + (k-1) \sum_{i=1}^{n} \ln t_i - n \ln (k-1)!.$$

Therefore

$$\frac{\partial \mathcal{L}}{\partial \phi} = \frac{nk}{\phi} - \sum_{i=1}^{n} t_i \quad \Rightarrow \quad \hat{\phi} = \frac{\hat{k}}{\bar{t}},$$

where, again, $\bar{t}$ is the sample mean of the data. Now to get the complete pair $(\hat{\phi}, \hat{k})$, consider k to be a continuous variable (say x) and proceed in the usual way to obtain the MLE $\hat{x}$ of x as the numerical solution to

$$\frac{\partial \mathcal{L}}{\partial x} = 0 = n \ln \hat{\phi} + \sum_{i=1}^{n} \ln t_i - n\psi(\hat{x})$$

$$= n(\ln \hat{x} - \ln \bar{t}) + \sum_{i=1}^{n} \ln t_i - n\psi(\hat{x}),$$

where $\psi(x)$ is the logarithmic derivative of the Γ function, that is, $\psi(x) \equiv d \ln \Gamma(x)/dx$. The function $\psi(x)$ is tabulated in Abramowitz and Stegun (1964), and a good approximation to it when x is not too small (say ≥ 3) is

$$\psi(x) \doteq \ln(x - \tfrac{1}{2}) + \frac{1}{24(x - \tfrac{1}{2})^2}.$$

Hence the MLE $\hat{k}$ of k is either $[\hat{x}]$ or $[\hat{x}] + 1$ (where $[x]$ is the greatest integer in x), depending on which pair, $([\hat{x}]/\bar{t}, [\hat{x}])$ or $(([\hat{x}] + 1)/\bar{t}, [\hat{x}] + 1)$, gives a higher value to the log likelihood.

In the case of the Erlang, the MOM answer follows much more easily. Here, since there are two parameters, two equations are needed, namely,

$$\bar{t} = \frac{\tilde{k}}{\tilde{\phi}} \quad \text{and} \quad s^2 = \frac{\tilde{k}}{\tilde{\phi}^2},$$

where s^2 is the sample variance. Thus, from the simultaneous solution of the above two equations, the moment estimates are found as

$$\tilde{\phi} = \frac{\bar{t}}{s^2} \quad \text{and} \quad \tilde{k} = \left[\frac{\bar{t}^2}{s^2}\right] \quad \text{or} \quad \left[\frac{\bar{t}^2}{s^2}\right] + 1.$$

For the Erlang case (and for most distributions), MOM and ML give different estimators for the parameters. If the mean and variance using the MLEs gave very different mean and variance values from those of the sample data, we would be hesitant in using the MLEs and would go with the MOM.

Distribution Selection
We now turn to the much more difficult, but very important topic of how to decide on which distribution family to choose to represent the input distributions (interarrival and service times). The choice of appropriate candidate probability distributions hinges upon knowing as much as possible about the characteristics of the potential distributions and the "physics" of the situation to be modeled. Generally, we have first to decide which probability functions are appropriate to use for the arrival and service processes. For example, we know the exponential distribution has the Markovian (memoryless) property. Is this a reasonable condition for the actual situation under study? Let us say we are looking to describe a service mechanism consisting of a single server. If the service for all customers is fairly repetitive, we might feel that the longer the customer is in service, the greater the probability of a completion in a given interval of time (nonmemoryless). In this case, the exponential distribution would not be a reasonable candidate for consideration. On the other hand, if the service is mostly diagnostic in nature (we must find the trouble in order to fix it), or there is a wide variation of service required from customer to customer, the exponential might indeed suffice.

The actual shape of the density function also gives quite a bit of information, as do its moments (see Law and Kelton, 1991, Chapter 6, for pictures of most standard distribution families). One particularly useful measure is the ratio of the standard deviation to the mean, called the coefficient of variation $(C = \sigma/\mu)$. The exponential distribution has $C = 1$, while E_k (Erlang type k), $k > 1$, has $C < 1$, and H_k (hyperexponential), $k > 1$, has

$C > 1$. Hence, choosing the appropriate distribution is a combination of knowing as much as possible about distribution characteristics, the "physics" of the situation to be modeled, and statistical analyses when data are available.

To help in characterizing probability distributions for consideration as candidates in describing interarrival or service times, we present a concept that emanates from the area of *reliability theory*, namely, the hazard rate (or as it is also called, the failure rate) function. We will relate this to the Markov property for the exponential distribution and point out its use as a way to gain general insight into probability distributions.

Suppose we desire to choose a probability distribution to describe a continuous random variable T (say interarrival or service time) with CDF $F(t)$. The density function, $f(t) = dF(t)/dt$, can be interpreted as

$$f(t)\, dt \doteq \Pr\{t \le T \le t + dt\},$$

that is, as the approximate probability that the random time will be in a neighborhood about a value t. The CDF $F(t)$ is, of course, the probability that the time will be less than or equal to the value t. We define a conditional type of probability as follows:

$$h(t)\, dt \doteq \Pr\{t \le T \le t + dt \mid T \ge t\},$$

which is the approximate probability that the time will be in a neighborhood about a value t, given that the time is already t. For example, if we are dealing with interarrival times, it is the approximate probability that an arrival occurs in an interval dt, given that it has been t since the last arrival. If we are dealing with service times, $h(t)$ is the approximate probability that a customer is completed in dt, given that the customer has already been in service for a time t.

From the law of conditional probability we have

$$h(t)\, dt \doteq \Pr\{t \le T \le t + dt \mid T \ge t\} = \frac{\Pr\{t \le T \le t + dt \text{ and } T \ge t\}}{\Pr\{T > t\}} = \frac{f(t)\, dt}{1 - F(t)}.$$

Therefore,

$$h(t) = \frac{f(t)}{1 - F(t)}. \tag{7.35}$$

This hazard or failure rate function, $h(t)$, can be increasing in t (called an increasing failure rate, or IFR), decreasing in t (DFR), constant (considered to be *both* IFR and DFR), or a combination. The constant case implies the memoryless or *ageless* property, and we shall shortly show this for the exponential distribution.

If we believe that service is consistent enough that the longer a customer has been in service, the more likely it is that the service is completed in the next dt, then we desire an $f(t)$ for which $h(t)$ is increasing in t, that is, an IFR distribution.

From (7.35) we can obtain $h(t)$ from $f(t)$. Thus, the hazard rate is another important source [as is the shape of $f(t)$ itself] for obtaining knowledge concerning candidate $f(t)$'s that may be considered for modeling arrival and service patterns.

Consider the exponential distribution, $f(t) = \theta e^{-\theta t}$. We desire to find $h(t)$. From (7.35),

$$ h(t) = \frac{\theta e^{-\theta t}}{e^{-\theta t}} = \theta. $$

Thus the exponential distribution has a constant failure (hazard) rate and is memoryless. Suppose we feel in a particular queueing situation that we need an IFR distribution for describing service times. It turns out that the Erlang $(k > 1)$ has this property. The density function, from Chapter 3, Section 3.3.1, where we now let $\theta = k\mu$, is

$$ f(t) = \frac{\theta^k t^{k-1} e^{-\theta t}}{(k-1)!}, $$

and (see Problem 7.14)

$$ F(t) = \frac{\theta^k}{(k-1)!} \int_0^t e^{-\theta x} x^{k-1}\, dx = 1 - \sum_{i=0}^{k-1} \frac{(\theta t)^i e^{-\theta t}}{i!}. \tag{7.36} $$

Thus,

$$ h(t) = \frac{\theta^k t^{k-1} e^{-\theta t}}{(k-1)!\, \Sigma_{i=0}^{k-1} (\theta t)^i e^{-\theta t}/i!} = \frac{\theta(\theta t)^{k-1}}{(k-1)!\, \Sigma_{i=0}^{k-1} (\theta t)^i/i!}. $$

Without doing numerical work, it is difficult to ascertain the direction of change of $h(t)$ with t. It can be shown, however (see Problem 7.15), that $h(t)$ can be written as

$$ h(t) = \frac{1}{\int_0^\infty (1 + u/t)^{k-1} e^{-\theta u}\, du}. \tag{7.37} $$

Now it is fairly easy to see that for $k > 1$, as t increases, the integrand in the denominator decreases, so that the integral decreases, and hence $h(t)$ is increasing with t (IFR). Further, it has an asymptote of θ as t goes to infinity, and $h(0) = 0$. Since $h(t)$ has an asymptote, even though $h(t)$ increases with t, it does so at an ever slower rate, and eventually approaches the constant θ.

Now suppose we were to desire the opposite IFR condition, that is, an accelerating rate of increase with t. There is a distribution called the Weibull [with $1 - F(t) = e^{-\theta t^{\alpha}}$] for which we can obtain this condition. In fact, depending on how we pick the shape parameter α of the Weibull, we can obtain an IFR with decreasing acceleration, constant acceleration (linear in t), or increasing acceleration, or even obtain a DFR or the constant-failure-rate exponential. Even though the Weibull does not lend itself to analytical treatment (except when it reduces to the exponential) in queueing situations, it still can be a candidate distribution for a simulation analysis. In many situations, more than one particular distribution may be a reasonable candidate. For example, if we decide we want an IFR with a deceleration, we could consider the Weibull or Erlang family of distributions.

We present one further example in the process of choosing an appropriate candidate distribution for modeling. Let us say we are satisfied with an IFR that has a deceleration effect, such as the Erlang, but we believe that the C might be greater than one. This latter condition eliminates the Erlang from consideration. But we know that a mixture of exponentials does have $C > 1$. It is also known (see Barlow and Proschan, 1975) that any mixture of exponentials is DFR. In fact, Barlow and Proschan prove that all IFR distributions have $C < 1$, while all DFR distributions have $C > 1$. However, the converse is not true, that is, $C < 1$ does not imply IFR, $C > 1$ does not imply DFR, and $C = 1$ does not imply a constant failure rate (CFR). An example of this is the lognormal family of distributions, which, depending on the value of its parameters, can yield Cs less than, equal to, or greater than one. Its hazard function is both IFR and DFR, that is, over a certain range it is IFR and over a certain range it is DFR. So, if we are convinced that we have an IFR situation, we must accept $C < 1$. Intuitively, this can be explained as follows. Situations that have $C > 1$ generally are ones where, say, service times are mixtures (say of exponentials). Thus, if a customer has been in service a long time, chances are it is of a type requiring a lot of service, so the chance of its being completed in the next dt diminishes. Situations for which we have an IFR condition indicate a more consistent service pattern among customers, thus yielding a $C < 1$.

In summary, we again make the point that choosing an appropriate probability model is a combination of knowing as much as possible about the characteristics of the probability distribution family being considered, and as much as possible about the actual situation being modeled. In most cases, data on the processes we are trying to model are available, but not always. Below we treat the case where data are available or can be collected and how these data can help us in choosing the appropriate family of distributions. But first we briefly comment that, in those cases where we have no data or cannot gather any (e.g., a new system we are modeling), the considerations we have mentioned above become paramount. Sometimes in the absence of data, the triangular distribution is chosen, where the modeler is asked to set the minimum, maximum, and most likely values, which will then yield the three parameters of this three parameter distribution. A more flexible distribution family is the beta, which can yield a large variety of shapes, and is employed along with "expert opinion" in deciding on the appropriate parameters (the beta is a two parameter distribution, with both parameters affecting the distribution shape). Further discussions of choosing distributions with no data available can be found in Law and Kelton (1991) and Banks et al. (1996).

We now turn to the case where we have observations available on interarrival and service times. We will also assume the data come from a homogeneous time period (i.e., the process from which the data were gathered was not changing in time—say a system during the peak traffic hours). It is very important to check the data to see that this is so. We also assume that all observations are independent. It is important to check the data for these conditions (IID), and there are some tests one can perform to do so (we refer the interested reader to Leemis, 1996, or Law and Kelton, 1991).

We now assume we have data which form an IID sample. One of the first things we can do in trying to decide which family of distributions might be appropriate is to calculate the sample mean, sample standard deviation, and sample C. A histogram plot of the data can also be very useful, although the shape of the resultant histogram depends on the length of intervals used to accumulate the frequencies (the number of observations falling in each interval). A rule of thumb is to choose interval lengths such that there are a minimum of five observations in each interval and that there are at least five intervals, but it is a good idea to try a variety of interval lengths. Suppose now, after looking at histogram plots and considering the physical characteristics of the system we are modeling, as well as the characteristics of distribution families, we choose a potential candidate distribution. There are a variety of statistical tests we can perform to see if our candidate distribution is a reasonable choice. We mention three general tests applicable for any family of distributions and one specific test if we believe our situation calls for choosing an exponential distribution.

The most commonly used (but not necessarily the best) test is the χ^2 goodness-of-fit test. It assumes we have our data in histogram form, with each block of the histogram (referred to as a frequency class) providing the number of observations in the interval that is covered by the block. Generally, the intervals are of equal length, spanning the range of the data, and the number of observations in each interval varies to yield a picture approximating a density function. Figure 7.3 shows a histogram of the 25 service-time observations given below:

27.6, 28.9, 3.8, 16.6, 13.3, 3.3, 7.8, 55.3, 12.6, 1.8, 12.9, 4.8, 12.6, 8.8, 3.3, 2.7, 0.6, 1.3, 1.1, 21.3, 11.3, 14.9, 15.7, 8.6, 9.6.

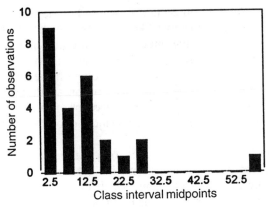

Fig. 7.3 Histogram of sample service-time data.

The mean, standard deviation, and C are, respectively, 12.02, 11.91, and 0.991. In reality, we should have considerably more than 25 observations, but we use this small data set in order to make the discussion below easier to follow.

Looking at the C, which is near one, we might consider the exponential family as a candidate distribution family. Both the ML and MOM estimators of the mean are 12.02. The histogram of Figure 7.3 has a roughly exponential distribution shape, and we first try the χ^2 test on these data to see if the exponential distribution is a reasonable choice. The χ^2 test statistic is

$$\chi_k^2 = \sum_{i=1}^{n} \frac{(o_i - e_i)^2}{e_i},$$

where o_i is the number observed in the ith frequency class, e_i the number expected in the ith frequency class if the hypothesized distribution were correct, and k the number of degrees of freedom, always equal to the total number of classes minus one and then minus one for each parameter estimated. Of course, the usual precautions must be taken to keep the number of observations in a class from being too small (a rule of thumb being five or more).

To get e_i, we integrate the theoretical distribution over the interval; for example, if i^- is the lower point of the ith interval and i^+ is the upper point, then for the exponential distribution,

$$e_i = n \int_{i^-}^{i^+} \theta e^{-\theta t}\, dt = n(e^{-\theta i^-} - e^{-\theta i^+}),$$

Table 7.7 χ^2 Goodness-of-Fit Tests

Interval	Upper Value	Sample Frequency	Theoretical Frequency	Contribution to Statistic
		(a) *Equal Intervals*		
1	5	9	8.51	0.028
2	10	4	5.61	0.462
3	15	6	3.70	1.429
4	20	2	2.44	0.079
5	∞	4	4.74	0.115
Total statistic				2.113
		(b) *Equal Probabilities*		
1	2.682	4	5	0.2
2	6.140	5	5	0
3	11.014	4	5	0.2
4	19.345	8	5	1.8
5	∞	4	5	0.2
Total statistic				2.4

where θ is replaced by the MLE $\hat{\theta}$ and n is the sample size. For our case, $\hat{\theta}$ is 1/12.02.

There are two variations of the χ^2 test: equal intervals or equal probabilities. The first has equal values for $i^+ - i^-$: in our case above, 5.0. The second has equal values for e_i, so that the $i^+ - i^-$ values vary. We perform both versions, and the results are given in Table 7.7. For the tail frequency, we combined the last eight intervals into one interval, >20 (20, ∞), for the equal-interval test. For the equal-probability version, we set the intervals so that the theoretical frequency in each would be 5.

Since there are three degrees of freedom, the critical value at the 5% level is 7.815 (see any statistical textbook), and neither test rejects the hypothesis that the data come from an exponential distribution.

Great care should always be exercised in doing χ^2 goodness-of-fit tests, and the analyst would, of course, be well advised to study for a definitive treatise on the subject in the statistical literature. The basic weaknesses of the χ^2 test are its requirement for large samples (our case of 25 is much too small), its heavy dependence upon the choice of the number and position of the time-axis intervals, and its possibly very high Type 2 error (i.e., the probability of accepting a false hypothesis) for some feasible alternative distributions.

Another popular goodness-of-fit test is the Kolmogorov–Smirnov (KS) test. The KS test compares deviations of the empirical CDF from the theoretical CDF, and uses as its test statistic a modified maximum absolute deviation, namely,

$$K = \max_j \max\left\{ \left| \frac{j}{n} - F(t_j) \right|, \left| \frac{j-1}{n} - F(t_j) \right| \right\}, \tag{7.38}$$

where t_j is the jth ordered (ascending) observation, and $F(t_j)$ is the value of the hypothesized distribution function at the jth observation. Unfortunately, as we see, the test presupposes that the CDF F is completely known. In case there is interest in performing a KS test, tables for critical values of the test statistic are widely available in the published literature. If, however, the parameters of the hypothesized CDF are unknown and are to be estimated from the data, then a special KS table must be established, or the test statistic must be modified for the particular family from which F came, or both. For example, see the work of Lilliefors (1967, 1969) on the normal and exponential respectively. Use of the general KS table for distributions with estimated means will give very conservative results in the sense that the actual significance level achieved will be much lower than that indicated. For our example (again we point out that 25 observations is too small a sample size to do any meaningful goodness-of-fit testing, but it serves our purpose for illustration of the technique), we find that the maximum deviation K obtained from applying Equation (7.38) is 0.11739. Modifying this according to Stephens (1974), who suggests using as the test statistic

$$\tilde{K} = \left(K - \frac{0.2}{n} \right)\left(\sqrt{n} + 0.26 + \frac{0.5}{\sqrt{n}} \right)$$

and using the modified tables (see Law and Kelton, 1991), the 5%-level critical value is 1.094, and since $\tilde{K}$ turns out to be $(0.117 - 0.2/25)(5 + 0.26 + 0.5/5) = 0.415$, we again have no reason to reject the hypothesis.

A variation on the KS test is the Anderson–Darling (AD) test, which instead of using only the maximum deviation as KS does, uses all the deviations (actually, a weighted average of the squared deviation, with the weights being the largest at the tails of the distribution). Again, special tables are available for certain distributions, the exponential being one of them. For our example, the test statistic turns out to be 0.30045, and the critical value at the 5% level is 1.29, so that again we cannot reject the hypothesis.

Although goodness-of-fit tests require large sample sizes to really discriminate, and many are limited if we have to estimate parameters from the sample, there are specific tests for the exponential distribution. Since the exponential distribution is so important in analytical queueing modeling, we would like to point out a specific test for exponentiality which is quite powerful (power meaning ability to discern false hypotheses) against almost any alternative hypothesis and will usually outperform these other tests. It is the F test.

To perform the F test, r ($\doteq n/2$) and $n - r$ of a set of n interoccurrence times t_i are randomly grouped. It follows that the quantity

$$F = \frac{\sum_{i=1}^{r} t_i / r}{\sum_{i=r+1}^{n} t_i / (n - r)} \tag{7.39}$$

is the ratio of two Erlangs and is distributed as an F distribution with $2r$ and $2(n - r)$ degrees of freedom when the hypothesis of exponentiality is true. Therefore a two-tailed F test will be performed on the F calculated from a set of data in order to determine whether the stream is indeed truly exponential. This argument for the F test can be extended easily to the case in which there are randomly occurring incomplete interoccurrence periods, as is common in repairlike problems. Tables of critical points for the F distribution are available in most standard statistics books. We again illustrate this test on our sample data above.

Since the data are in random order, we sum the first 13 observations to get 201.3 and the last 12 to get 99.2. Calculating the F statistic according to (7.39) gives $F = (201.3/13)/(99.2/12) = 1.873$. The 95% critical values for $F_{26,24}$ are $1/2.23 \doteq 0.45$ and 2.26, respectively, so we accept the hypothesis that the data are exponential.

To close this discussion on input modeling, we mention that there are software packages which will run data and recommend the distribution which "best" represents the data (e.g., ExpertFit by Law and Associates, 1995). One note of caution is that most of these packages use ML to estimate parameters, and have multiple criteria (besides goodness-of-fit testing) to select their recommendations. Often the model selected as best has considerable differences in the second and higher moments from those of the sample. We know from certain analytical theory (e.g., PK formula and heavy-traffic approximations) that the first and second moments are very important, and for these cases they are the *only* thing that matters—the actual distribution does not even enter into the formulas. Juttijudata (1996) and Gross and Juttijudata (1997) have shown how important moments are, and we caution the reader against choosing a distribution for which the theoretical moments (especially the first three or four) differ significantly from those of the sample.

Many simulation modelers recommend that, instead of trying to choose a theoretical probability distribution, one simply use the empirical distribution, that is, the sample histogram (this is very closely akin to what is referred to in statistical circles as *bootstrapping*). For a discussion on the pros and cons of this debate, see Fox (1981) and Kelton (1984).

Generation of Random Variates

Once the appropriate probability distributions are selected for representing the input processes (interarrival and service times), it is necessary to generate typical observations from them for "running" the simulated system. Determining how many to generate (i.e., how long to observe the simulated system) will be treated in a later section.

The procedure for generating IID random variates from a given specified probability distribution, say $f(x)$, generally consists of two phases: (1) generation of pseudorandom nubers distributed uniformly on (0, 1) and (2) using the pseudorandom numbers to obtain variates (observations) from $f(x)$. We first discuss generation of uniform (0, 1) pseudorandom numbers and then how to use these to generate random variates from a specified $f(x)$.

Most computer programming languages contain a pseudorandom-number generator. These are "pseudo" in that they are completely reproducible by a mathematical algorithm, but "random" in the sense that they have passed statistical tests, which basically test for equal probability of all values and statistical independence. Most computer routines are based on linear-congruential methods which involve modulo arithmetic. It is a recursive algorithm of the form

$$r_{n+1} = (kr_n + a) \bmod m, \qquad (7.40)$$

where k, a, and m are positive integers ($k < m$, $a < m$). That is, r_{n+1} is the remainder when $kr_n + a$ is divided by m. We must choose an initial value, r_0, which is called the *seed*, and this should be less than m.

For example, if $k = 4$, $a = 0$, and $m = 9$, and we choose r_0 initially as 1, we generate the numbers.

$$1, 4, 7, 1, 4, 7, 1, 4, 7, \ldots.$$

Since the smallest number (remainder on division by 9) could be 0 and the largest number could be 8, the range is [0–8]. To normalize to (0, 1), all numbers are divided by $m = 9$—note that since 0 is a possibility, we are really normalizing to [0, 1). The results are then

$$0.111, 0.444, 0.778, 0.111, 0.444, 0.778, \ldots.$$

It is clear that this sequence is not acceptable. First of all, only three of the possible nine numbers appear. Second we see that the sequence is cyclic with a cycle length of three. If one desired more than three random numbers, this sequence would be unusable. So let us instead change k to 5, a to 3, and m to 16 with an r_0 of 7 (an example from Law and Kelton, 1991), which yields

$$7, 6, 1, 8, 11, 10, 5, 12, 15, 14, 9, 0, 3, 2, 13, 4, 7, 6, 1, 8, \ldots.$$

Here all possible numbers [0–15], are generated, so that we have a *full* cycle generator, but the cycle length is only 16 [the maximum cycle length of any stream using Equation (7.40) is $m - 1$]. To normalize on $[0, 1)$, we divide by 16 [if we truly desire uniform random numbers on $(0, 1)$, then we can discard the 0's generated].

Thus we see that very careful consideration must be given to selecting k, a, and m (and to some extent r_0 also). Some values of m which which seem to work well and appear in simulation software packages are $2^{31} - 1$ and 2^{48}, which work well with 32- and 64-bit machines in accomplishing the modulo arithmetic. For a more detailed discussion of random-number generation, the interested reader is referred to Fishman (1978), Law and Kelton (1991), and Rubinstein (1981).

We desire now to generate representative observations from any specified probability distribution, with CDF (say) $F(x)$. Again, there are several methods of achieving this. We present the most popular and refer the reader to the aforementioned references for further detail, if desired.

The main method we offer is sometimes referred to as the inverse or probability transformation method, or generation by inversion. It can best be described graphically by considering a plot of the CDF from which we desire to generate random variates. Such a plot is shown in Figure 7.4. The procedure is to first generate uniform $(0, 1)$ random variates, say $r_1, r_2, \ldots$. To obtain x_1, the first random variate corresponding to $F(x)$, we simply enter the ordinate with r_1 and project over and down, as shown in Figure 7.4; the resulting value from the abscissa is x_1. Repeating the procedure with r_2, $r_3 \ldots$ will yield $x_2, x_3, \ldots$.

To prove that this procedure works, we would like to show that a random variate (say X_i) generated by this procedure obeys the relation $\Pr\{X_i \leq x\} =$

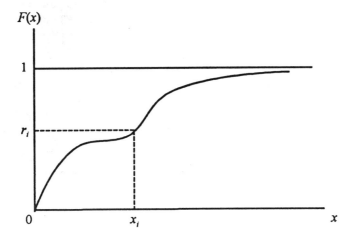

Fig. 7.4 Inversion technique for generating random variates.

$F(x)$. We have, considering Figure 7.4, that $\Pr\{X_i \le x\} = \Pr\{R_i \le F(x)\}$. Since R_i is uniform $(0, 1)$, we can write $\Pr\{R_i \le F(x)\} = F(x)$; hence $\Pr\{X_i \le x\} = F(x)$. Note that this procedure holds for discrete distributions as well. For this case, the CDF will be a step function, thus yielding only discrete values of X.

For some theoretical distributions, the inversion can be obtained analytically in closed form. For example, consider the exponential distribution with parameter θ. Its CDF is given by $F(x) = 1 - e^{-\theta x}, x \ge 0$. Entering the ordinate with a uniform $(0, 1)$ random number r and finding the resulting x after the projection procedure amounts to solving the following equation for x:

$$r = 1 - e^{-\theta x} \quad \Rightarrow \quad e^{-\theta x} = 1 - r.$$

Since r is uniform $(0, 1)$, it is immaterial whether we use r or $1 - r$ as our random number, and hence we can write $e^{-\theta x} = r$. Taking natural logarithms of both sides finally gives

$$x = \frac{-\ln r}{\theta}. \tag{7.41}$$

Unfortunately, analytical inversion is not possible for all probability distributions. In the case of some continuous distributions, we can find alternative ways to generate the variates, sometimes using numerical integration techniques, or more often making use of the fundamentals of probability theory to aid in the random-variate generation procedure. As an example of how we may take advantage of statistical theory, we examine the Erlang type k. Instead of attempting inversion on the Erlang CDF, we can merely take sums of k exponential random variates, which are quite easy to generate by inversion, as we have shown above. This was actually done in generating the data for Problem 3.18 of Chapter 3. Thus if we desired to generate random variates from an Erlang type k with mean $1/\mu$, we could obtain this type of random variate, say x, from the uniform $(0, 1)$ random variates $r_1, r_2, \ldots, r_k$ by

$$x = \sum_{i=1}^{k} \left(-\frac{\ln r_i}{k\mu} \right) = -\frac{\ln \Pi_{i=1}^{k} r_i}{k\mu}.$$

Another useful continuous distribution in the modeling of queues is the mixed exponential. To generate mixed exponential variates according to the density

$$f(t) = \sum_{i=1}^{n} p_i \lambda_i e^{-\lambda_i t},$$

we recognize that each observation is essentially the result of two independent probabilistic events. That is, we first select the relevant exponential subpopulation using the discrete mixing probabilities $\{p_i\}$, and then, given that the jth population is indeed selected, an exponential variate is generated from (7.41) using mean $1/\lambda_j$. For procedures for generating other random variates (e.g., normal, gamma with general shape parameters, lognormals, etc.), we refer the reader to Law and Kelton (1991, Chapter 8).

It is possible, therefore, to generate random variates from any probability distribution using procedures such as those given above, although in some cases it may be time-consuming and/or approximate. Nevertheless, in most cases it can be done without too much difficulty, and most modern simulation software has procedures for doing so.

We next turn our attention to consideration of the bookkeeping phase of a simulation analysis.

7.4.3   Bookkeeping Aspects of Simulation Analysis

As mentioned earlier, the bookkeeping phase of a simulation model must keep track of the transactions moving around the system, and set up counters on ongoing processes in order to calculate various measures of system performance. The simulation modeler has a large variety of languages and packages from which to choose. These can be categorized into three main types: general-purpose languages, simulation languages, and simulators. General-purpose languages such as FORTRAN, C++, BASIC, etc., allow the most flexibility in modeling, but require the most effort to program. One can get a feel from the very simple example given in Section 7.4.1 of the programming effort involved in generating variates from the input probability distributions, keeping track of which transactions are at what places at what times, and performing the statistical calculations needed for obtaining output measures of performance.

Requirements for bookkeeping, random-variate generation, and data collection necessary for statistical analyses of output are similar for large classes of simulation models, and this situation has given rise to the development of a variety of simulation language packages. These simulation language packages (e.g., GPSS/H, SIMAN, SIMSCRIPT II.5, and SLAM, to mention a few) make programming a simulation model much, much easier. However, some flexibility in modeling is sacrificed, since the model must fit into the general language environment. For most cases, this is not a problem. As a general rule, one can expect that the easier the programming becomes, the less flexibility there is in deviating from the language environment.

Most of the popular simulation languages use a next-event approach to bookkeeping (as opposed to fixed time increments), in that the master clock is advanced, as in Example 7.2, to the next event scheduled to occur, rather than the clock being advanced in fixed increments of time, where, for many of the increments, nothing may have happened (see Tables 1.4 and 7.6). Further, most of the event-oriented routines employ a transaction-process technique which keeps track of the entire experience of a transaction as it proceeds along its way in the system.

Even easier to use than simulation languages are packages called *simulators*. These require very little, if any, programming, and the model can usually be built by choosing among icons, filling out tables, etc. However, these offer the least flexibility in modeling in that the modeling environment is quite fixed—for example, a manufacturing plant, or a communications network. Some examples of simulators are SIMFACTORY, ProModel, and WITNESS for manufacturing environments and NETWORK and OPNET for communications environments. See INFORMS (1997, pp. 38–46) for a comprehensive listing of simulation software.

7.4.4 Output Analysis

Reaching reliable conclusions from simulation output requires a great deal of thought and care. When simulating stochastic systems, a single run yields output values which are statistical in nature, so that sound experimental design and sound statistical analyses are required for valid conclusions. Unlike sampling from a population in the classical sense, where great effort is made to have random samples with independent observations, we often purposely induce correlation in simulation modeling as a variance reduction technique, so that classical, off-the-shelf statistical techniques for analyzing sample data are often not appropriate. We present next some basic procedures for analyzing simulation output.

There are two major types of simulation models: terminating and continuing (nonterminating). A terminating model has a natural start and stop time, for example, a bank opens its doors at 9:00 A.M. and closes its doors at 3:00 P.M. (Example 7.2 is also a terminating model). On the other hand, a continuing model does not have a start and stop time, for example, a manufacturing process where, at the beginning of a shift, things are picked up exactly as they were left at the end of the previous shift, so that, in a sense, the process runs continuously. In these latter cases, steady-state results are usually of interest, and in simulating such a system, a determination must be made as to when the initial transients are damped out and the simulation is in steady state.

Considering first a terminating simulation such as Example 7.2, from a single run we cannot make any statistical statements. For example, the maximum waiting time is a single observation, that is, a sample of one. What can be done is to *replicate* the experiment (repeated runs), using different random number streams for the order arrival and processing times for each run, and thus generate a sample of independent observations to which classical statistics can be applied. Assuming we replicate n times, we then will have n values for the maximum waiting time, say, $w_1, w_2, \ldots, w_n$. Assuming n is large enough to employ the central limit theorem, we can get a

$100(1 - \alpha)\%$ confidence interval (CI) by first calculating the mean and sample standard deviation of the maximum waiting time by

$$\bar{w} = \frac{\sum_{i=1}^{n} w_i}{n}$$

and

$$s_w = \sqrt{\frac{\sum_{i=1}^{n} (w_i - \bar{w})^2}{n - 1}},$$

and then obtaining the CI as

$$\left[\bar{w} - \frac{t(n - 1, 1 - \alpha/2)s_w}{\sqrt{n}}, \quad \bar{w} + \frac{t(n - 1, 1 - \alpha/2)s_w}{\sqrt{n}} \right],$$

where $t(n - 1, 1 - \alpha/2)$ is the upper $1 - \alpha/2$ critical value for the t distribution with $n - 1$ degrees of freedom.

For continuing simulations for which we are interested in steady-state results, we know from the ergodic theory of stochastic processes that

$$\lim_{T \to \infty} \frac{1}{T} \int_0^T X^n(t) \, dt = E[X^n],$$

so that if we run long enough, we shall get close to the limiting average value. But it is not clear how long is long enough, and we often wish to be able to obtain a CI statement. Thus, we have two additional problems: determining when we reach steady state and deciding when to terminate the simulation run. Assuming for the moment that these problems are solved and we decide to run the simulation for n transactions after reaching steady state and measure the time a customer spends waiting in a particular queue for service, we can obtain n queue wait values, which we shall again denote by w_i (now these are actual waits, not maximum waits). It might be tempting to calculate the average and standard deviation of these n values and proceed as above to form a CI. However, these w_i are correlated, and using the above formula for s_w greatly underestimates the true variance. There are procedures for estimating the required correlations and obtaining an estimate of the standard deviation from these correlated data, but this requires a great deal of estimation from this single data set, which has its drawbacks in statistical precision.

To get around the correlation problem, we may again replicate the run m times, using a different random number seed each time, as we did in the terminating case. For each run, we still calculate the mean of the w_i, denoting the mean for the jth replication by, $\bar{w}_j$, that is,

$$\bar{w}_j = \frac{\sum_{i=1}^{n} w_{ij}}{n},$$

where w_{ij} is the waiting time for transaction i on replication j, $i = 1, 2, \ldots, n$ and $j = 1, 2, \ldots, m$. The $\bar{w}_j$ are now independent and in an analogous fashion to that of the terminating simulation, we can form a $100(1 - \alpha)\%$ CI by calculating

$$\bar{w} = \frac{\sum_{j=1}^{m} \bar{w}_j}{m}$$

and

$$s_{\bar{w}_j} = \sqrt{\frac{\sum_{j=1}^{m} (\bar{w}_j - \bar{w})^2}{m - 1}},$$

and then the CI becomes

$$\left[\bar{w} - \frac{t(m - 1, 1 - \alpha/2)s_{\bar{w}_j}}{\sqrt{m}}, \quad \bar{w} + \frac{t(m - 1, 1 - \alpha/2)s_{\bar{w}_j}}{\sqrt{m}} \right].$$

Returning to the previous two problems mentioned for continuing simulations (namely, when steady state is reached and when to stop each run), we first discuss the latter and then the former. The run length (n) and the number of replications (m) will both influence the size of the standard error $(s_{\bar{w}_j}/\sqrt{m})$ required for making the CI above. The smaller the standard error, the more precise is the CI (narrower limits) for a given confidence $(1 - \alpha)$. We know that the standard error goes down by the square root of m, so that the more replications made, the more precise the CI. Also, we might expect that as the run length n is increased, the computed value of $s_{\bar{w}_j}$ itself will be smaller for a given number of replications, so that longer run lengths will also increase the precision of the CI. Thus for a fixed amount of computer running time, we can trade off size of n versus size of m. Setting n and m is still something of an art, and trial runs can be made to evaluate the tradeoffs.

The warmup period (the initial amount of time required to bring the process near steady-state conditions) is also not an easy thing to determine. The basic idea is that if we could compute the time or number of transactions required so that the process is near steady state, we could simply not start recording data for calculating output measures until the master clock passed that point. A variety of procedures have been developed, and the reader is referred to the basic simulation texts referenced previously.

One particular method that seems to work fairly well (again, the warmup period analysis is still more of an art than a science) is that due to Welch (1981, 1983). The procedure involves choosing one of the output performance measures (e.g., average waiting time for service), calculating means of this measure over the replications for *each* transaction (i.e., if we have m replications of run length n, we average the m values we obtain for transaction i from each replication, for $i = 1, 2, \ldots, n$), taking moving averages of neighboring values of these transaction averages, plotting these, and visually determining when the graph appears to be stabilizing. It is recommended to try various moving average windows (the number of adjacent points in the moving average). The size of the moving average window and the point at which the graph settles down are again judgment calls.

Another approach using the transaction averages calculated as described above is a regression approach suggested by Kelton and Law (1983). The transaction average data stream of n values is segmented into b batches. A regression line is fitted for the last batch, and if the slope of the line is not significantly different from zero, steady state is assumed. Then the next to last batch of data is included in the regression, the test made again, and if not significantly different from zero, the last $2b$ values are said to be in steady state. This procedure is continued, and when the test for slope becomes significantly different from zero, that batch and all preceeding ones are considered to be in the transient region. This approach is predicated on the assumption that the performance measure is monotonic in time, which should be the case when the initial conditions assume the system starts empty and idle. Again, decisions on the values for n, m, and b must be made subjectively.

A third approach differs from the two previous approaches in that, rather than attempting to find a point at which the process enters steady state, the bias of the transient effects is estimated in order to determine if the data do show an initial conditions bias (Schruben, 1982). This procedure could be used in conjunction with one of the two above to check whether the warmup period chosen was adequate to remove the initial-conditions bias.

In order to avoid throwing away the initial warmup period for each of the m replications in a nonterminating simulation experiment, the procedure of *batch means* has been suggested (Law, 1977; Schmeiser, 1982). Rather than replicating, a single long run (say mn transactions) is made and then broken up into m segments (batches) of n each. The performance measures for the segments are assumed to be approximately independent (if the segments are long enough, the correlation between segments should be small), so that the classical estimate of the standard deviation can be employed, that is, the segments act as if they were independent replications. The methodology for determining the CIs is identical to that for m independent replications. But now, one only has to discard a single warmup period instead of m as before.

Other methods have been suggested, such as the regenerative method (Crane and Iglehart, 1975; Fishman, 1973a, b), and time series analyses (Fishman, 1971; Schruben, 1982).

In comparing two alternative system designs, the technique most commonly used is a paired t CI on the difference of a given performance measure for each design. For example, if we are simulating a queueing system, and one design has two servers serving at a particular rate at a service station in the system and a competing design replaces the two servers with automatic machines, we may be interested in the average holding time of a transaction.

We make a run for design 1, calculate the average holding time, make a run for design 2, calculate the average holding time, and then compute the difference between the two average holding times. Then we replicate the pair of runs m times, obtaining m differences, d_i, $i = 1, 2, \ldots, m$. The mean and standard deviation of the d_i are calculated and the t distribution used to form a $100(1 - \alpha)\%$ CI on the mean difference in a manner analogous to that described above, yielding

$$\left[\bar{d} - \frac{t(m - 1, 1 - \alpha/2)s_{d_i}}{\sqrt{m}}, \quad \bar{d} + \frac{t(m - 1, 1 - \alpha/2)s_{d_i}}{\sqrt{m}} \right].$$

Whenever possible, the same random number stream(s) should be used for each design *within* a replication, so that the difference observed depends only on the design-parameter change and not the variation due to the randomness of the random variates generated. Of course, different random numbers streams are used *between* the replications. This is a *variance reduction technique* (VRT) called *common random numbers* (discussed in more detail below) and is quite effective in narrowing the CI limits.

Often, we wish to compare more than two designs, which necessitates using multiple comparison techniques. There are a variety of procedures which can be of help. All systems could be compared in a pairwise fashion using the methodology for comparing only two systems. However, if the confidence level of a CI for single pair is $1 - \alpha$, and we have k pairs, the confidence associated with a statement concerning all the pairs simultaneously drops to $1 - k\alpha$ (Bonferroni inequality). Therefore, if we desire the overall confidence to be $1 - \alpha$, then it is necessary to have the confidence for each pair be $1 - \alpha/k$.

Also available are a variety of ranking and selection procedures, such as selecting the best of k systems, selecting a subset of size r containing the best of the k systems, or selecting the r best of k systems. These are treated in some detail in Law and Kelton (1991).

We now turn our attention to some variance reduction techniques (VRTs). Unlike sampling from the real world, the simulation modeler has control over the randomness generated in the system. Often, purposely introducing correlation among certain of the random variates in a simulation run can reduce variance and provide narrower CIs. One example of this was mentioned above in forming a paired t CI by using *common random numbers* (CRNs) within a replication, which introduces positive correlation between the two performance measures within a replication, yielding a smaller variance of the mean difference over the replications.

Another technique, called *antithetic variates* (AVs), introduces negative correlation between two successive replications of a given design with the idea that a large random value in one of the pairs will be offset by a small random value in the other. The performance measures for the pairs are averaged to give a single "observed" performance measure. Hence, if m

replications are run, only $m/2$ independent values end up being averaged for the CI calculation, but the variance of these values should be considerably lower than for m independent observations.

Among other variance reduction techniques are *indirect estimation*, *conditioning*, *importance sampling*, and *control variates*, and again we refer the reader to one of the basic simulation texts.

One drawback mentioned at the beginning of this simulation section was the difficulty in finding an optimal system design. Multiple comparison procedures will help us in finding the best of those tried, but finding the true optimal design is quite difficult. There has been quite a bit of attention paid lately to sensitivity analyses and optimization, and we refer the reader to Rubinstein and Melamed (1997). In spite of many excellent researchers working in this area, it is still a very complex and difficult task, much more so than for analytical modeling.

We end this discussion with the very important topic of *model validation*.

7.4.5   Model Validation

Model validation is a very important step in a simulation study which often is glossed over by modelers. Prior to embarking on developing a simulation model, it behooves the simulation analyst to become very familiar with the system being studied, to involve the managers and operating personnel of the system, and thus to agree on the level of detail required to achieve the goal of the study. The appropriate level of detail is always the coarsest that can still provide the answers required. One problem with simulation modeling is that since any level of detail can be modeled, models are often developed in more detail than necessary and this can be very inefficient and counterproductive.

Validity is closely associated with *verification* and *credibility*. Verification has to do with program debugging to make sure the computer program does what is intended. This is generally the most straightforward of the three goals to accomplish, as there are well-known and established methods for debugging computer programs.

Validation deals with how accurate a representation of reality the model provides, and credibility deals with how believable the model is to the users. To establish validity and credibility, users must be involved in the study early and often. Goals of the study, appropriate system performance measures,and level of detail must be agreed upon and kept as simple as possible. A log book of assumptions should be kept, updated frequently, and signed off periodically by the model builders and users.

When possible, simulation model output should be checked against actual system performance, if the system being modeled is in operation. If the model can duplicate (in a statistical sense) *actual* data, both validity and credibility are advanced. If no system currently exists, then if the model can be run under conditions where theoretical results are known (e.g., in studying a queueing system, one can compare the simulation results with known queueing-theoretic results), and if the simulation results duplicate theoretical results, then verification is comfirmed. The model can be run under a variety of conditions, and results examined by the users for plausibility, thus providing some validity and credibility checks. Most simulation texts have at least one chapter devoted to this important topic. Other references include Carson (1986), Gass and Thompson (1980), Sargent (1988), and Schruben (1980).

PROBLEMS

Whenever a problem is best solved by use of this book's accompanying software, we have added a boldface **C** *to the right of the problem number.*

7.1.   The Bearing Straight Corporation (of Example 5.2 and Problem 6.4) is now in bad straits because it has found that an estimated machine-breakdown rate as low as 5/hr was optimistic and that a more realistic estimate would be 6/hr. It is then observed that the service rate is also 6/hr, based on a time of 9 min two out of three times and 12 min one-third of the time. Use the heavy-traffic approximation of Section 7.2.5 to determine what decrease below the 9-min figure would guarantee Bearing machines an average wait twice what it was before, namely, now equal to 72 min.

7.2.   Use Equation (7.5) to derive the mean line wait for a stationary $E_j/E_k/1$ queue.

7.3.   Derive the inequality given by (7.6) from Equation (7.5).

7.4C. Assemblies come to an inspection station with a single inspector from two production lines. The randomly merged interarrival times coming into the inspection station are found to be hyperexponential [see Problem 5.28(a)] with $q = 0.25$ and $1/\lambda = 6.5$ min. Approximately one-fourth of the assemblies come from line 1, and three-fourths from line 2. The inspection time for an assembly from line 1 is found to be exponential with a mean time to inspect of 9 min, and for an assembly from line 2 is found to be exponential with a mean time to inspect of 5 min. The inspector chooses items to inspect on a FCFS basis, so that the probability that the next item chosen is a line-1 item is $\frac{1}{4}$, and the probability that it is a line 2-item is $\frac{3}{4}$. The production supervisor is interested to know what the worst-case wait (total time at the inspection station) might be.

7.5.   Employ the bounds on W_q derived in Section 7.1 for $D/M/1$.

7.6.   Do a Markov-chain exact analysis to find the stationary delay distribution for a $D/G/1$ with interarrival times of 2 min and two equiprobable service times of 0 and 3 min.

7.7.   Find upper and lower bounds for the W_q of the $D/G/1$ queue of Problem 7.6, and compare with the exact answer.

7.8.   Considering Problem 7.1, what decrease below 9 min would guarantee Bearing machines waits of more than 400 min less than 5% of the time?

7.9.   Our moonlighting graduate assistant of Problem 2.11 now finds in the new semester that the arrival rate to his counter has increased to 20/hr. If his service times remain exponential with mean 4 min, use the result of Section 7.2 to find the approximate probability that his Nth customer (N large) will have to wait. Then compare this result with what you would have obtained via the Chebyshev approximation.

7.10. Show that the density function $p(x, t \mid x_0)$ given by (7.20) satisfies the diffusion partial differential equation (7.19).

7.11. Apply the central limit theorem to the one-dimensional random walk which moves left with probability q, moves right with probability p, and stands still with probability r. Then use the limiting procedures of Section 7.3 which led to Equation (7.20) to show that the continuous problem is a Wiener process, and that this Wiener density satisfies an equation of the same form as (7.19).

7.12. For a machine-repair problem, $M = Y = c = 1$, $\lambda = 1$, $\mu = 1.5$, find $p_2(t)$, $p_1(t)$, and $p_0(t)$ for $t = \frac{1}{2}$ and 1, where $p_0(0) = 1$, $p_2(0) = p_1(0) = 0$:

(a) exactly by using the Laplace transform,
(b) by Euler's method using $\Delta t = 0.10$, and
(c) by the randomization procedure using $\epsilon = 0.01$.

7.13. For the previous problem, find the approximate steady-state probability distribution by:

(a) Jacobi stepping on the uniformized imbedded discrete-parameter Markov chain with transition probability matrix $\tilde{P}$, and
(b) Gauss–Seidel stepping using $\tilde{P}$.
(c) In view of the results of the randomization technique used in Problem 7.12(c), comment on the merit, in this case, of using it to obtain $p(t)$ for t suitably large.

7.14. Show that $F(t)$ for the Erlang can be written as Equation (7.36).

7.15. Show that $h(t)$ for the Erlang can be written as Equation (7.37).

7.16. Use a pseudorandom-number generator from your spreadsheet software to create five observations from the following distributions:
(a) Uniform between 5 and 15.
(b) Exponential, mean 5.
(c) Erlang type 3, mean 5.

7.17. Use a pseudorandom-number generator from your spreadsheet software to create six observations from the following distributions:
(a) Mixed exponential, with mixing probabilities $(\frac{1}{3}, \frac{2}{3})$ and subpopulation means of 5 and 10.
(b) Gamma, with shape parameter 2.5 and mean 5.

7.18. Use a pseudorandom-number generator from your spreadsheet software to create five observations from the following distributions:
(a) The triangular distribution, where

$$f(x) = \begin{cases} 2x/3 & (0 \le x \le 1), \\ 1 - x/3 & (1 \le x \le 3). \end{cases}$$

(b) Poisson, with mean 2.
(c) The distribution given by

x:	0	1	2	3	4	5
$f(x)$:	0	$\frac{1}{10}$	0	$\frac{3}{10}$	$\frac{2}{10}$	$\frac{4}{10}$

7.19C. Reformulate the SIM1 simulator module in QTS to replace the Erlang interarrival and service times with the distributions given below and simulate this single-server queue:

Interarrival time, min:	4	5	6
Probability of occurrence:	.1	.3	.6

Service time, min:	4	5	6
Probability of occurrence:	.5	.3	.2

Now estimate L, L_q, W, and W_q.

7.20. Rewrite the SIM1 simulator module, which is based on the iterative equation for the waiting time of the $(n + 1)$st customer in terms of the nth (see Section 6.2), in a computer language of your choice, so that run length is not as restrictive as that of the QTS spreadsheet. Determine waiting times for an $M/G/1$ queue. You may use any service-time distribution you please as long as you remember to keep ρ less than one. Start with a run size of 10,000 customers, and compare your answer with that expected from the PK formula. Then increase your run size as befits your computer and continue comparisons. What is your conclusion?

7.21. Write an event-oriented simulation program (as opposed to the recursion of SIM1), in the general programming language of your choice, giving the expected system size and waiting time in queue for a single-channel queueing model where interarrival times and service times are to be generated from subroutines to be specified by the user. Check out the program by using exponential interarrival and service times and comparing results with the known formula for $M/M/1/\infty$ given in Chapter 2.

7.22. Use a simulation language (e.g., GPSS/H, SIMCRIPT II.5, SLAM, SIMON, etc.) to program the model of Problem 7.21. Compare programming effort.

7.23. Write a simulator in the language of your choice for a single-server machine-repair problem, with exponential lifetimes of mean 2, exponential service times also with mean 2, and four machines.

7.24. Suppose you have programmed a general single-channel queueing simulator which allows for any input and service patterns. To validate the model, you decide to make runs with exponential interarrival times (mean 10) and exponential service times (mean 8). The following are results of average system-size calculations for 20 replications:

5.21, 3.63, 4.18, 2.10, 4.05, 3.17, 4.42, 4.91, 3.79, 3.01, 3.71, 2.98, 4.31, 3.27, 3.82, 3.41, 5.00, 3.26, 3.19, 3.63.

Based on these values, what can you conclude?

7.25. Autos are attempting to merge onto a high-speed highway. Cars in the highway merge lane proceed along with separations, which are IID exponential variables with mean 2.5 sec. The lead car on the ramp is accepted whenever a gap passes which is ≥ 1.5 sec. Cars arrive on the ramp as a Poisson process with mean interarrival time of 7 sec. Write a simulation program to estimate the mean waiting time in the ramp queue.

7.26. Write a simulation program for estimating the mean stationary waiting time of a $G/G/1$ queue with interarrival and service times distributed as follows. The interarrival CDF is the mixed exponential

$$A(t) = 1 - \frac{e^{-t}}{2} - \frac{e^{-2t}}{2},$$

while service times are distributed as

$$B(t) = 1 - e^{-(4n/3)t},$$

where n is the number in the system at the instant service begins.

7.27. In comparing two alternative system designs via a queueing simulation model, the results in Table 7.8 on mean waiting times under each design for 15 replications are obtained. For each replication, the two designs were compared on the same random number stream. Does it appear that one design may be preferable?

7.28. Suppose that in Problem 7.27, design 2 is the current design. Another alternative design, which we shall call design 3, is proposed. Results for 15 replications (using the same random number streams as before) are presented in Table 7.9. Is Design 3 any better than current

Table 7.8   Data for Problem 7.27

Replication Number	Mean Waiting Time	
	Design 1	Design 2
1	23.02	23.97
2	25.16	24.98
3	19.47	21.63
4	19.06	20.41
5	22.19	21.93
6	18.47	20.38
7	19.00	21.97
8	20.57	21.31
9	24.63	23.17
10	23.91	23.09
11	27.19	26.93
12	24.61	24.82
13	21.22	22.18
14	21.37	21.99
15	18.78	20.61

practice? Comment from both a statistical and a practical point of view.

Table 7.9

Replications 1–5:	23.91, 24.95, 21.52, 20.37, 21.90,
Replications 6–10:	20.17, 21.90, 21.26, 23.10, 23.02,
Replications 11–15:	26.90, 24.67, 22.09, 21.91, 20.60.

7.29. The manager of your local supermarket is having problems in determining the number of checkers (and baggers) to have on Saturday morning, one of her busiest periods. Because your spouse is always complaining about the wait and challenges you to put some of your queueing-theory education to use, you undertake the task of aiding the manager. You decide—because of the complexities of jockeying and reneging, of the ability to add and subtract baggers (which decreases or increases the μ of a channel), and of other complexities, to model the system by simulation. Develop the simulation model, and program it in any language you desire. Try to estimate, by actual observation over several Saturday mornings, the arrival and service patterns as well as the queue discipline. Use part of the observations to develop empirical distributions for the simulation and validate the simulator using the remaining observations. Then attempt to determine the solutions to the manager's problems.

7.30. Take a system that gives you great annoyance (registration for college courses, barber or beauty shop, local cafeterias, getting a free computer terminal, etc.), and build a simulation model of the system with an eye toward determining a better system design.

REFERENCES

Abramowitz, M., and Stegun, I. A. (1964). *Handbook of Mathematical Functions.* Applied Mathematics Series 55. Washington, DC: National Bureau of Standards.

Allen, A. O. (1990). *Probability, Statistics, and Queueing Theory with Computer Science Applications*, 2nd ed. New York: Academic Press.

Ash, R. B. (1972). *Real Analysis and Probability.* New York: Academic Press.

Ashour, S., and Jha, R. D. (1973). "Numerical Transient-State Solutions of Queueing Systems." *Simulation* **21**, 117–122.

Bailey, N. T. J. (1952). "Study of Queues and Appointment Systems in Out-patient Departments with Special Reference to Waiting Times." *J. Roy. Statist. Soc. Ser. B.* **14**, 185–199.

Bailey, N. T. J. (1954). "A Continuous Time Treatment of a Single Queue Using Generating Functions." *J. Roy. Statist. Soc. Ser. B.* **16**, 288–291.

Banks, J., Carson, J. S., and Nelson, B. L. (1996). *Discrete-Event System Simulation*, 2nd ed. Upper Saddle River, NJ: Prentice-Hall.

Barbour, A. D. (1976). "Networks of Queues and the Methods of Stages." *Adv. Appl. Prob.* **8**, 584–591.

Barlow, R. E., and Proschan, F. (1975). *Statistical Theory of Reliability and Life Testing.* New York: Holt, Rinehart and Winston.

Barrer, D. Y. (1957a). "Queuing with Impatient Customers and Indifferent Clerks." *Oper. Res.* **5**, 644–649.

Barrer, D. Y. (1957b). "Queuing with Impatient Customers and Ordered Service." *Oper. Res.* **5**, 650–656.

Baskett, F., Chandy, K. M., Muntz, R. R., and Palacios, F. G. (1975). "Open, Closed and Mixed Networks of Queues with Different Classes of Customers." *J. Assoc. Comp. Mach.* **22**, 248–260.

Beneš, V. E. (1957). "A Sufficient Set of Statistics for a Simple Telephone Exchange Model." *Bell. Syst. Tech. J.* **36**, 939–964.

Beneš, V. E. (1963). *General Stochastic Processes in the Theory of Queues.* Reading, MA: Addison-Wesley.

Beneš, V. E. (1965). *Mathematical Theory of Connecting Networks and Telephone Traffic.* New York: Academic Press.

Bengtsson, B. (1983). "On Some Control Problems for Queues." Linköping Studies in Science and Technology, Dissertation No. 87.

Bertsekas, D., and Gallager, R. (1992). *Data Networks*, 2nd ed. Upper Saddle River, NJ: Prentice-Hall.

Bhat, U. N. (1964). "Imbedded Markov Chain Analysis of Single-Server Bulk Queues." *J. Aust. Math. Soc.* **4**, 244–263.

Bhat, U. N. (1967). "Some Explicit Results for the Queue $GI/M/1$ with Group Service." *Sankhyā Ser. A.* **29**, 199–206.

Bhat, U. N. (1968a). "Transient Behavior of Multi-server Queues with Recurrent Input and Exponential Service Times." *J. Appl. Prob.* **5**, 158–168.

Bhat, U. N. (1968b). *A Study of the Queueing Systems M/G/1 and GI/M/1.* Lecture Notes in Operations Research and Mathematical Economics, No. 2. Berlin: Springer-Verlag.

Bhat, U. N. (1969a). "Queueing Systems with First-Order Dependence." *Opsearch* **6**, 1–24.

Bhat, U. N. (1969b). "Sixty Years of Queueing Theory." *Manage. Sci.* **15**, B280–B292.

Bhat, U. N. (1984). *Elements of Applied Stochastic Processes*, 2nd ed. New York: Wiley.

Bhat, U. N., and Rao, S. S. (1972). "A Statistical Technique for the Control of Traffic Intensity in the Queuing Systems M/G/1 and GI/M/1." *Oper. Res.* **20**, 955–966.

Bhat, U. N., Shalaby, M., and Fischer, M. J. (1979). "Approximation Techniques in the Solution of Queueing Problems." *Nav. Res. Log. Quart.* **26**, 311–326.

Billingsley, P. (1961). *Statistical Inference for Markov Processes*. Chicago: University of Chicago Press.

Blackburn, J. D. (1972). "Optimal Control of a Single-Server Queue with Balking and Reneging." *Manage. Sci.* **19**, 297–313.

Bookbinder, J. H., and Martell, P. L. (1979). "Time-Dependent Queueing Approach to Helicopter Allocation for Forest Fire Initial Attack." *Infor.* **17**, 58–70.

Borofsky, S. (1950). *Elementary Theory of Equations*. New York: Macmillan.

Botta, R. F., and Harris, C. M. (1980). "Approximation with Generalized Hyperexponential Distribution: Weak Convergence Results." *Queueing Syst.* **1**, 169–190.

Brigham, G. (1955). "On a Congestion Problem in an Aircraft Factory." *J. Oper. Res. Soc. Amer.* **3**, 412–428.

Brill, P. H., and Posner, M. J. M. (1977). "Level Crossings in Point Processes Applied to Queues: The Single-Server Case." *Oper. Res.* **25**, 662–674.

Brosh, I. (1970). "The Policy Space Structure of Markovian Systems with Two Types of Service." *Manage. Sci.* **16**, 607–621.

Bruell, S. C., and Balbo, G. (1980). *Computational Algorithm for Closed Queueing Networks*. Operating and Programming Systems Series, P. J. Denning (Ed.). New York, Oxford: North Holland.

Brumelle, S. L. (1971). "Some Inequalities for Parallel-Server Queues." *Oper. Res.* **19**, 402–413.

Buffa, E. S. (1966). *Readings in Production and Operations Management*. New York: Wiley.

Bunday, B. D., and Scraton, R. E. (1980). "The $G/M/r$ Machine Interference Model." *Euro. J. Oper. Res.* **4**, 399–402.

Burke, P. J. (1956)." The Output of a Queueing System." *Oper. Res.* **4**, 699–714.

Burke, P. J. (1964). "The Dependence of Delay in Tandem Queues." *Ann. Math. Statist.* **35**, 874–875.

Burke, P. J. (1969). "The Dependence of Service in Tandem $M/M/s$ Queues." *Oper. Res.* **17**, 754–755.

Buzen, J. P. (1973). "Computational Algorithms for Closed Queueing Networks with Exponential Servers." *Comm. ACM* **16**, 527–531.

Carson, J. S. (1986). "Convincing Users of Model's Validity Is Challenging Aspect of Modeler's Job." *Ind. Eng.* **18**, 74–85.

Champernowne, D. G. (1956). "An Elementary Method of Solution of the Queueing Problem with a Single Server and a Constant Parameter." *J. Roy. Statist. Soc. Ser. B* **18**, 125–128.

Chaudhry, M. L., and Templeton, J. G. C. (1983). *A First Course in Bulk Queues*. New York: Wiley.

Chaudhry, M. L., Harris, C. M., and Marchal, W. G. (1990). "Robustness of Rootfinding in Single-Server Queueing Models." *ORSA J. Comp.* **2**, 273–286.

Çinlar, E. (1975). *Introduction to Stochastic Processes*. Englewood Cliffs, NJ: Prentice-Hall.

Clarke, A. B. (1957). "Maximum Likelihood Estimates in a Simple Queue." *Ann. Math. Stat.* **28**, 1036–1040.

Cobham, A. (1954). "Priority Assignment in Waiting Line Problems." *Oper. Res.* **2**, 70–76; correction, **3**, 547.

Cochran, W. G., and Cox, G. M. (1957). *Experimental Designs*, 2nd ed. New York: Wiley.

Cohen, J. W. (1982). *The Single Server Queue*, 2nd ed. New York: North Holland.

Conway, R. W., and Maxwell, W. L. (1962). "A Queueing Model with State Dependent Service Rates." *J. Ind. Eng.* **12**, 132–136.

Conway, R. W., Maxwell, W. L., and Miller, L. W. (1967). *Theory of Scheduling*. Reading, MA: Addison-Wesley.

Cooper, R. B. (1981). *Introduction to Queueing Theory*, 2nd ed. New York: North Holland.

Cooper, R. B., and Gross, D. (1991). "On the Convergence of Jacobi and Gauss–Seidel Iteration for Steady-State Probabilities of Finite-State Continuous-Time Markov Chains." *Stochastic Models* **7**, 185–189.

Cox, D. R. (1955a). "A Use of Complex Probabilities in the Theory of Stochastic Processes." *Proc. Camb. Phil. Soc.* **51**, 313–319.

Cox, D. R. (1955b). "The Analysis of Non-Markovian Stochastic Processes by the Inclusion of Supplementary Variables." *Proc. Camb. Phil. Soc.* **51**, 443–441.

Cox, D. R. (1965). "Some Problems of Statistical Analysis Connected with Congestion." In *Proceedings of the Symposium on Congestion Theory*, W. L. Smith and W. E. Wilkinson (Eds). Chapel Hill, NC: University of North Carolina Press.

Crabill, T. B. (1968). "Sufficient Conditions for Positive Recurrence of Specially Structured Markov Chains." *Oper. Res.* **16**, 858–867.

Crabill, T. B. (1972). "Optimal Control of a Service Facility with Variable Exponential Service Time and Constant Arrival Rate." *Manage. Sci.* **18**, 560–566.

Crabill, T. B., and Maxwell, W. L. (1969). "Single Machine Sequences with Random Processing Times and Random Due-Dates." *Nav. Res. Log. Quart.* **16**, 549–554.

Crabill, T. B., Gross, D., and Magazine, M. (1977). "A Classified Bibliography of Research on Optimal Design and Control of Queues." *Oper. Res.* **28**, 219–232.

Crane, M. A., and Iglehart, D. L. (1974a). "Simulating Stable Stochastic Systems I: General Multi-server Queues." *J. Assoc. Comput. Mach.* **21**, 103–113.

Crane, M. A., and Iglehart, D. L. (1974b). "Simulating Stable Stochastic Systems, II: Markov Chains." *J. Assoc. Comput. Mach.* **21**, 114–123.

Crane, M. A., and Iglehart, D. L. (1975). "Simulating Stable Stochastic Systems, III: Regenerative Processes and Discrete-Event Simulations." *Oper. Res.* **23**, 33–45.

Crommelin, C. D. (1932). "Delay Probability Formulae When the Holding Times Are Constant." *P. O. Elec. Eng. J.* **25**, 41–50.

Davenport, W. B., Jr., and Root, W. L. (1958). *An Introduction to the Theory of Random Signals and Noise.* New York: McGraw-Hill.

Davis, O. L. (1956). *The Design and Analysis of Industrial Experiments.* London: Oliver and Boyd.

De Smit, J. H. A. (1973). "Some General Results for Many Server Queues." *Adv. Appl. Prob.* **5**, 153–169.

Disney, R. L. (1981). "Queueing Networks." *Amer. Math. Soc. Proc. Symp. Appl. Math.* **25**, 53–83.

Disney, R. L. (1996). "Networks of Queues" In *Encyclopedia of Operations Research & Managment Science*, S. I. Gass and C. M. Harris (Eds). Boston: Kluwer Academic.

Disney, R. L., and Franken, P. (1982). "Further Comments on Some Queueing Inequalities." *EIK* **18**, 595–602.

Disney, R. L., McNickle, D. C., and Simon, B. (1980). "The *M/G/1* Queue with Instantaneous Bernoulli Feedback." *Nav. Res. Log. Quart.* **27**, 635–644.

Eilon, S. (1969). "A Simpler Proof of $L = \lambda W$." *Oper. Res.* **17**, 915–917.

Emshoff, J. R., and Sisson, R. L. (1970). *Design and Use of Computer Simulation Models.* New York: Macmillan.

Erlang, A. K. (1909). "The Theory of Probabilities and Telephone Conversations." *Nyt Tidsskrift Mat. B* **20**, 33–39.

Erlang, A. K. (1917). "Solution of Some Problems in the Theory of Probabilities of Significance in Automatic Telephone Exchanges." *Electroteknikeren* (Danish) **13**, 5–13. [English translation, *P. O. Elec. Eng. J.* **10**, 189–197 (1917–1918).]

Evans, R. V. (1971). "Programming Problems and Changes in the Stable Behavior of a Class of Markov Chains." *J. Appl. Prob.* **8**, 543–550.

Fabens, A. T. (1961). "The Solution of Queueing and Inventory Models by Semi-Markov Processes." *J. Roy. Statist. Soc. Ser. B* **23**, 113–127.

Feller, W. (1957). *An Introduction to Probability Theory and Its Applications*, vol. 1. New York: Wiley.

Feller, W. (1971). *An Introduction to Probability Theory and Its Applications*, vol. 2, 2nd ed. New York: Wiley.

Ferziger, J. H. (1981). *Numerical Methods for Engineering Applications*. New York: Wiley.

Fishman, G. S. (1967). "Problems in the Statistical Analysis of Simulation Experiments: The Comparison of Means and the Length of Sample Records." *Comm. ACM* **10**, 94–99.

Fishman, G. S. (1968). "The Allocation of Computer Time in Comparing Simulation Experiments." *Oper. Res.* **16**, 280–295.

Fishman, G. S. (1971). "Estimating Sample Size in Computer Simulation Experiments." *Manage. Sci.* **18**, 21–38.

Fishman, G. S. (1972). "Bias Considerations in Simulation Experiments." *Oper. Res.* **20**, 785–790.

Fishman, G. S. (1973a). "Statistical Analysis for Queueing Simulations." *Manage. Sci.* **20**, 363–369.

Fishman, G. S. (1973b). *Concepts and Methods in Discrete Event Digital Simulation*. New York: Wiley.

Fishman, G. S. (1974). "Estimation in Multi-server Queueing Simulations." *Oper. Res.* **22**, 72–78.

Fishman, G. S. (1978). *Principles of Discrete Event Simulation*. New York: Wiley.

Fisz, M. (1963). *Probability Theory and Mathematical Statistics*, 3rd ed. New York: Wiley.

Foster, F. G. (1953). "On Stochastic Matrices Associated with Certain Queuing Processes." *Ann. Math. Stat.* **24**, 355–360.

Fox, B. L. (1981). "Fitting 'Standard' Distributions to Data is Necessarily Good: Dogma or Myth?" In *Proceedings of the 1981 Winter Simulation Conference*. Institute of Electrical and Electronic Engineers, Piscataway, New Jersey, 305–307.

Fry, T. C. (1928). *Probability and Its Engineering Uses*. Princeton, NJ: Van Nostrand.

Galliher, H. P., and Wheeler, R. C. (1958). "Nonstationary Queuing Probabilities for Landing Congestion of Aircraft." *Oper. Res.* **6**, 264–275.

Gass, S. I., and Thompson, B. W. (1980). "Guidelines for Model Evaluation." *Oper. Res.* **28**, 431–439.

Gaver, D. P., Jr. (1959). "Imbedded Markov Chain Analysis of a Waiting Line Process in Continuous Time." *Ann. Math. Statist.* **30**, 698–720.

Gaver, D. P., Jr. (1966), "Observing Stochastic Processes and Approximate Transform Inversion." *Oper. Res.* **14**, 444–459.

Gaver, D. P., Jr. (1968). "Diffusion Approximations and Models for Certain Congestion Problems." *J. Appl. Prob.* **5**, 607–623.

Gear, C. W. (1969). "The Automatic Integration of Stiff Ordinary Differential

Equations." In *Information Processing 68*, A. J. H. Morrell (Ed). Amsterdam: North Holland, 187–193.

Gear, C. W. (1971). "DIFSUB for Solution of Ordinary Differential Equations." *Comm. ACM* **14**, 185–190.

Gebhard, R. F. (1967). "A Queueing Process with Bilevel Hysteretic Service-Rate Control." *Nav. Res. Log. Quart.* **14**, 55–68.

Gnedenko, B. V., and Kovalenko, I. N. (1989). *Introduction to Queueing Theory*, 2nd ed. Translated by S. Kotz. Boston: Birkhäuser.

Gordon, G. (1978). *System Simulation*, 2nd ed. Englewood Cliffs, NJ: Prentice-Hall.

Gordon, W. J., and Newell, G. F. (1967). "Closed Queuing Systems with Exponential Servers." *Oper. Res.* **15**, 254–265.

Grassmann, W. (1977). "Transient Solutions in Markovian Queueing Systems." *Comput. & Oper. Res.* **4**, 47–56.

Greenberg, I. (1973). "Distribution-Free Analysis of $M/G/1$ and $G/M/1$ Queues." *Oper. Res.* **21**, 629–635.

Greenberg, I. (1977). "Single Server Queues with Hyperexponential Service Times." *Nav. Res. Log. Quart.* **24**, 451–455.

Gross, D., and Harris, C. M. (1971). "On One-for-One Ordering Inventory Policies with State-Dependent Leadtimes." *Oper. Res.* **19**, 735–760.

Gross, D., and Harris, C. M. (1985). *Fundamentals of Queueing Theory*, 2nd ed. New York: Wiley.

Gross, D., and Ince, J. (1981). "The Machine Repair Problem with Heterogeneous Populations." *Oper. Res.* **29**, 532–549.

Gross, D., and Juttijudata, M. (1997). "Sensitivity of Output Measures to Input Distributions in Queueing Simulation Modeling." In *Proceedings of the 1977 Winter Simulation Conference*, Institute of Electrical and Electronics Engineers, Piscataway, New Jersey.

Gross, D., and Miller, D. R. (1984). "The Randomization Technique as a Modeling Tool and Solution Procedure for Transient Markov Processes." *Oper. Res.* **32**, 343–361.

Gross, D., Miller, D. R., and Soland, R. M. (1983). "A Closed Queueing Network Model for Multi-echelon Repairable Item Provisioning." *IIE Trans.* **15**, 344–352.

Gross, D., Kioussis, L. C., Miller, D. R., and Soland, R. M. (1984). "Computational Aspects of Determining Steady-State Availability for Markovian Multi-echelon Repairable Item Inventory Models." T-483/84, IMSE, The George Washington University, Washington, DC.

Haight, F. A. (1957). "Queueing with Balking, I." *Biometrika* **46**, 360–369.

Haight, F. A. (1959). "Queueing with Reneging." *Metrika* **2**,186–197.

Harris, C. M. (1967)." Queues with State-Dependent Stochastic Service Rates." *Oper. Res.* **15**, 117–130.

Harris, C. M. (1974). "Some New Results in the Statistical Analysis of Queues." In *Mathematical Methods in Queueing Theory*, A. B. Clarke (Ed). Berlin: Springer-Verlag.

Harris, C. M. (1985). "A Note on Mixed Exponential Approximations for $GI/G/1$ Queues." *Comput. & Oper. Res.* **12**, 285–289.

Harris, C. M., and Marchal, W. G. (1988). "State Dependence in $M/G/1$ Server-Vacation Models." *Oper. Res.* **36**, 560–565.

Heffer, J. C. (1969). "Steady State Solution of the $M/E_k/c$ (∞, FIFO) Queueing System." *CORS J.* **7**, 16–30.

Henrici, P. (1962). *Discrete Variable Methods in Ordinary Differential Equations.* New York: Wiley.

Heyman, D. P. (1968). "Optimal Operating Policies for $M/G/1$ Queuing Systems." *Oper. Res.* **16**, 362–382.

Heyman, D. P., and Sobel, M. J. (1982). *Stochastic Models in Operations Research,* vol. I. New York: McGraw-Hill.

Heyman, D. P., and Sobel, M. J. (1984). *Stochastic Models in Operations Research,* vol. II. New York: McGraw-Hill.

Heyman, D. P., and Sobel, M. J. (Eds). (1990). *Stochastic Models.* Handbooks in Oper. Res., vol. 2. Amsterdam: North-Holland.

Hildebrand, F. B. (1949). *Advanced Calculus for Engineers.* Englewood Cliffs, NJ: Prentice-Hall.

Hildebrand, F. B. (1952). *Methods of Applied Mathematics.* Englewood Cliffs, NJ: Prentice-Hall.

Hillier, F. S., and Lieberman, G. J. (1995). *Introduction to Operations Research,* 6th ed. New York: McGraw-Hill.

Hoover, S. V., and Perry, R. F. (1989). *Simulation: A Problem-Solving Approach.* Reading, MA: Addison-Wesley.

Hunt, G. C. (1956). "Sequential Arrays of Waiting Lines." *Oper. Res.* **4**, 674–683.

Iglehart, D. L. (1972). "Extreme Values in the $GI/G/1$." *Ann. Math. Statist.* **49**, 627–635.

INFORMS (1997). "Simulation Software Survey." *OR/MS Today*, October. Linthicum, MD: INFORMS.

Jackson, J. R. (1957). "Networks of Waiting Lines." *Oper. Res.* **5**, 518–521.

Jackson, J. R. (1963). "Jobshop-like Queueing Systems." *Manage. Sci.* **10**, 131–142.

Jaiswal, N. K. (1968). *Priority Queues.* New York: Academic Press.

Jewell, W. S. (1967). "A Simple Proof of $L = \lambda W$." *Oper. Res.* **15**, 1109–1116.

Juttijudata, M. (1996). "Sensitivity of Output Performance Measures to Input Distributions in Queueing Simulation Modeling." D.Sc. Dissertation, Department of Operations Research, The George Washington University, Washington, DC.

Kao, E. P. C. (1991). "Using State Reduction for Computing Steady State Probabilities of $GI/PH/1$ Types." *ORSA J. Comput.* **3**, 231–240.

Karlin, S. (1966). *A First Course in Stochastic Processes.* New York: Academic Press.

Karlin, S., and McGregor, J. L. (1957). "The Classification of Birth and Death Processes." *Trans. Amer. Math. Soc.* **86**, 366–400.

Karlin, S., and Taylor, H. M. (1975). *A First Course on Stochastic Processes.* New York: Academic Press.

Kelly, F. P. (1975). "Networks of Queues with Customers of Different Types." *J. Appl. Prob.* **12**, 542–554

Kelly, F. P. (1976). "Networks of Queues." *Adv. Appl. Prob.* **8**, 416–432.

Kelly, F. P. (1979). *Reversibility and Stochastic Networks*. New York: Wiley.

Kelton, W. D. (1984). "Input Data Collection and Analysis." In *Proceedings of the 1984 Winter Simulation Conference*, Institute of Electrical and Electronics Engineers, Piscataway, New Jersey, 305–307.

Kelton, W. D., and Law, A. M. (1983). "A New Approach for Dealing with the Startup Problem in Discrete Event Simulation." *Nav. Res. Log. Quart.* **30**, 641–658.

Kendall, D. G. (1951). "Some Problems in the Theory of Queues." *J. Roy. Statist. Soc. Ser. B* **13**, 151–185.

Kendall, D. G. (1953). "Stochastic Processes Occurring in the Theory of Queues and Their Analysis by the Method of Imbedded Markov Chains." *Ann. Math. Statist.* **24**, 338–354.

Kendall, D. G. (1964). "Some Recent Work and Further Problems in the Theory of Queues." *Theory Prob. Appl.* **9**, 1–15.

Kennedy, D. P. (1972). "The Continuity of the Single-Server Queue." *J. Appl. Prob.* **9**, 370–381.

Kesten, H., and Runnenburg, J. T. (1957). "Priority Waiting Line Problems I, II." *Koninkz. Ned. Akad. Wetenschap. Proc. Ser. A* **60**, 312–336.

Khintchine, A. Y. (1932). "Mathematisches über die Erwortung vor einemöffenthchen Schalter." *Mat. Sb.* **39**, 73–84.

Kingman, J. F. C. (1962). "Some Inequalities for the Queue $GI/G/1$." *Biometrika* **49**, 315–324.

Kleinrock, L. (1975). *Queueing Systems, Vol. 1: Theory*. New York: Wiley.

Kleinrock, L. (1976). *Queueing Systems, Vol. 2: Computer Applications*. New York: Wiley.

Koenigsberg, E. (1966). "On Jockeying in Queues." *Manage. Sci.* **12**, 412–436.

Kolmogorov, A. N. (1931). "Sur le problème d'attente." *Mat. Sb.* **8**,101–106.

Kosten, L. (1948–1949). "On the Validity of the Erlang and Engset Loss Formulae." *Het P.T.T. Bedriff* **2**, 22–45.

Krakowski, M. (1973). "Conservation Methods in Queueing Theory." *RAIRO* **7**(V-1), 63–84.

Krakowski, M. (1974). "Arrival and Departure Processes in Queues. Pollaczek–Khintchine Formulas for Bulk Arrivals and Bounded Systems." *RAIRO* **8**(V-1), 45–56.

Lavenberg, S. S., and Reiser, M. (1979). "Stationary State Probabilities at Arrival Instants for Closed Queueing Networks with Multiple Types of Customers." Research Report RC 759. IBM T. J. Watson Research Center, Yorktown Heights, NY.

Law, A. M. (1977). "Confidence Intervals in Discrete Event Simulation: A Comparison of Replication and Batch Means." *Nav. Res. Log. Quart.* **27**, 667–678.

Law, A. M., and Kelton, W. D. (1991). *Simulation Modeling and Analysis*, 2nd ed. New York: McGraw-Hill.

Ledermann, W., and Reuter, G. E. (1954). "Spectral Theory for the Differential Equations of Simple Birth and Death Process." *Phil. Trans. Roy. Soc. London Ser. A* **246**, 321–369.

Lee, A. M. (1966). *Applied Queueing Theory*. Montréal: St. Martin's Press.

Leemis, L. M. (1996). "Discrete-Event Simulation Input Process Modeling." In *Proceedings 1996 Winter Simulation Conference*, San Diego, CA, Institute of Electrical and Electronics Engineers, Piscataway, New Jersey, 39–46.

Lemoine, A. J. (1977). "Networks of Queues—A Survey of Equilibrium Analysis." *Manage. Sci.* **24**, 464–481.

Liitschwager, J., and Ames, W. F. (1975). "On Transient Queues—Practice and Pedagogy." In *Proceedings 8th Symposium on the Interface*, Los Angeles, 206.

Lilliefors, H. W. (1966). "Some Confidence Intervals for Queues." *Oper. Res.* **14**, 723–727.

Lilliefors, H. W. (1967). "On the Kolmogorov–Smirnov Statistic for Normality with Mean and Variance Unknown." *J. Amer. Statist. Assoc.* **62**, 399–402.

Lilliefors, H. W. (1969). "On the Kolmogorov–Smirnov Statistic for the Exponential Distribution with Mean Unknown." *J. Amer. Statist. Assoc.* **64**, 387–389.

Lindley, D. V. (1952). "The Theory of Queues with a Single Server." *Proc. Camb. Phil. Soc.* **48**, 277–289.

Little, J. D. C. (1961). "A Proof for the Queuing Formula $L = \lambda W$." *Oper. Res.* **9**, 383–387.

Marchal, W. G. (1974). "Some Simple Bounds and Approximations in Queueing." D.Sc. Dissertation, Department of Operations Research, The George Washington University, Washington, DC.

Marchal, W. G. (1978). "Some Simpler Bounds on the Mean Queuing Time." *Oper. Res.* **26**, 1083–1088.

Maron, M. J. (1982). *Numerical Analysis, A Practical Approach*. New York: Macmillan.

Marshall, K. T. (1968). "Some Inequalities in Queuing." *Oper. Res.* **16**, 651–665.

Mayhugh, J. O., and McCormick, R. E. (1968). "Steady-State Solution of the Queue $M/E_k/r$." *Manage. Sci.* **14**, 692–712.

Mehdi, J. (1991). *Stochastic Models in Queueing Theory*. Boston: Academic Press.

Melamed, B. (1979). "Characterization of Poisson Traffic Streams in Jackson Queueing Networks." *Adv. Appl. Prob.* **11**, 422–438.

Miller, B. I. (1969). "A Queueing Reward System with Several Customer Classes." *Manage. Sci.* **16**, 234–245.

Miller, D. R. (1981). "Computation of the Steady-State Probabilities for $M/M/1$ Priority Queues." *Oper. Res.* **29**, 945–958.

Moder, J. J., and Phillips, C. R., Jr. (1962). "Queuing with Fixed and Variable Channels." *Oper. Res.* **10**, 218–231.

Molina, E. C. (1927). "Application of the Theory of Probability to Telephone Trunking Problems." *Bell Syst. Tech. J.* **6**, 461–494.

Montgomery, D. C. (1976). *Design and Analysis of Experiments*. New York: Wiley.

Montgomery, D. C., and Bettencourt, Jr., V. M. (1977). "Multiple Response Surface Methods in Computer Simulation." *Simulation* **29**, 113–121.

Morse, P. M. (1958). *Queues, Inventories and Maintenance*. New York: Wiley.

Nelson, R. (1995). *Probability, Stochastic Processes, and Queueing Theory*. New York: Springer-Verlag.

Neuts, M. F. (1965). "The Busy Period of a Queue with Batch Service." *Oper. Res.* **13**, 815–819.

Neuts, M. F. (1966). "An Alternative Proof of a Theorem of Takács on the *GI/M/*1 Queue." *Oper. Res.* **14**, 313–316.

Neuts, M. F. (1973). "The Single Server Queue in Discrete Time—Numerical Analysis, I." *Nav. Res. Log. Quart.* **20**, 297–304.

Neuts, M. F. (1981). *Matrix-Geometric Solutions in Stochastic Models*. Baltimore: Johns Hopkins University Press.

Newell, G. F. (1972). *Applications of Queueing Theory*. London: Chapman and Hall.

Olmsted, J. M. H. (1959). *Real Variables*. New York: Appleton-Century-Crofts.

Palm, C. (1938). "Analysis of the Erlang Traffic Formulae for Busy-Signal Arrangements." *Ericsson Tech.* **6**, 39–58.

Panico, J. (1969). *Queuing Theory: A Study of Waiting Lines for Business, Economics, and Science*. Englewood Cliffs, NJ: Prentice-Hall.

Papoulis, A. (1991). *Probability, Random Variables and Stochastic Processes*, 2nd ed. New York: McGraw-Hill.

Parzen, E. (1960). *Modern Probability and Its Applications*. New York: Wiley.

Parzen, E. (1962). *Stochastic Processes*. San Francisco: Holden-Day.

Perros, H. (1994). *Queueing Networks with Blocking*. New York: Oxford University Press.

Phipps, T. E., Jr. (1956). "Machine Repair As a Priority Waiting-Line Problem." *Oper. Res.* **4**, 76–85. (Comments by W. R. Van Voorhis, **4**, 86.)

Pollaczek, F. (1932). "Lösung eines Geometrischen Wahrscheinlichkeits-problems." *Math. Z.* **35**, 230–278.

Pollaczek, F. (1934). "Über das Warteproblem." *Math. Z.* **38**, 492–537.

Posner, M., and Bernholtz, B. (1968). "Closed Finite Queueing Networks with Time Lags." *Oper. Res.* **16**, 962–976.

Prabhu, N. U. (1965a). *Stochastic Processes*. New York: Macmillan.

Prabhu, N. U. (1965b). *Queues and Inventories*. New York: Wiley.

Prabhu, N. U. (1974). "Stochastic Control of Queueing Systems." *Nav. Res. Log. Quart.* **21**, 411–418.

Prabhu, N. U. (1980). *Stochastic Storage Processes*. New York: Springer-Verlag.

Prabhu, N. U. (1997). *Foundations of Queueing Theory*. Boston: Kluwer Academic.

Prabhu, N. U., and Bhat, U. N. (1963a). "Further Results for the Queue with Poisson Arrivals." *Oper. Res.* **11**, 380–386.

Prabhu, N. U., and Bhat, U. N. (1963b). "Some First Passage Problems and Their Application to Queues." *Sankhyā Ser. A* **25**, 281–292.

Pritsker, A. A. B., and Pegden, C. D. (1979). *Introduction to Simulation and SLAM*. West Lafayette, IN: Systems Publ.

Puterman, M. L. (1990). "Markov Decision Processes." In *Stochastic Models*, D. P. Heyman, and M. J. Sobel (Eds). Handbooks in Oper. Res., vol. 2. Amsterdam: North Holland.

Puterman, M. L. (1991). *Markov Decision Processes*. New York: Wiley.

Pyke, R. (1961). "Markov Renewal Processes." *Ann. Math. Statist.* **32**, 1231–1242.

Rainville, E. D., and Bedient, P E. (1969). *A Short Course in Differential Equations*. New York: Macmillan.

Rao, S. S. (1968). "Queueing with Balking and Reneging in M/G/1 Systems." *Metrika* **12**, 173–188.

Reich, E. (1957). "Waiting Times When Queues Are in Tandem." *Ann. Math. Statist.* **28**, 768–773.

Romani, J. (1957). "Un Modelo de la Teoria de Colas con Número Variable de Canales." *Trabajos Estadestica* **8**, 175–189.

Rosenshine, M. (1967). "Queues with State-Dependent Service Times." *Transp. Res.* **1**, 97–104.

Rosenshine, M. (1968). "Operations Research in the Solution of Air Traffic Control Problems." *J. Ind. Eng.* **19**, 122–128.

Ross, S. M. (1970). *Applied Probability Models with Optimization Examples*. San Francisco: Holden-Day.

Ross, S. M. (1996). *Stochastic Processes*, 2nd ed. New York: Wiley.

Rubinstein, R. Y. (1981). *Simulation and the Monte Carlo Method*. New York: Wiley.

Rubinstein, R. Y. (1986). *Monte Carlo Optimization, Simulation and Sensitivity of Queueing Networks*. New York: Wiley.

Rubinstein, R. Y., and Melamed, B. (1997). *Modern Simulation and Modeling*. New York: Wiley.

Rue, R. C., and Rosenshine, M. (1981). "Some Properties of Optimal Control Policies for Entries to an *M/M/*1 Queue." *Nav. Res. Log. Quart.* **28**, 525–532.

Saaty, T. L. (1961). *Elements of Queueing Theory with Applications*. New York: McGraw Hill.

Saaty, T. L. (1966). "Seven More Years of Queues: A Lament and a Bibliography." *Nav. Res. Log. Quart.* **13**, 447–476.

Sargent, R. G. (1988). "A Tutorial on Validation and Verification of Simulation Models." In *Proceedings of the 1988 Winter Simulation Conference*, Institute of Electrical and Electronics Engineers, Piscataway, New Jersey, 33–39.

Schmeiser, B. W. (1982). "Batch Size Effects in the Analysis of Simulation Output." *Oper. Res.* **30**, 556–568.

Schrage, L. E., and Miller, L. W. (1966). "The Queue *M/G/*1 with the Shortest Remaining Processing Time Discipline." *Oper. Res.* **14**, 670–684.

Schriber, T. J. (1991). *Introduction to Simulation*. New York: Wiley.

Schruben, L. W. (1980). "Establishing the Credibility of Simulations." *Simulation* **34**, 101–105.

Schruben, L. W. (1982). "Detecting Initialization Bias in Simulation Output." *Oper. Res.* **30**, 569–590.

Serfozo, R. F. (1981). "Optimal Control of Random Walks, Birth and Death Processes, and Queues." *Adv. Appl. Prob.* **13**, 61–83,

Serfozo, R. F., and Lu, F. V. (1984). "*M/M/*1 Queueing Decision Processes with Monotone Hysteretic Optimal Policies." *Oper. Res.* **32**, 1116–1132.

Sevick, K. C., and Mitrani, I. (1979). "The Distribution of Queueing Network States at Input and Output Instants." In *Proceedings of the 4th International Symposium on Modelling and Performance Evaluation of Computer Systems*, Vienna.

Simon, B., and Foley, R. D. (1979). "Some Results on Sojourn Times in Cyclic Jackson Networks." *Manage. Sci.* **25**, 1027–1034.

Smith, W. L. (1953). "On the Distribution of Queueing Times." *Proc. Camb. Phil. Soc.* **49**, 449–461.

Smith, W. L., and Wilkinson, W. E. (Eds). (1965). *Proceedings Symposium on Congestion Theory*. Chapel Hill, NC: University of North Carolina Press.

Sobel, M. J. (1969). "Optimal Average Cost Policy for a Queue with Start-up and Shut-down Costs." *Oper. Res.* **17**, 145–162.

Sobel, M. J. (1974). "Optimal Operation of Queues." In *Mathematical Methods in Queueing Theory*, A. B. Clarke (Ed). Lecture Notes in Economics and Mathematical Systems 98. Berlin: Springer-Verlag, 231–261

Stephens, M. A. (1974). "EDF Statistics for Goodness of Fit and Some Comparisons." *J. Amer. Statist. Assoc.* **69**, 730–737.

Stidham, S., Jr. (1970). "On the Optimality of Single-Server Queuing Systems." *Oper. Res.* **18**, 708–732.

Stidham, S., Jr. (1972). "$L = \lambda W$: A Discounted Analogue and a New Proof." *Oper. Res.* **20**, 1115–1126.

Stidham, S., Jr. (1982). "Optimal Control of Arrivals to Queues and Network of Queues." Paper presented at the 21st IEEE Conference on Decision and Control, Orlando, FL.

Stidham, S., Jr., and Prabhu, N. U. (1974). "Optimal Control of Queueing Systems." In *Mathematical Methods in Queueing Theory*, A. B. Clarke (Ed). Lecture Notes in Economics and Mathematical Systems 98. Berlin: Springer-Verlag, 263–294.

Syski, R. (1960). *Introduction to Congestion Theory in Telephone Systems*. London: Oliver and Boyd.

Takács, L. (1955). "Investigations of Waiting Time Problems by Reduction to Markov Processes." *Acta Math. Acad. Sci. Hung.* **6**, 101–129.

Takács, L. (1962). *Introduction to the Theory of Queues*. Oxford, England: Oxford University Press.

Takács, L. (1967). *Combinatorial Methods in the Theory of Stochastic Processes*. New York: Wiley.

Takács, L. (1969). "On Erlang's Formula." *Ann. Math. Statist.* **40**, 71–78.

Taylor, H., and Karlin, S. (1984). *An Introduction to Stochastic Modeling*. Orlando, FL: Academic Press.

Thorndyke, F. (1926). "Application of Poisson's Probability Summation." *Bell Syst. Tech. J.* **5**, 604–624.

Vandergraft, J. S. (1983). "Fluid Flow of Networks of Queues." *Manage. Sci.* **29**, 1198–1208.

van Dijk, N. M. (1993). *Queueing Networks and Product Forms*: *A Systems Approach*. New York: Wiley.

Vaulot, A. E. (1927). "Extension des Formules d'Erlang au Cas où les Durées des

Conversations Suivent une Loi Quelconque." *Rev. Gén. Electricité* **22**, 1164–1171.

Walrand, J. (1988). *An Introduction to Queueing Networks*. Englewood Cliffs, NJ: Prentice-Hall.

Welch, P. D. (1981). "On the Problem of the Initial Transient in Steady-State Simulation." Technical Report, IBM T. J. Watson Research Center, Yorktown Heights, NY.

Welch, P. D. (1983). "The Statistical Analysis of Simulation Results." In *The Computer Performance Modeling Handbook*, S. S. Lavenberg (Ed). New York: Academic Press.

Whitt, W. (1974). "The Continuity of Queues." *Adv. Appl. Prob.* **6**, 175–183.

Wolff, R. W. (1965). "Problems of Statistical Inference for Birth and Death Queuing Models." *Oper. Res.* **13**, 343–357.

Wolff, R. W. (1989). *Stochastic Modeling and the Theory of Queues*. Englewood Cliffs, NJ: Prentice-Hall.

Yadin, M., and Naor, P. (1967). "On Queueing Systems with Variable Service Capacities." *Nav. Res. Log. Quart.* **14**, 43–54.

APPENDIX 1

Symbols and Abbreviations

This appendix contains definitions of common symbols and abbreviations used frequently and consistently throughout the text. Symbols that are used only occasionally in isolated sections of the text are not always included here. The symbols are listed in alphabetical order. Greek symbols are filed according to their English names; for example, λ (lambda) is found under L. Within an alphabetical category, Latin symbols precede Greek. Listed at the end are nonliteral symbols such as primes and asterisks.

$A/B/X/Y/Z$	Notation for describing queueing models, where A indicates interarrival pattern B indicates service pattern, X indicates number of channels Y, indicates system capacity limit, Z indicates queue discipline
a.s.	Almost surely
$A(t)$	Cumulative distribution function of interarrival times
$a(t)$	Probability density of interarrival times
$B(t)$	Cumulative distribution function of service times
$b(t)$	Probability density of service times
b_n	(1) Probability of n services during an interarrival time; (2) discouragement function in queueing models with balking
C	(1) Arbitrary constant; (2) cost per unit time, often used in functional notation as $C(\cdot)$; (3) coefficient of variation ($\equiv \sigma/\mu$)
c	Number of parallel channels (servers)
CDF	Cumulative distribution function
CK	Chapman–Kolmogorov
CTMC	Continuous-time Markov chain
CV	Coefficient of variation ($\equiv \sigma/\mu$)
C_s	Marginal cost of a server per unit time

c_n	Probability that the batch size is n
C_W	Cost of customer wait per unit time
$C(t)$	CDF of the interdeparture process
$C(z)$	Probability generating function of $\{c_n\}$
$c(t)$	Probability density of the interdeparture process
D	(1) Deterministic interarrival or service times; (2) linear difference operator, $Dx_n = X_{n+1}$; (3) linear differential operator, $Dy(x) = dy/dx$
$\bar{d}$	Mean observed interdeparture time of a queueing system
DFR	Decreasing failure rate
DTMC	Discrete-time Markov chain
df	Distribution function
Δy_n	First finite difference; that is, $\Delta y_n = y_{n+1} - y_n$
E_k	Erlang type-k distributed interarrival or service times
$E[\cdot]$	Expected value
η_{ij}	Mean time spent in state i before going to j
FCFS	First-come, first-served queue discipline
$F_n(t)$	Joint probability that n are in the system at time t after the last departure and t is less than the interdeparture time
$F_{ij}(t)$	Conditional probability that, given that a process begins in state i and next goes to state j, the transition time is $\leq t$ (a conditional CDF)
f_{ij}	Probability that state j of a process is ever reached from state i
$f_{ij}^{(n)}$	Probability that the first passage of a process from state i to state j occurs in exactly n steps
G	General distribution for service and/or interarrival times
GCD	Greatest common divisor
GD	General queue discipline
$G(N)$	Normalizing constant in a closed network
$G(t)$	Cumulative distribution function of the busy period for $M/G/1$ and $G/M/1$ models
$G(z)$	Generating function associated with Erlang-service steady-state probabilities $\{p_{n,i}\}$
$G_j(t)$	Conditional probability that, given that a process starts in state i, the time to the next transition is $\leq t$ (a conditional CDF)
γ_i	External flow rate to node i of a network
H_k	Mixture of k exponentials used as distribution for interarrival and/or service times

H	Hyperexponential (a balanced H_2) distribution for service and/or interarrival times
$H(z, y)$	(1) Probability generating function for $\{p_{n,i}\}$; (2) joint generating function for a two-priority queueing model
$H_{ij}(t)$	Cumulative distribution function of time until first transition of a process into state j beginning at state i
$H_r(y, z)$	Generating function associated with $P_{mr}(z)$ for a two-priority queueing model
$h(u)$	Failure or hazard rate of a probability distribution
I	Phase of service the customer is in for Erlang service models (a random variable)
IFR	Increasing failure rate
IID	Independent and identically distributed
I_u	Expected useful server idle time
$I_n(\cdot)$	Modified Bessel function of the first kind
$\tilde{I}(t)$	Probability of a server being idle for a time $>t$ (a complementary CDF)
$\tilde{I}_n(t)$	Conditional probability that one of the $c - n$ idle servers remains idle for a time $>t$ (a conditional complementary CDF)
$i(t)$	Probability density of idle time
$J_n(\cdot)$	Regular Bessel function
K	System capacity limit (truncation point of system size)
K_q	Greatest queue length at which an arrival would balk (a random variable)
$K(z)$	Probability generating function of $\{k_n\}$
$K_i(z)$	Probability generating function of $\{k_{n,i}\}$
k_n	Probability of n arrivals during a service time
$k_{n,i}$	Probability of n arrivals during a service time, given i in the system when service began
L	Expected system size
LCFS	Last-come, first-served queue discipline
LST	Laplace–Stielties transform
LT	Laplace transform
$L^{(D)}$	Expected system size at departure points
$L^{(P)}$	Expected number of phases in the system of an Erlang queueing model
$L^{(n)}$	(1) Expected number of customers of type n in system; (2) expected system size at station n in a series or cyclic queue

$L_{(k)}$	The kth factorial moment of system size
L_q	Expected queue size
L_q'	Expected queue size of nonempty queues
$L_q^{(D)}$	Expected queue size at departure points
$L_q^{(P)}$	Expected number of phases in the queue of an Erlang queueing model
$L_q^{(n)}$	(1) Expected queue size for customers of type n; (2) expected queue size in front of station n in a series or cyclic queue
$L_{q(k)}^{(D)}$	The kth factorial moment of the departure-point queue size
$L(\cdot)$	Likelihood function
$\mathscr{L}(\cdot)$	(1) Log-likelihood function; (2) Laplace transform
Λ	Minimum diagonal element of Q
λ	Mean arrival rate (independent of system size)
λ_n	(1) Mean arrival rate when there are n in the system; (2) mean arrival rate of customers of type n
M	(1) Poisson arrival or service process (or equivalently exponential interarrival or service times); (2) finite population size
MC	Markov chain
MGF	Moment generating function
MLE	Maximum-likelihood estimator
MOM	Method-of-moments (estimator)
$M_x(t)$	Moment generating function of the random variable X
m_i	Mean time a process spends in state i during a visit
m_{ij}	Mean first passage time of a process from state i to state j
m_{jj}	Mean recurrence time of a process to state j
μ	Mean service rate (independent of system size)
$\mu^{(B)}$	Mean service rate for a bulk queueing model
μ_n	(1) Mean service rate when there are n in the system; (2) mean service rate of server n; (3) mean service rate for customers of type n
N	Steady-state number in the system (a random variable)
N_q	Steady-state number in the queue (a random variable)
$N(t)$	Number in the system at time t (a random variable)
$N_q(t)$	Number in the queue at time t (a random variable)
n_a, n_{ae}, n_b	Number of observed arrivals to a system, to an empty system, and to a busy system, respectively ($n_a = n_{ae} + n_b$)

$o(\Delta t)$	Order Δt; that is, $\lim_{\Delta t \to 0} o(\Delta t)/\Delta t = 0$
ω	Expected remaining work
P	Single-step transition probability matrix of a DTMC
PDE	Partial differential equation
PK	Pollaczek–Khintchine formula
PR	Priority queue discipline
$P(z)$, $P(z, t)$	Probability generating function of $\{p_n\}$ and $\{p_n(t)\}$, respectively
$P_{mr}(z)$	Probability generating function of priority steady-state probabilities $\{P_{mnr}\}$
p	Steady-state probability vector of a CTMC
p_n	(1) Steady-state probability of n in the system; (2) steady-state probability that a CTMC is in state n
$p_n^{(B)}$	Steady-state probability of n in a bulk queueing system
$p_n^{(P)}$	Steady-state probability of n phases in an Erlang queueing system
$p_n(t)$	Probability of n in the system at time t
p_{ij}	Single-step transition probability of going from state i to state j
$p_{n,i}$	Steady-state probabilities for Erlang models of n in the system and the customer in service (if service is Erlangian) or next to arrive (if arrivals are Erlangian) in phase i
$p_{ij}^{(n)}$	Transition probability of going from state i to state j in n steps
$p_{n,i}(t)$	Probability that, in an Erlang queueing model at time t, n are in the system and the customer in service (if the service is Erlangian) or next to arrive (if arrivals are Erlangian) is in phase i
$p_{mnr}(t)$	Probability at time t of m units of priority 1, n units of priority 2 in the system, and a unit of priority r in service ($r = 1$ or 2)
$p_{i,j}(u, s)$	Transition probability of moving from state i to state j in time beginning at u and ending at s
$p_{n_1,n_2,\dots,n_k}(t)$	Probability of n_1 customers at station 1, n_2 at station 2, $\dots$, n_k at station k in a series queue at time t
$p_{n_1,n_2,\dots,n_k}$	Steady-state probability of $p_{n_1,n_2,\dots,n_k}(t)$
p–c	Predictor–corrector
$\Pi(z)$	Probability generating function of $\{\pi_n\}$
π	Steady-state probability vector of a DTMC

π_n	(1) Steady-state probability of n in the system at a departure point; (2) steady-state probability that a DTMC is in state n
Q	Infinitesimal generator matrix of a CTMC
$Q_{ij}(t)$	Joint conditional probability that, given that a process begins in state i, the next transition will be to state j in an amount of time t (a conditional CDF)
q_n	Steady-state probability that an arriving customer finds n in the system
R	Network routing probability matrix
$R(t)$	Distribution function of remaining service time
Re	Real portion of a complex number
RK	Runge–Kutta
RSS	Random selection for service
RV	Random variable
r	Defined as λ/μ for multichannel models; defined as $\lambda/k\mu$ for Erlang service models
r_i	The ith root of a polynominal equation (if there is only one root, r_0 is used)
r_n	The nth uniform $(0, 1)$ random number
r_{ij}	Routing probability in a queueing network of a customer going to station j after being served at station i
$r(n)$	Reneging function
ρ	Traffic intensity $(=\lambda/\mu$ for single-channel and *all* network models, and $=\lambda/c\mu$ for other multichannel models)
S	Steady-state service time (a random variable)
SMP	Semi-Markov process
$S^{(n)}$	Service time of the nth arriving customer (a random variable)
$S_k^{(n)}$	Service time of the nth arriving customer of type k
$S_k (S_k')$	Time it takes to serve $n_k(n_k')$ waiting customers of type k (a random variable)
S_0	Time required to finish customer in service (remaining time of service; a random variable)
s_n	Probability that n servers are busy $(c - n$ idle) in a multichannel system
s_X	Sample standard deviation of the random variable X
σ^2, σ_B^2	Variance of the service-time distribution
σ_A^2	Variance of the interarrival-time distribution

σ_k	Sum of traffic intensities in a priority queueing model; that is, $\sigma_k = \Sigma \lambda_i / \mu_i$
T	(1) Time spent in system (a random variable), with expected value W; (2) steady-state interarrival time (a random variable); (3) steady-state interdeparture time (a random variable)
$T^{(n)}$	Interarrival time between the nth and $(n + 1)$st customers (a random variable)
T_A	Instant of arrival
T_i	Length of time a stochastic process spends in state i (a random variable)
T_S	Instant of service completion
T_{busy}	Length of a busy period (a random variable)
T_q	Time spent in queue (a random variable), with expected value W_q
$T_{b,i}$	Length of i channel busy period for $M/M/c$ (a random variable)
t_b, t_e, t	Observed time a system is busy, observed time a system is empty, and total observed time, respectively ($t = t_b + t_e$)
τ_i	Time at which the ith arrival to a Poisson process occurred
U	Steady-state difference between service time and interarrival time, $U = S - T$ (a random variable)
$U^{(n)}$	Service time of nth customer minus interarrival time between customer $n + 1$ and n; that is $U^{(n)} = S^{(n)} - T^{(n)}$ (a random variable)
$U(t)$	(1) Cumulative distribution function of $U = S - T$; (2) cumulative distribution function of the time back to the most recent transition
$U^{(n)}(t)$	Cumulative distribution function of $U^{(n)}$
$U_i(t)$	Cumulative distribution function of the time back to the most recent transition, given the process starts in state i
$\text{Var}[\cdot]$	Variance
V	Expected virtual wait
$V(t)$	Virtual waiting-time function
v_j	(1) Steady-state probability of a semi-Markov process being in state j; (2) relative throughput in a closed network
W	Expected waiting time in system
$W^{(n)}$	Waiting time including service at station n of a series or cyclic queue

W_k Ordinary kth moment of waiting time in system

W_q Expected waiting time in queue

$W_{q,k}$ Regular kth moment of waiting time in queue

$W(t)$ Cumulative distribution function of waiting time in system

$W_q^{(H)}$ Expected time in queue for a system in heavy traffic

$W_q^{(n)}$ (1) Waiting time in queue for the nth arriving customer (a random variable); (2) expected wait in queue for customers of priority class n; (3) expected time in queue at station n of a network of queues

$W_q(t)$ Cumulative distribution function of waiting time in queue

$\tilde{W}_q(t \mid j)$ Probability in $M/M/c$ model that the delay undergone by an arbitrary arrival who joined when $c + j$ were in the system is more than t

$X(t)$ Stochastic process with state space X and parameter t

x, x_i Observed interval of a queueing system, and observed interval of type i (busy, empty, etc.) of a queueing system, respectively

$[x]$ Greatest integer value $\leq x$

$\doteq$ Approximately equal to

$\sim$ Asymptotic to

$\in$ Set membership

$*$ (1) LST; (2) used for various other purposes as specifically defined in text

$^{-}$ Laplace transform

$\binom{n}{c}$ Binomial coefficient, $n!/[(n - c)!c!]$

$[\cdot]$ Batch queueing model

$(\cdot)$ (1) Order of convolution; (2) order of differentiation, (3) number of steps (transitions) in a discrete-parameter Markov chain.

$'$ (1) Differentiation; (2) conditional; for example, p_n' is a conditional probability distribution of n in the system given system not empty; (3) used for various other purposes as specifically defined in text

$\sim$ (1) Complementary CDF; (2) used for various other purposes as specifically defined in text

Summary of Models Treated and Types of Results

This appendix provides a table showing the models treated in the book and the types of results obtained. Results are indicated as analytical, numerical, results in terms of generating functions or transforms only, or results that are mentioned with references given. The table's notation is provided at the end.

Model (Notation Explained in Table 1.1, Section 1.3)	$\{p_n\}$	Expected-Value Measures (L, L_q, W, W_q)	Waiting-Time Distribution	Section	QTS Module
		(a) Steady State			
FCFS					
$M/M/1$	a	a	a	2.1	✓
$M/M/1/K$	a	a	a	2.3	✓
$M/M/c$	a	a	a	2.2	✓
$M/M/c/K$	a	a	a^b	2.3	✓
$M/M/c/c$	a	a	—	2.5	✓
$M/M/\infty$	a	a	—	2.6	✓
Finite-source $M/M/c$	a	a	a	2.7	✓
State-dep. serv. $M/M/1$	a, n^c	a, n^c	0	2.8	✓
Impatience $M/M/c$	a, n^c	a, n^c	a, n^c	2.9	—
$M^{[X]}/M/1$	g	a	0	3.1	✓
$M/M^{[Y]}/1$	a^d	a^d	0	3.2	✓
$M/M^{[Y]}/c$		Results indicated	0	3.2	—
$M/E_k/1$ $(M/D/1)$	$g, (a)$	a	0	3.3.3	✓
$M/D/c$	g	a	0	6.3	✓
$E_k/M/1$ $(D/M/1)$	a^d	n	n	3.3.4	✓
$E_l/E_k/1$		Some numerical results possible—reference given		3.3.4	✓
$M/G/1$	n	a	n	5.1–5.1.6	—
$M/G/1/K$	a	a	0	5.1.7	—
Impatience $M/G/1$		Results indicated—reference given		5.1.8	—
Finite source $M/G/1$		Reference given		5.1.8	—
$M^{[X]}/G/1$	g, n^c	a, n^c	0	5.1.9	—
$M/G^{[K]}/1$		Reference given		5.1.8	—
State-dep. serv. $M/G/1$	n	n	0	5.1.10	—
$M/G/c$		Little's relation on higher moments		5.2.1	—

Model				Section	
$M/G/c/c$	a	a	—	5.2.2	✓
$M/G/\infty$	a	a	—	5.2.2	✓
$G/M/1$	a^{d}	a^{d}	a^{d}	5.3.1, 6.4	✓
$G/M/c$	a^{d}	n	a^{d}	5.3.2	✓
$G/M/1/K$		Results indicated—reference given		5.3.2	—
$G/M/c/K$		Results indicated—reference given		5.3.2	—
Impatience $G/M/1$		Results indicated—reference given		5.3.2	—
Impatience $G/M/c$		Results indicated—reference given		5.3.2	—
$G/M^{[x]}/1$		Results indicated—reference given		5.3.2	—
$G/M^{[x]}/c$		Results indicated—reference given		5.3.2	—
$G/E_k/1$		Results indicated—reference given		6.1.1	—
$G^{[K]}/M/1$				6.1	—
$G/PH_k/1$	0	Results indicated—reference given	n	6.1.1	—
$G/G/1$	0	n, bounds, approx.	0	6.2, 7.1.1, 7.2	✓
$G/G/c$	0	bounds	0	7.1.2	—
PR					
$M/M/1$, two priorities	g	a	0	3.4.1	✓
$M/M/1$, many priorities	0	a	0	3.4.2	✓
$M/M/1$ preemptive	g	a	0	3.4.3	✓
$M/G/1$, many priorities	0	0	a	5.1.8	✓
$G/G/1$, many priorities		Results discussed—reference given		5.3.2	—
$M/M/c$, many priorities	0	a	0	3.4.2	✓
Series					
$M/M/c$	a	a	a	4.1.1	✓
$M/M/1$ with blocking		Partial results, depending on the model		4.1.2	—
Cyclic					
$M/M/1$, $M/M/c$	a	a	0	4.4	✓
Networks					
"$M/M/1$," "$M/M/c$"	a	a	0	4.2, 4.3	✓

Types of Results^a for:

Model (Notation Explained in Table 1.1, Section 1.3)	$\{p_n\}$	Expected-Value Measures (L, L_q, W, W_q)	Waiting-Time Distribution	Section	QTS Module
		(b) Transient			
M/M/1/1	a	n^b	—	2.10.1	—
M/M/1	a, n	n^b	0	2.10.2	√
M/M/∞	a, n	a, n^b	—	2.10.3, 7.3.2	√
M/M/c	n	n^b	0	7.3.2	—
M/M/c/k	n	n^b	0	7.3.2	—
Finite source M/M/c	n	n^b	0	7.3.2	—
General birth–death	n	n^b	0	7.3.2	—
M/G/1	Results indicated—reference given			5.1.8	—
M/G/∞	Results indicated—reference given			5.2.2	—
G/M/1	Results indicated—reference given			5.3.2	—

^{a}Notation: a, analytical results; n, numerical results; g, results in form of generating function or Laplace transform; 0, no results; —, not applicable.
^{b}Indicated but not presented.
^{c}Depends on particular model.
^{d}Analytical results follow after a root to a nonlinear equation is found; finding the root may require numerical analysis.

APPENDIX 3

QTS Software

The QTS software is available for Quattro Pro 6 for Windows 3.x and Excel for Windows 95. Instructions for downloading the software from Wiley's ftp site are found at the end of the appendix.

The software generally follows the text and is organized into seven "chapters," as is the text, with a correspondence between a software chapter and

CHAPTER 2:						
m-m-1.wb2	M/M/1: Single-Server/Unlimited Markovian Queue					
m-m-1-k.wb2	M/M/1/K: Single-Server/Space-Limited Queue					
m-m-c.wb2	M/M/c: Multiple-Servers/Unlimited Markovian Queue					
m-m-c-k.wb2	M/M/c/K: Multi-Server/Space-Limited Queue					
m-g-c-c.wb2	M/G/c/c: Pure Overflow Model					
overflow.wb2	M/G/c/c: Optimal Number of Servers					
m-g-inf.wb2	M/G/inf: Poisson Input, Gen'l Service, Unlimited Servers					
m-m-inf.wb2	M/M/inf: Poisson Input, Expo. Service, Unlimited Servers					
mr-m-1.wb2	Markov Single-Server, Finite-Source Queue w/o Spares					
mrs-m-1.wb2	Markov Single-Server, Finite-Source Queue with Spares					
mrs-m-c.wb2	Markov Multi-Server, Finite-Source Queue with Spares					
ssdm-m-1.wb2	Simple Markov Queue with State-Dependent Service					
csdm-m-1.wb2	General State-Dependent Birth-Death Queue					
mm1_trans.wb2	Transient M/M/1 Solution					

Fig. A3.1.

a text chapter. When loaded properly, a menu comes up which gives the modules included in a particular chapter. Figure A3.1 is an example of Chapter 2 in the Quattro Pro version (the Excel version is virtually identical).

There are fourteen modules in Chapter 2, corresponding to most of the models treated in the text. The user would click on a module "button" to select the model. A sample problem (with explanation and/or instructions at the top, if necessary) appears. Illustrated in Figure A3.2 is the *M/G/c/c* Pure Overflow Model module.

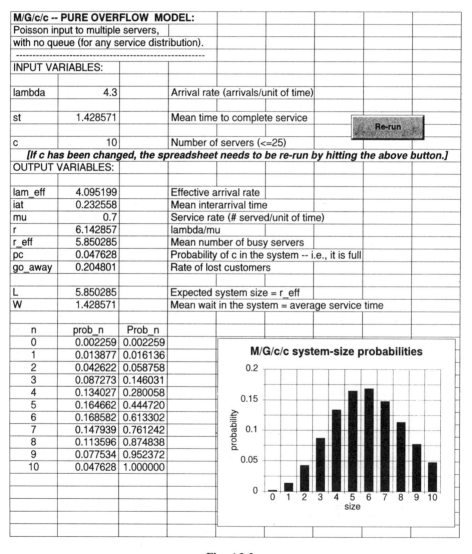

Fig. A3.2.

To run another problem, one would change the INPUT VARIABLES as desired, and the spreadsheet automatically calculates new OUTPUT VARIABLES (output measures of performance). As an illustration, suppose we desire to run a case with $\lambda = 4$, $1/\mu = 2$, and $c = 5$. We enter these new input values by simply writing over the current ones, and the output values are changed accordingly as shown in Figure A3.3. Note that there is a "Re-run" button to be hit in order to plot the smaller set of probabilities.

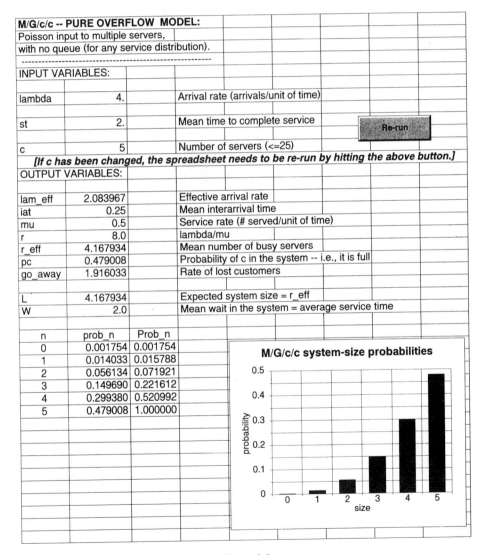

M/G/c/c -- PURE OVERFLOW MODEL:							
Poisson input to multiple servers,							
with no queue (for any service distribution).							

INPUT VARIABLES:							
lambda	4.		Arrival rate (arrivals/unit of time)				
st	2.		Mean time to complete service			Re-run	
c	5		Number of servers (<=25)				
[If c has been changed, the spreadsheet needs to be re-run by hitting the above button.]							
OUTPUT VARIABLES:							
lam_eff	2.083967		Effective arrival rate				
iat	0.25		Mean interarrival time				
mu	0.5		Service rate (# served/unit of time)				
r	8.0		lambda/mu				
r_eff	4.167934		Mean number of busy servers				
pc	0.479008		Probability of c in the system -- i.e., it is full				
go_away	1.916033		Rate of lost customers				
L	4.167934		Expected system size = r_eff				
W	2.0		Mean wait in the system = average service time				
n	prob_n	Prob_n					
0	0.001754	0.001754					
1	0.014033	0.015788					
2	0.056134	0.071921					
3	0.149690	0.221612					
4	0.299380	0.520992					
5	0.479008	1.000000					

Fig. A3.3.

SAMPLE SINGLE-SERVER QUEUE:						
This program does a basic waiting-time analysis for a sample G/G/1 queue						
constructed from an input sequence of interarrival and service times. To expand						
the size of the data table, copy (cell formulas) its first three columns as far down						
as needed, put in the extra service and interarrival times, and finally change N.						
(Note that the first customer is assumed to arrive to an idle system.)						

INPUT:						
N	10		Number of observed customers (input)			
OUTPUT SUMMARY:						
T	30		Total time horizon = sum(IAT) + final waiting time			
iat	3		Mean interrarrival time computed over T			
st	2.6		Mean service time of input			
lambda	0.333333		Computed as N/T			
mu	0.384615		Computed from ST data			
rho	0.866667		Empirical utilization rate (can be > 1)			
Wq	0.913		Average line delay			
W	3.513		Average waiting time in system			
OUTPUT DATA:				INPUT DATA:		
	line delays	system waits		service times		interarrival times
	Wq(n)	W(n)		S		T
1	0	3		3		4
2	0	5		5		3
3	2	7		5		3
4	2	5		3		5
5	0	3		3		5
6	1.9	3.9		2		1.1
7	0.6	1.6		1		3.3
8	0.72	1.72		1		0.88
9	0.95	1.95		1		0.77
10	0.96	2.96		2		0.99

Fig. A3.4.

In Figure A3.4 we present one additional module, Sample Single-Server Queue from Chapter 1, which is one with self-contained instructions given at the top. If we desire to run ten different service and interarrival times, we merely enter new values. If we wish to run more than ten, we need to copy the cell formulas of the first three columns as far down as desired and enter the new values. A twelve-value problem is illustrated in Figure A3.5.

Note that in the $N = 12$ example, as you change each interarrival and service-time value, the spreadsheet automatically calculates new output values for $W_q(n)$ and $W(n)$. Also note that in this example, it is assumed that the clock does not start until the first customer arrives; hence the 0 in the first row for interarrival time.

SAMPLE SINGLE-SERVER QUEUE:							
This program does a basic waiting-time analysis for a sample G/G/1 queue							
constructed from an input sequence of interarrival and service times. To expand							
the size of the data table, copy (cell formulas) its first three columns as far down							
as needed, put in the extra service and interarrival times, and finally change N.							
(Note that the first customer is assumed to arrive to an idle system.)							

INPUT:							
N	12		Number of observed customers (input)				

OUTPUT SUMMARY:							
T	31		Total time horizon = sum(IAT) + final waiting time				
iat	2.583333		Mean interarrival time computed over T				
st	2.5		Mean service time of input				
lambda	0.387097		Computed as N/T				
mu	0.4		Computed from ST data				
rho	0.967742		Empirical utilization rate (can be > 1)				
Wq	3.333333		Average line delay				
W	5.833333		Average waiting time in system				

OUTPUT DATA:				INPUT DATA:			
	line delays	system waits		service times		interarrival times	
	Wq(n)	W(n)		S		T	
1	0	1		1		0	
2	0	3		3		2	
3	2	8		6		1	
4	5	7		2		3	
5	6	7		1		1	
6	6	7		1		1	
7	3	7		4		4	
8	5	7		2		2	
9	2	7		5		5	
10	6	7		1	·	1	
11	3	4		1		4	
12	2	5		3		2	

Fig. A3.5.

A3.1 INSTRUCTIONS FOR DOWNLOADING QTS SOFTWARE

Self-extracting compressed files for installing either the Quattro Pro or Excel versions of the QTS software can be obtained from the Wiley public ftp site. Both software versions require approximately 6 MB of hard drive space. To access the files, type the following location address in your ftp or web browsing program:

ftp://ftp.wiley.com/public/sci_tech_med/queueing_theory/

From the queueing_theory area, you can download either of the two versions by double-clicking on the preferred file and saving the file to your hard drive. To install the software, double-click on the saved file in your Windows File Manager or Explorer. The target location for the Quattro Pro software needs to be the C:\qts_qpw directory, and C:\qts_xcel for the Excel version. Changing the drive or directory from the default choice will require extensive macro editing.

Once the software is installed, you can load QTS by opening the QTS-MENU.WB2 file of the Quattro Pro version or the QTS-MENU.XLS file of the Excel version. You can access the software directly from your desktop by creating application icons. For the Quattro Pro version, use the File/New/-Program Item selection in Windows 3.x, name the icon QTS, and browse in the Command box to select the QTS-MENU.WB2 file. For the Excel version, go into Windows Explorer and drag QTS-MENU to your desktop to create a shortcut icon.

The QTS menu will open to Chapter 1's page. Other chapters can be reached by selecting worksheet page tabs.

To learn more about related titles from Wiley, visit the Publisher's web site http://www.wiley.com.

Subject Index

Author Index

440

WILEY SERIES IN PROBABILITY AND STATISTICS
ESTABLISHED BY WALTER A. SHEWHART AND SAMUEL S. WILKS

Editors
*Vic Barnett, Ralph A. Bradley, Noel A. C. Cressie, Nicholas I. Fisher,
Iain M. Johnstone, J. B. Kadane, David G. Kendall, David W. Scott,
Bernard W. Silverman, Adrian F. M. Smith, Jozef L. Teugels,
Geoffrey S. Watson; J. Stuart Hunter, Emeritus*

Probability and Statistics Section

*Now available in a lower priced paperback edition in the Wiley Classics Library.

*Now available in a lower priced paperback edition in the Wiley Classics Library.

*Now available in a lower priced paperback edition in the Wiley Classics Library.

*Now available in a lower priced paperback edition in the Wiley Classics Library.

*Now available in a lower priced paperback edition in the Wiley Classics Library.

Texts and References Section

AGRESTI · An Introduction to Categorical Data Analysis

ANDERSON · An Introduction to Multivariate Statistical Analysis, *Second Edition*

ANDERSON and LOYNES · The Teaching of Practical Statistics

ARMITAGE and COLTON · Encyclopedia of Biostatistics: Volumes 1 to 6 with Index

BARTOSZYNSKI and NIEWIADOMSKA-BUGAJ · Probability and Statistical Inference

BERRY, CHALONER, and GEWEKE · Bayesian Analysis in Statistics and Econometrics: Essays in Honor of Arnold Zellner

BHATTACHARYA and JOHNSON · Statistical Concepts and Methods

BILLINGSLEY · Probability and Measure, *Second Edition*

BOX · R. A. Fisher, the Life of a Scientist

BOX, HUNTER, and HUNTER · Statistics for Experimenters: An Introduction to Design, Data Analysis, and Model Building

BOX and LUCEÑO · Statistical Control by Monitoring and Feedback Adjustment

BROWN and HOLLANDER · Statistics: A Biomedical Introduction

CHATTERJEE and PRICE · Regression Analysis by Example, *Second Edition*

COOK and WEISBERG · An Introduction to Regression Graphics

COX · A Handbook of Introductory Statistical Methods

DILLON and GOLDSTEIN · Multivariate Analysis: Methods and Applications

DODGE and ROMIG · Sampling Inspection Tables, *Second Edition*

DRAPER and SMITH · Applied Regression Analysis, *Third Edition*

DUDEWICZ and MISHRA · Modern Mathematical Statistics

DUNN · Basic Statistics: A Primer for the Biomedical Sciences, *Second Edition*

FISHER and VAN BELLE · Biostatistics: A Methodology for the Health Sciences

FREEMAN and SMITH · Aspects of Uncertainty: A Tribute to D. V. Lindley

GROSS and HARRIS · Fundamentals of Queueing Theory, *Third Edition*

HALD · A History of Probability and Statistics and their Applications Before 1750

HALD · A History of Mathematical Statistics from 1750 to 1930

HELLER · MACSYMA for Statisticians

HOEL · Introduction to Mathematical Statistics, *Fifth Edition*

JOHNSON and BALAKRISHNAN · Advances in the Theory and Practice of Statistics: A Volume in Honor of Samuel Kotz

JOHNSON and KOTZ (editors) · Leading Personalities in Statistical Sciences: From the Seventeenth Century to the Present

JUDGE, GRIFFITHS, HILL, LÜTKEPOHL, and LEE · The Theory and Practice of Econometrics, *Second Edition*

KHURI · Advanced Calculus with Applications in Statistics

KOTZ and JOHNSON (editors) · Encyclopedia of Statistical Sciences: Volumes 1 to 9 wtih Index

KOTZ and JOHNSON (editors) · Encyclopedia of Statistical Sciences: Supplement Volume

KOTZ, REED, and BANKS (editors) · Encyclopedia of Statistical Sciences: Update Volume 1

KOTZ, REED, and BANKS (editors) · Encyclopedia of Statistical Sciences: Update Volume 2

LAMPERTI · Probability: A Survey of the Mathematical Theory, *Second Edition*

LARSON · Introduction to Probability Theory and Statistical Inference, *Third Edition*

LE · Applied Survival Analysis

MALLOWS · Design, Data, and Analysis by Some Friends of Cuthbert Daniel

MARDIA · The Art of Statistical Science: A Tribute to G. S. Watson

MASON, GUNST, and HESS · Statistical Design and Analysis of Experiments with Applications to Engineering and Science

MURRAY · X-STAT 2.0 Statistical Experimentation, Design Data Analysis, and Nonlinear Optimization

*Now available in a lower priced paperback edition in the Wiley Classics Library.

Texts amd References (Continued)

PURI, VILAPLANA, and WERTZ · New Perspectives in Theoretical and Applied Statistics

RENCHER · Methods of Multivariate Analysis

RENCHER · Multivariate Statistical Inference with Applications

ROSS · Introduction to Probability and Statistics for Engineers and Scientists

ROHATGI · An Introduction to Probability Theory and Mathematical Statistics

RYAN · Modern Regression Methods

SCHOTT · Matrix Analysis for Statistics

SEARLE · Matrix Algebra Useful for Statistics

STYAN · The Collected Papers of T. W. Anderson: 1943–1985

TIERNEY · LISP-STAT: An Object-Oriented Environment for Statistical Computing and Dynamic Graphics

WONNACOTT and WONNACOTT · Econometrics, *Second Edition*

WILEY SERIES IN PROBABILITY AND STATISTICS

ESTABLISHED BY WALTER A. SHEWHART AND SAMUEL S. WILKS

Editors
Robert M. Groves, Graham Kalton, J. N. K. Rao, Norbert Schwarz, Christopher Skinner

Survey Methodology Section

BIEMER, GROVES, LYBERG, MATHIOWETZ, and SUDMAN · Measurement Errors in Surveys

COCHRAN · Sampling Techniques, *Third Edition*

COX, BINDER, CHINNAPPA, CHRISTIANSON, COLLEDGE, and KOTT (editors) · Business Survey Methods

*DEMING · Sample Design in Business Research

DILLMAN · Mail and Telephone Surveys: The Total Design Method

GROVES · Survey Errors and Survey Costs

GROVES, BIEMER, LYBERG, MASSEY, NICHOLLS, and WAKSBERG · Telephone Survey Methodology

*HANSEN, HURWITZ, and MADOW · Sample Survey Methods and Theory, Volume 1: Methods and Applications

*HANSEN, HURWITZ, and MADOW · Sample Survey Methods and Theory, Volume II: Theory

KASPRZYK, DUNCAN, KALTON, and SINGH · Panel Surveys

KISH · Statistical Design for Research

*KISH · Survey Sampling

LESSLER and KALSBEEK · Nonsampling Error in Surveys

LEVY and LEMESHOW · Sampling of Populations: Methods and Applications

LYBERG, BIEMER, COLLINS, de LEEUW, DIPPO, SCHWARZ, TREWIN (editors) · Survey Measurement and Process Quality

SKINNER, HOLT, and SMITH · Analysis of Complex Surveys

*Now available in a lower priced paperback edition in the Wiley Classics Library.